Automotive Milestones

The Technological Development of the Automobile

Who, What, When, Where, and How It All Works

Robert L. Norton

Industrial Press, Inc.

Industrial Press, Inc.

32 Haviland Street, Unit 2C
South Norwalk, Connecticut 06854
Tel: 203-956-5593, Toll-Free: 888-528-7852
E-mail: info@industrialpress.com

Library of Congress Cataloging-in-Publication Data

Norton, Robert L., author.
Automotive milestones: the technological development of the automobile / Robert L. Norton.
  320 pages 28 cm
Includes bibliographical references and index.
ISBN 978-0-8311-3520-1 (softcover)
1. Automobiles—Popular works. I. Title.
TL146.5.N67 2015
629.222—dc23   2015020769

ISBN print: 978-0-8311-3520-1
ISBN ePDF: 978-0-8311-9340-9
ISBN ePUB: 978-0-8311-9341-6
ISBN eMOBI: 978-0-8311-9342-3

Copyright © 2016 by Industrial Press, Inc.
All rights reserved.
Printed in the United States of America.

This book, or any parts thereof, with the exception of those figures in the public domain or presented under a Creative Commons license as listed in Appendix D, may not be reproduced, stored in a retrieval system, or transmitted in any form without the permission of the copyright holders.

Sponsoring Editor: John Carleo
Developmental Editor: Robert Weinstein
Interior Text Designer: Robert L. Norton
Cover Designer: Robert L. Norton

Key to front-cover photographs:

                            1967 Jaguar E-Type
2002 BMW 325ci   2005 Porsche Boxster S
                    2007 Maserati Quattroporte
1983 BMW 633csi   1997 BMW 540i Sport
                        2008 Corvette C6

All cover photographs are of automobiles in the author's personal collection. All text, drawings, and equations in this book were prepared and typeset electronically by the author on a Macintosh® computer using Freehand®, Illustrator®, MathType®, and InDesign® desktop publishing software. The body text was set in Times, and headings were set in Syntax. The book was printed directly from the author's files.

Some figures contain photographs from Wikipedia under the Creative Commons license. Use of these images does not imply endorsement of any views or opinions expressed by their originators on Wikipedia.

industrialpress.com
ebooks.industrialpress.com

10 9 8 7 6 5 4 3 2 1

This book is dedicated to the memory of:

John W. Lothrop
1918–2011

*A gentleman, scholar, master machinist,
restorer of antique automobiles extraordinaire,
and a good friend and mentor.
He is greatly missed.*

Table of Contents

Foreword XIII
Acknowledgments XV
About the Author XVI
A Caveat XVI
Introduction XVII

Chapter 1
Motive Power 1

External Combustion Engines (ECE) 2
Steam 2
Steam Engine Terminology 3
Steam-Powered Machines 5
- Cugnot (1769) 5
- Early Steam Vehicles in Europe 6
- Oliver Evans' Oruktor Amphibolos 7
- Stanley Steamer (1897–1924) 7
- Locomobile Steamer (1899–1903) 8
- White Steamer (1901–1911) 9
- Doble Steamer (1909–1931) 10
- Besler Steam Machines 12
- Stirling Engine 12

Internal Combustion Engines (ICE) 14
Stroke Cycles 15
- Otto Four-Stroke Cycle 15
- Clerk Two-Stroke Cycle 18
- Diesel Cycle (Compression Ignition) 20
- Atkinson Cycle 21
- Miller Cycle 24
- Brayton Cycle 24
- Wankel Cycle 25

Electric Vehicles 26
Pure Electric 26
Hybrids 28
- Electric/ICE 28
- Hydraulic/ICE 30
- Flywheel Hybrids 32

Chapter 2

Chassis Layouts and Drivelines 33

The First Production Automobiles 33
Benz 33
Daimler 34
Flocken Elektrowagen 34
Panhard-Levassor 35
Other Early European Makes 35
United States 35
Oldsmobile 35
Ford 36

Rear-Wheel Drive (RWD) 37
Front-Wheel Drive (FWD) 41
Transverse Engine FWD 43
Constant Velocity Joints 45
Four-Wheel Drive (4WD) and All Wheel Drive (AWD) 46

Chapter 3

Engine Configurations 51

Engine Construction 51
Cylinder Arrangements 53
Multi-Cylinder Inline Engines 54
Two-Cylinder (Twin) Inline Engines 54
Four-Cylinder Inline Engines 57
Five-Cylinder Inline Engines 60
Six-Cylinder Inline Engines 61
Three-Cylinder Inline Engines 63
Eight-Cylinder Inline Engines 65
Twelve-Cylinder Inline Engines 67
Multi-Cylinder V-Engines 69
V-Twin Engines 71
V4 Engines 72
V8 Engines 72
V6 Engines 77
V10 Engines 78
V12 Engines 78
V16 Engines 80
Multi-Cylinder, Horizontally Opposed Boxer Engines 81
Opposed Boxer Twins 81
Opposed Boxer Four 82
Opposed Boxer Six 83

  W-Engines..83
  Rotary and Radial Engines..84
 Engine Balancing..87
  Crankshaft Counterweights..88
  Balance Shafts..89
  Crankshaft Torsional Dampers..90
  Floating Power..90
 Summary..91

CHAPTER 4
VALVE TRAINS, INDUCTION, AND SUPERCHARGING................................. 92

 Valve Actuation 93
 Valve Cam Functions 95
 Valve Arrangements 95
  Valve in Block...95
   T-head... 96
   L-head.. 97
   F-head.. 97
  Valve in Head...97
   Overhead Valves (OHV) 97
   Overhead Camshaft.. 99
  Sleeve Valves..101
   Dual-Sleeve Valves—The Knight Engine 101
   Single-Sleeve Valves—The Burt-McCollum Engine...................... 105
 Variable Valve Timing (VVT).. 105
 Cylinder Deactivation 109
  Engine Start-Stop..110
 Fuel Control 111
  Carburetors 111
  Fuel Injection..113
   Injection Methods 114
    Mechanical Fuel Injection 114
    Electronic Fuel Injection..114
 Forced Induction... 115
  Types of Superchargers..116
   POSITIVE DISPLACEMENT 116
   POSITIVE PRESSURE 117
  Driving the Supercharger 117
   MECHANICAL SUPERCHARGERS 117
   TURBOCHARGERS 119
 Summary 120

Photo Gallery of Notable Automobiles

Chapter 5

Transmissions and Differentials 121

Gearboxes 123
Sliding-Gear Transmissions 124
Synchromesh Transmissions 129

Friction Drive 131
CVTs 132

Planetary Transmissions 134

Overdrive Transmissions 137

Semi-Automatic Gearboxes 139
Clutches Versus Fluid Couplings 139
Mechanical Clutches 140
Fluid Couplings 141
Preselector Gearboxes 143
Dry-Clutched, Semi-Automatics 144
Reo's Self-Shifter 144
Oldsmobile's Safety Transmission 144
Hudson's Drive-Master 145
Fluid-Drive, Semi-Automatics 146
Daimler (England) 146
Chrysler's Fluid Drive 146
Chrysler's "Vacamatic" Transmission 146
Other Semi-Automatics 148

Automatic Transmissions 148
The Sturtevant Automatic Transmission 148
Other Attempts 149
GM's Hydramatic Transmission 149
Torque Converters 152
Buick's Dynaflow Transmission 153
Chevrolet Powerglide 154
GM Turbo Hydramatic 155
Everybody Wants In On The Act! 156
Modern Automatic Transmissions 156
Manumatic Transmissions 157

Automated Manual Transmissions (AMT) 157
Single-Clutch AMT (SC-AMT) 157
Dual-Clutch AMT (DC-AMT) 158

Differentials 159

  Viscous Couplings 160
  Spiral Bevel Vs. Hypoid Differentials 161
  Torsen Differentials 162
 Two-Speed Rear Ends 162
 Summary 163

Chapter 6

Suspension and Steering 164

 Springs 165
  Leaf Springs 165
  Torsion Bars 166
  Coil Springs 167
  Air Springs 167
   Hydro-Pneumatic Suspension 168
  Tires 169
 Sprung Versus Unsprung Weight 170
 Non-Independent Suspension 170
  Non-Independent Front Suspension 170
   Caster, Camber, and Toe 171
  Shimmy, Wobble, and Bounce 172
  Non-Independent Rear Suspension 174
   Dead Axles 174
   Live Axles 174
    Full-Floating Versus Semi-Floating Axles 175
   De Dion Axles 175
   Axle Control 176
    Hotchkiss Suspension 177
    Torque-Tube Suspension 177
    Multi-Link Suspension 178
 Independent Suspension 179
  Wheel Control 179
  Independent Front Suspension (IFS) 181
   Sliding Pillar 181
   Leaf Spring Linkages 182
   Trailing Arm Suspension 183
   Swing Axle Suspension 184
   Unequal Arm Linkages 184
   MacPherson Strut Suspension 186
   Modern IFS Practice 187
  Independent Rear Suspension (IRS) 187
   Swing-Axle IRS 188
   Other IRS Types 188

- Spatial-Linkage Suspension 188
- Modern Rear Suspension Practice 189

Dampers 189
- MagnetoRheological Dampers 190

Active Suspension 191
- Bose Active Suspension System 192

Steering 193
- Ackerman Steering 193
- Steering Mechanisms 194
  - Parallelogram Steering Mechanisms 194
  - Rack and Pinion Steering 195

Power Steering 195
- Hydraulic Power Assist 195
- Electric Power Assist 197

Handling 197
- Weight Distribution 197
- Mass-Center Height 198
- Roll Center and Roll Axis 198
- Slip Angle 199
- Understeer and Oversteer 199
- Sway Bars 199
- Anti-Dive and Anti-Squat 200
- Bump Steer 201

Ride Comfort 201

CHAPTER 7
BRAKES 203

Four-Wheel Brakes 204

Drum Brakes 204
- Hydraulic Brakes 204
- Self-Energizing Brakes 205

Disk Brakes 206

Power Brakes 209
- Mechanical Power Brakes 209
- Vacuum Power Brakes 210
- Other Power-Brake Systems 210

Anti-Lock Braking Systems 210
- Stability, Cornering, and Traction Control 211

Air Brakes 212
Parking Brakes 212

Chapter 8
Body Design 213

Early Body Designs 214
Closed Bodies 216
- Fabric Bodies 217
- Woodies 217

Streamlining 218
- Drag Coefficient (C_d) 218
- The Chrysler Airflow 219
- European Streamliners 221

Unit–Bodies 222
- Ruler Frameless—1915 222
- Lancia Lambda—1922–1931 222
- Adler Trumpf—1932–1939 223
- Citroën Traction Avant—1934–1957 223
- Cord 810/812—1934–1937 223
- Lincoln Zephyr—1936–1941 224
- Tatra T87—1937–1948 224
- Volkswagen Beetle—1938–2011 225
- Nash Ambassador 600—1941–1949 225
- The Unit-Body Revolution—1960–on 226

Aluminum Bodies 226
- Pierce-Arrow—1904–1928 226
- Peugeot Darl'Mart—1930–1953 227
- Amilcar—1938–1941 228
- Hotchkiss-Gregoire—1950–1954 228
- Jaguar D-Type—1954–1957 228
- Mercedes 300S—1955–1957 229
- Acura NSX—1990–2005 and 2015–? 229
- Audi A8—1994–Present 230
- Audi A2—1999–2005 230
- Morgan Aero 8—2001–2010 230
- Jaguar XJ (X350)—2004–2007 231
- Ford F150—2015–? 231

Plastic Bodies 231
- Fiberglass and Carbon Fiber 231

  Chevrolet Corvette—1953–Present 232
  Lotus Elite—1957–1963 233
  Lotus Elan—1962–1975 and 1989–1995 234
  Pontiac Fiero—1984–1988 234
  Saturn—1985–2010 234
  Dodge Viper—1992–Present 234
  Bugatti Veyron—2005–2014 235
  Lexus LFA—2010–2012 235
  Lamborgini Aventador—2011–Present 236
  McLaren MP4-12C—2011–2014 and McLaren P1—2014–Present 236
  Mercedes-Benz SLR McLaren—2003–2010 237
  BMW i3—2014–Present 237

Crushable Bodies 237

Pedestrian-Friendly Cars 238

NVH 239

Summary 239

Chapter 9
Summary 240

Appendix A - Glossary of Terms 241

Appendix B - Milestones by Year 247

Appendix C - Milestones by Category 251

Appendix D - List of Figures 255

Appendix E - Bibliography 267

Index 277

Foreword

Neither a history lesson nor an engineering text, *Automotive Milestones* is a bit of both. Samuel Johnson said, *The two most engaging powers of an author are to make new things familiar and familiar things new*. Professor Bob Norton certainly has succeeded in this. From Nicholas Cugnot's steam-powered vehicle of 1769 and Dr. Ferdinand Porsche's 1901 Lohner P1 hybrid, to modern dual-clutch, automatic-manual transmissions, Bob explains automotive engineering and progress in a way that is uniquely understandable, yet not oversimplified, and without requiring deep math skills. Well-chosen diagrams and photos clearly illustrate concepts. Although easily readable front-to-back, its logical organization makes it an excellent reference work. Bob provides Internet links and references for those desiring more extensive information.

Many automotive books depend on opinions and secondary sources for information, and may therefore contain inaccuracies. One of the most significant aspects of this book is its extensive use of original sources. Bob visited many museums and collections and read many patents and original technical papers. Many of the photos were taken specifically for the book, and many facts were verified by examining the actual vehicles. I was blessed with the opportunity of making available the many unique resources of the collection at the Tampa Bay Automobile Museum.

What will be the automotive milestones of the future? It is obvious that we will see significantly more dependence on digital technology, requiring more capable electrical systems. So one milestone will surely be the introduction of 48-volt systems in a production vehicle, already planned by Audi. Another will be fully active suspension—Mercedes' "Magic Ride Control", already in production, is a predecessor. I'm not so sure about autonomous vehicles—will they be a milestone or a millstone? They'll certainly be a boon for the lawyers.

Shakespeare wrote, *What's past is prologue*. This book proves it, at least as far as automobiles are concerned. Many of today's automotive milestones had their origins in the distant past, and to enter production only required advances in supporting technologies, often prompted by economic and social pressures. Regardless of their impetus, *Automotive Milestones* covers them in an engaging and readable manner.

John Perodeau

Acknowledgments

The author would like to thank the people who reviewed the book and the organizations who helped with its research. Several individuals were kind enough to read and review many draft chapters and provide useful feedback. The Revs Institute of Naples, FL, very generously gave the author full access to their extensive library of automotive literature spanning the late 19th century to the present, and allowed the author to photograph their Collier Collection of automobiles. The Revs Institute's Managing Librarian, Trina Purcell, helped the author find many obscure references, and their Volunteer Coordinator, Susan Kuehne, supported the author's efforts. The research librarians of Worcester Polytechnic Institute (WPI) also found many publications needed by the author.

John Perodeau of the Tampa Bay Automobile Museum spent copious time with the author showing the cars in their collection and describing them in detail. He reviewed every chapter in draft form and provided valuable feedback on each. He also kindly agreed to write the Foreword for the book and provided some of the photos of cars in their collection. Gregory Aviza, a mechanical engineer and graduate of WPI with extensive experience restoring automobiles, also reviewed every chapter and gave valuable feedback. Dr. Robert Dyke of the Steam Car Club of Great Britain had helpful comments on the steam section of Chapter 1. Other reviewers who offered useful comments were:

- Dr. William Crochetiere, Professor Emeritus, Tufts University
- Richard Curley of Barnes and Noble
- Roland Gaucher, model maker
- Jack Hall, Professor, WPI
- Christopher Morss, retired English teacher and car collector

The author also visited a number of museums in addition to those mentioned above to photograph their collections and talk with some of their curators. These included:

- The AACA (Antique Automobile Club of America) Museum, Hershey, PA
- The Bahre Collection, Paris, ME
- Boothbay Railroad Village Museum, Boothbay, ME
- The Bristol Museum, Bristol, England
- Carolinas Aviation Museum, Charlotte, NC
- Deutsches-Museum, Munich, Germany
- Florida Air Museum, Lakeland, FL
- Fort Lauderdale Antique Car Museum, Fort Lauderdale, FL
- Henry Ford Museum, Dearborn, MI
- The Heritage Museum, Sandwich, MA
- Larz Anderson Auto Museum, Brookline, MA
- Musée National de l'Automobile de Mulhouse, France
- Owls Head Transportation Museum, Owls Head, ME
- Pensacola Air Museum, Pensacola, FL
- The Ralph Lauren Collection, at the Museum of Fine Arts, Boston, MA
- Sarasota Classic Car Museum, Sarasota, FL
- Seal Cove Auto Museum, Seal Cove, ME

About the Author

Robert L. Norton, M.S., D.Eng. (h.c.), P.E., taught mechanical engineering for 45 years, first at Northeastern University, followed by Tufts University, and then for 31 years at Worcester Polytechnic Institute, where he is now the Milton P. Higgins II Distinguished Professor Emeritus. His teaching career was devoted to the machine design and automotive aspects of mechanical engineering. He is a licensed Professional Engineer in Florida and Massachusetts, has been president of Norton Associates Engineering LLC, Engineering Consultants for over 45 years, and holds 13 patents. In that role, he has designed and analyzed several engines and many valve trains, as well as machinery of all types for companies around the world. His linkage- and cam-design software programs are used by many companies worldwide.

He has published five college-level engineering textbooks on Kinematics, Dynamics of Machinery, Cam Design, and Machine Design topics. These books have been translated into six languages and are used worldwide. He has also published dozens of technical papers in journals of the American Society of Mechanical Engineers (ASME), Society of Automotive Engineers (SAE), and elsewhere. He is a Life Fellow of the ASME and was a member of SAE for many years.

This book is the result of a lifelong interest in all things automotive. Norton is a "certified car nut." As a teen in the 1950s, he built and raced hot rods. Later, he restored a number of antique cars and is still an avid collector of automobiles and motorcycles, especially ones that go fast.

A Caveat

The world of automobiles is very broad, and we must limit its scope to keep this book to a reasonable number of pages. Therefore, we have had to establish some criteria as to which makes and models to cover and which to ignore. So, we have focused our discussion on automobiles that actually entered production and sold a significant number to the public.

This book thus excludes all prototypes, experimental automobiles, and show cars made as "one-offs." It also excludes some cars for which a serious effort was made to produce in quantity, but for various reasons did not succeed. We realize this eliminates some very interesting automobiles such as the Tucker, Buckminster Fuller's Dymaxion, the DeLorean, and the Chrysler Turbine car.

Moreover, our focus is on automotive technology rather than on makes and models of automobiles. We have selected what we believe to be the most significant examples of automobiles and automotive systems that have pushed the state of the art forward. Then we have attempted to explain how these systems work and how they have improved the breed.

If any reader is disappointed because we did not mention his or her "favorite car," then we apologize in advance. Nevertheless, we hope that all readers find the collection of automobiles and technology presented to be interesting and informative. Your comments and feedback are most welcome and can be directed to **norton@designofmachinery.com**.

Introduction

This is a book about all things automotive and how they work. The root of the word automotive is "self-moving" or self-propelled. That covers every non-living thing that moves under its own power. The word automobile was coined in 1897. We will focus on the technology of the automobile in particular, but sometimes will stray into related areas such as aircraft or locomotives. The explanations attempt to be technically correct without resorting to the often complicated mathematics that underlie them. We will address the major systems and subsystems that make up an automobile or light truck, which are very similar, and **explain how they work**. Combustion cycles, chassis layouts, engine types and designs, valve trains, transmissions, drive lines, suspension, steering, brakes, and body design—all will be discussed in some detail. We must necessarily use a number of technical terms in these discussions, some of which may not be familiar to you. We provide a glossary with definitions of these terms, and on first use, a word that is in the glossary is shown in **bold** and in **color**.

Sidebars
When a technical topic needs more detailed explanation, it will be discussed in a sidebar with a colored background. These can be skipped if the reader already knows the information or is not really interested in this level of detail. Some simple mathematics may sneak into a sidebar from time to time.

We start with the earliest vehicles. The saying, *There is nothing new under the sun* is quite applicable to automotive technology. Even now, well over a century since the first self-propelled vehicles appeared, inventors are proposing novel engine designs. If one checks, one can usually find that these novel ideas (or ones very similar) were proposed in the early 20th century, tried and abandoned. Nevertheless, just because an idea was abandoned so long ago, does not mean it is a bad idea today. Technology has advanced so much in that time, particularly in respect to material science and digital electronics, that something which could not be made to work back then, now can be. Several companies are actively working to perfect variants on existing engines or totally new engines to achieve the holy grails of better economy and better power-to-weight ratio. Venture capitalists and the U.S. government have put multiple millions of dollars on the table, betting that these new ideas will work.

An example: electric cars, now all the rage, were first invented over a century ago but were abandoned because battery technology of the day could not provide sufficient range. Now, with new types of batteries available, you can buy a new, pure-electric car (not a hybrid) capable of traveling more than 250 miles on a single charge (the Tesla S). Hybrid automobiles, which typically combine gasoline engines and electric motors, reappeared on the scene quite recently (e.g., Prius: 1997 in Japan, 2000 in the U.S.), and all are variants of a transportation system that has been around a long time. The diesel-electric locomotive uses diesel engines turning generators to supply electric current to electric traction motors at the wheels. WWI submarines were diesel-electric. On the surface, diesels charged batteries used for electric propulsion when submerged. Modern luxury liners also use diesel-engine generators to drive electric motors on the propeller shafts. Electric motors offer speed control and quick reversing. The generators also power "the hotel," as the ship's engineers call the systems used by the passengers. And, Ferdinand Porsche helped develop a hybrid automobile in 1898 that used a 2-cylinder gasoline engine to drive a generator that powered electric motors mounted in the wheel hubs. *Nothing new under the sun, indeed*! The history of the technological development of the many systems that comprise a modern automobile and how they work is quite interesting. Let's explore it.

Chapter 1
Motive Power

Something has to make our self-propelled vehicle go. We call that something *motive power*. It comes in many forms, among which are **external combustion**, **internal combustion**, **electric**, and **hybrid** systems. One of the most difficult tasks that automotive pioneers and inventors faced in the late 19th century was finding a practical means to power a vehicle. They experimented with and developed working examples of all the systems listed above before the 20th century arrived.

Steam power was already well established, first as a means to pump water from mines and later to power railway locomotives, run mills and factories, and perform other tasks. These applications, with the exception of the locomotive, were stationary and so their engines could be large and heavy with no penalty. Even locomotive steam engines were large and heavy, but the size and weight of a locomotive could accommodate this. Developing a practical power plant for a small, light vehicle like an automobile was their goal. They ultimately achieved it and their work provided the foundation for improvements in engine technology that continue to this day.

There were many proponents of steam power and electric propulsion for automobiles. Steam and electric vehicles dominated the U.S. market into the start of the first decade of the 20th century. Many inventors were determined to replace steam and electricity in automobiles with an entirely new technology—the internal combustion engine—and they eventually succeeded.

This chapter describes how these various *motive power* methods were developed and applied as the automotive century began. To understand where these pioneers were in a technological sense, we need to first understand the technology and technical environment in which they found themselves in the late 19th century. We start by looking back at what had transpired in the century leading up to the automobile revolution.

The opening photograph is of a steam locomotive in The Deutsches-Museum, Munich, Gerrmany, taken by the author.

External Combustion Engines (ECE)

By external combustion is meant that the fuel is burned outside the engine. Internal combustion engines (**ICE**) such as gasoline and diesel engines burn their fuel inside the cylinders. **ECE** fuel can be anything that is combustible: wood, coal, natural gas, propane, gasoline, kerosene, diesel, or even peat or dung. The heat produced is used to raise the temperature of a working **fluid**: water, oil, air, or any fluid—either liquid or gas. Water heated to steam is the most common working fluid and powers steam engines and steam turbines. Air also can be the working fluid as it is in a Stirling engine, described in a later section.

Steam

The power of steam was explored as early as the 1st century AD when Hero of Alexandria devised the *aeolipile*, which was simply a rotating ball spun by steam jets, in essence a primitive steam **turbine**. But the practical application of steam power had to wait until the early 1700s when Newcomen developed the first practical steam-driven piston engine, used to pump water from coal mines. James Watt made several improvements to Newcomen's design during the second half of the 18th century, culminating with the first Watt engine, which used a *sun and planet drive* (see Figure 1-1) to convert the reciprocating motion of the piston into rotation of a **flywheel**. Watt had to invent this planetary gear train to get around James Pinkney's 1780 patent of the **crankshaft** and **connecting rod**.

No one in the 18th century really understood how steam worked, as the *laws of thermodynamics* were not discovered until the 19th century. As a result, these early steam engines were very inefficient and had to be quite large to get enough power to perform a task like pumping water out of a deep mine. Large engines required large **bores** and long **strokes** (both measured in feet) to get the needed **displacement** (the volume displaced by the piston's bore during its stroke). (See Figure 1-3.) Watt needed a way to guide the long straight-line motion of the piston. But the machinery needed to cut a

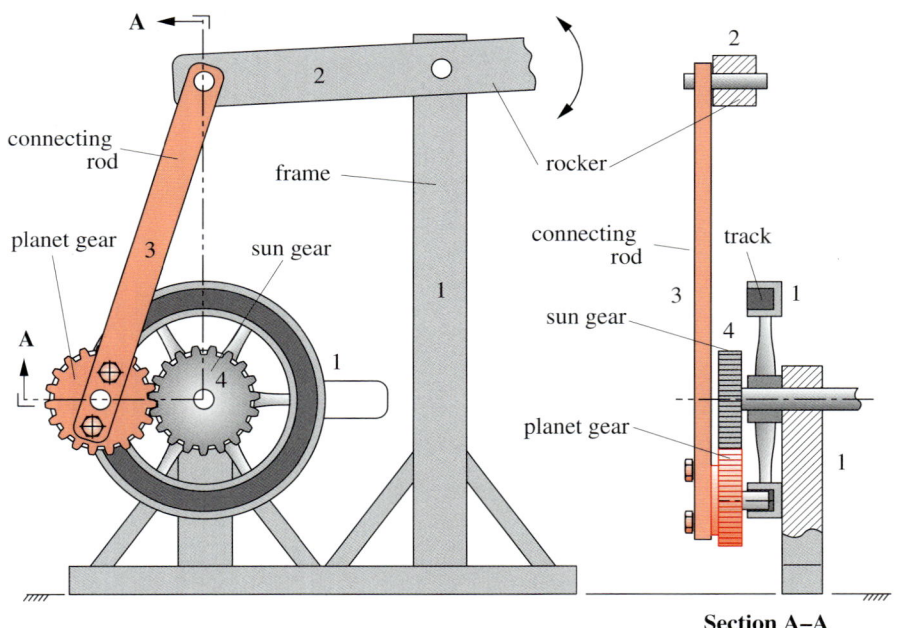

FIGURE 1-1 James Watt's sun and planet drive

CHAPTER 1 MOTIVE POWER

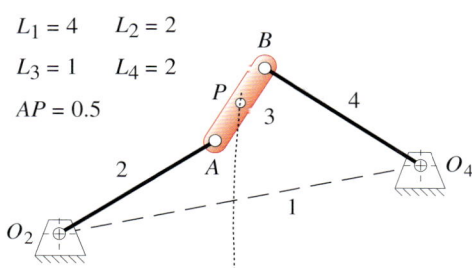

FIGURE 1-2 Watt's straight-line linkage

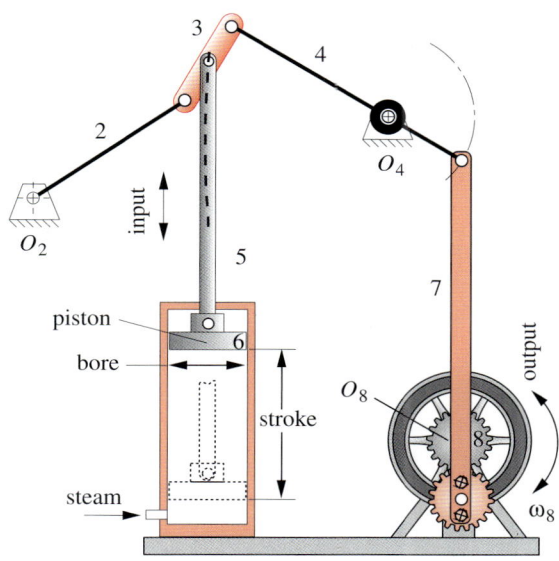

FIGURE 1-3 Watt's steam engine

long straight guide in metal (the planer) had not been invented yet. So Watt invented his straight-line linkage (see Figure 1-2), which uses simple rotating links, is easily made, and generates a nearly straight-line motion at one point (P) on link 3. What we now call *straight-line motion* was then called *parallel motion* by Watt and his contemporaries.

Watt is considered by many to be the *father of the steam engine*. Nevertheless, when his son interviewed him late in life while writing a biography of his father, the son quotes him as saying, *Though I am not over anxious after fame, yet I am more proud of the parallel motion than of any other mechanical invention I have made.*[1] This from the father of the steam engine! Watt's *parallel motion* linkage is shown in Figure 1-2. This linkage is used to guide the suspension motion of rear axles on many vehicles today. Watt's engine with his *parallel motion* and *sun and planet drive* are shown in Figure 1-3.

Steam Engine Terminology

The most common type of steam engine used in early automobiles was the piston steam engine. These reached a very high state of development both in automotive applications and in railroad locomotives, where they were first used extensively. But, the earliest mobile use of a steam engine was in a road-going vehicle by Cugnot as described below. A steam engine of this type has a **cylinder**, **piston**, connecting rod, and crank that are essentially similar to those of an internal combustion engine. Figure 1-4 shows a schematic of a locomotive steam engine and **linkwork**. The piston drives a **crosshead** that

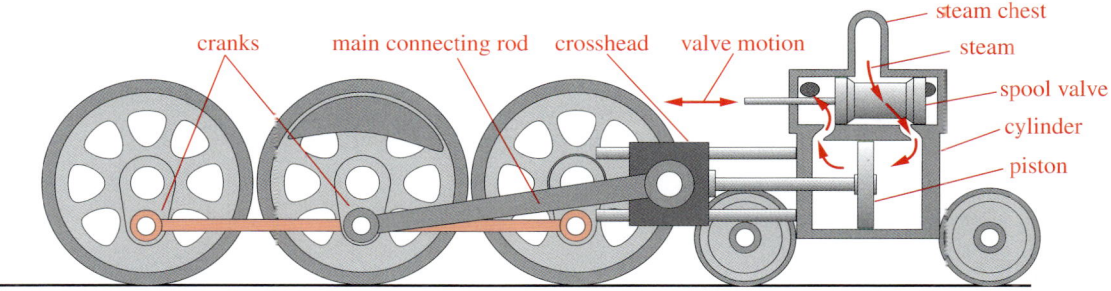

FIGURE 1-4 Locomotive steam engine, linkwork, and valve (valve mechanism not shown)

---

[1] Quoted in Muirhead, J. P. (1854). *The Origin and Progress of the Mechanical Inventions of James Watt*, Vol. 3, London, p.89.

3

translates on guides. The main connecting rod is pivoted to the crosshead at one end and to the main drive crank on the center driving wheel at its other end. Additional connecting rods take the motion to the other drive wheels. It also has a **valve** or valves that are actuated from the crank and piston motion through linkwork.

Steam at high pressure enters through the **intake port**, pushes the piston through its **stroke**, expanding the steam to a lower pressure and larger volume. **As it expands, it gives up most of its energy to drive the piston**. It exits the **exhaust port** and is either dumped to atmosphere (referred to as an **open-circuit system**), or collected to be fed to additional, lower-pressure cylinders, and/or routed to a **condenser** where it cools to liquid water. The water is reheated to steam to continue the cycle. This is called a **closed-circuit system**.

A **single-acting** engine is one in which the steam is introduced to only one side of the piston. This has one power stroke per revolution. A **double-acting** engine introduces steam to both sides of the piston in turn to give two power strokes per revolution, driving the piston both left and right as shown in Figure 1-4.

A **simple expansion** engine uses the steam for only one expansion in a cylinder and then either dumps it to atmosphere (very wasteful) or returns it to a condenser to be reheated. Watt's main contribution was to add a condenser to its engine. A **compound expansion** engine expands the steam multiple times in different cylinders linked in series. The first cylinder, or high-pressure stage, extracts the bulk of the **energy** by expanding against a relatively small diameter piston. The steam exhausted from the first stage is still above atmospheric pressure and is directed to another cylinder of larger diameter to drive a second piston and extract more energy from the steam. Both cylinders drive the same crankshaft and can act on cranks that are **out of phase**, meaning the pistons do not all reach top dead center simultaneously. An engine with two cylinders in series is called a **double-expansion** engine; with more than two cylinders, a **multiple-expansion** engine. Ideally, the steam should be expanded to be close to atmospheric pressure in order to extract most of its energy before returning to the condenser.

Multiple cylinders (or multiple sets of multiple-expansion cylinders) arranged in parallel are also used, as is common in **IC** engines. Then, the multiple cylinders are typically arranged with cranks that are out of phase. This has the advantage of carrying the crankshaft through the dead points of top and bottom dead center (**TDC** and **BDC**). If a single-cylinder piston steam engine (or IC engine) happens to stop at either top or bottom dead center, it will not be able to restart unless it is moved off dead center by some external means. That is one reason you have to spin a lawn-mower engine (or any IC engine) to start it. Once spinning, the **flywheel** carries it through the dead-center positions. Steam locomotives have cylinders on each side of the locomotive, arranged with their cranks 90° out of **phase**. If it happens to stop in a station with one of the pistons at TDC or BDC, the piston on the other side will get it moving.

However, one major difference between a steam engine and an IC engine is that the former has maximum piston **force** (and thus also some **torque** depending on the crank angle) available at **stall,** or zero speed. The IC engine has zero torque available at stall. It must be kept running at idle speed to be ready to do **work**. When steam is introduced to the ECE cylinder at stall, it immediately begins driving the wheels with maximum torque as long as the crank is not at TDC or BDC. An IC engine, on the other hand, has the least torque and **power** available at idle speed. Its torque typically peaks at some speed between idle and its **red-line speed**. The steam engine then needs no clutch, **fluid coupling**, or transmission and is thus ideal for starting and driving large loads. An electric or **hydraulic motor** also develops maximum torque at stall and so is also well suited to driving large loads, as in the diesel-electric locomotive and heavy trucks such as those used in open-pit mines. Both of those vehicles are driven by electric motors at the wheels with diesel engines providing the electricity through generators.

Steam-Powered Machines

This section will focus on steam-powered machinery intended to operate on roads as opposed to rails. The history of this type of vehicle is older than that of steam-powered railroad locomotives. William Murdoch is generally credited with building the first prototype steam locomotive in 1784 in Scotland. It was not until 1804 that the first working steam locomotive designed by Richard Trethiwick hauled a train of wagons loaded with 10 tons of iron in Wales. But even earlier, an intrepid French engineer and inventor named Nicolas Cugnot built the first road-going self-propelled vehicle. It was steam-powered.

Cugnot (1769)

As is the case with much of technology, this machine was invented to serve a military need. Monsieur Cugnot designed a vehicle in 1769 to haul heavy artillery for the French army. It was built at the Paris arsenal. It had three wheels and could carry four passengers at a maximum speed of 2.5 mph. It was an ungainly contraption as can be seen in Figure 1-5. Its wood-fired boiler was cantilevered off the front, and it was an open-circuit system, which dumped exhaust steam to the atmosphere. Consequently, it had to stop every ten or fifteen minutes to allow the boiler to build up more steam.

Motion was transmitted to the drive wheel by a **ratchet and pawl** mechanism, shown schematically in Figure 1-6. In the Cugnot, instead of a rotating arm driving the pawl, the translating piston motion drove it. Cugnot is considered by some to have invented the first

FIGURE 1-5   The *Fardier de Cugnot* of 1769—the first self-propelled road vehicle (*Courtesy Tampa Bay Automobile Museum*)

AUTOMOTIVE MILESTONES

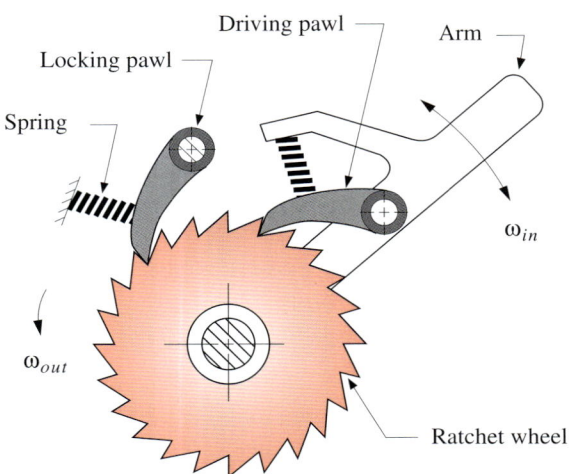

FIGURE 1-6   Ratchet and pawl mechanism

automobile and it contains some technology that is now quite current—it had front-wheel drive, did not use fossil fuel, and had **rack and pinion steering**. It was reversible in the same way a common **ratchet wrench** is. The pawl can be flipped over to drive the ratchet wheel in the opposite direction. The image in Figure 1-5 is of a working replica that resides in the *Tampa Bay Automobile Museum* in Florida. The original *Fardier* (wheeled cart) *De Cugnot* has been in the collection of *Le Conservatoire Des Arts et Metiers*, Paris, France since 1801. A video of the Tampa Museum's replica in operation can be seen at http://www.polypack.com/videos/tbautocugnot.wmv. The ratchet and pawl action can be clearly seen in this film.

Cugnot was the first to use steam pressure to drive the piston. Newcomen's engine was called an "**atmospheric engine**" because the steam was introduced to the cylinder below the piston when the piston was at the top as shown in Figure 1-3. Then the cylinder was cooled to make the steam condense. The cylinder was open to atmosphere at the top. The resulting vacuum allowed atmospheric pressure to push the piston down.

This is very inefficient as the largest pressure differential that can be achieved this way is 14.7 psi. Cugnot put steam pressure directly on the piston. Watt also put low-pressure steam on the piston and added a condenser to improve efficiency of the Newcomen engine.

Early Steam Vehicles in Europe

England and France had a head start on technology; their industrial revolution started almost a century earlier than in the U.S. As a result, steam power developed there before it did in the United States. The Cugnot effort has already been described.

In England, Richard Trevithick was servicing Watt engines in the mines and saw opportunities for their improvement. Around the same time as Oliver Evans (see next section), he began work on a high-pressure steam engine. He built one in 1800 but did not patent it until 1802. (Evans patented his in 1790 but did not build it until 1801.) They apparently knew nothing of each other's work.

Trevithick nevertheless gets credit for first using a steam engine in a vehicle intended for passenger transport. After an experiment in 1801 in which he built a "steam road locomotive" called the "Puffing Devil" that successfully climbed "Camborne Hill" with six passengers aboard and inspired a song, he built his London Carriage intended to carry passengers. He also is credited with the first steam railroad locomotive, which was reported in a previous section.

In 1825, Sir Goldsworthy Gurney built a steam-powered tractor that pulled carriages with paying passengers at lower fares than horse-drawn carriages. In 1829, Walter Hancock built a ten-passenger, steam-powered bus that later carried passengers between London and Brighton.

In France, Bollee began making steam carriages in 1873. Leon Serpollet built a steam car

CHAPTER 1   MOTIVE POWER

with a flash boiler in 1894, five years before White in the U.S. They also broke the world land speed record in 1902 with a 75.06 mph run. The Stanley brothers would smash this record in 1906.

Oliver Evans' Oruktor Amphibolos

Oliver Evans was a prolific American inventor and engineer who lived between 1755 and 1819. He invented and built the first automated mills for production of flour and became wealthy in the process. He then turned his attention to steam and steam engines and patented the first high-pressure steam engine in the U.S. in 1790 and built it in 1801. There is no indication that he knew of Cugnot's or of Trevithick's work.

He used high pressure in the cylinder and designed an engine that was smaller and more efficient than Watt's. His steam engines were successful as stationary power plants for mills and factories, and his company had great success in manufacturing and selling them.

Evans had been talking of a steam-powered wagon since 1789 and finally built one in 1805, called the *Oruktor Amphibolos* or *Amphibious Digger*. This vehicle was distinctive for two reasons: it was the first self-propelled vehicle in the U.S. and also the first **amphibious** vehicle in the world.

The Philadelphia Board of Health wanted to dredge the river around the dockyards and hired Evans to design a device to do so. He built a steam-powered, flat-bottomed **scow** with a paddle wheel for water travel and four wheels to get it to the water. The steam engine powered both wheels and paddle. It was completed and driven through the city to the water. But it was not very successful in dredging the bottom. Most considered the overall effort a failure, but the vehicle did move over land under its own power. It was eventually scrapped.

Stanley Steamer (1897–1924)

When most people think about steam automobiles, they think of the Stanley Steamer. The Stanley brothers produced their first Stanley Steamer in 1897. It had two simple-expansion, open-circuit, double-acting cylinders with **slide valves**. The engine drove the rear **differential** via a chain. There was no **clutch**, **driveshaft**, or **transmission**. They were not needed because the steam engine could be stopped completely and restarted immediately as long as steam pressure was available. Early models burned vaporized gasoline. Later models used kerosene. From 1898–1899 they produced and sold more than 200 cars, exceeding the production of any other car maker in the U.S. in that year. They sold the rights to their design to Locomobile in 1899 and worked for that company until 1902, when they formed a new company to build an improved steamer. Then they had to avoid their own patents, which Locomobile still owned.

Steam was a significant mode of automotive power in the U.S. during the early 1900s. In 1902, 485 of 909 new car registrations in the U.S. were steamers. The early steamers were more powerful than most of their ICE contemporaries, especially in the 19th century and the first decade of the 20th.

The new Stanley Steamer had two cylinders geared to the rear axle rather than using a chain drive. A Stanley Steamer set the world speed record in 1906, traversing a mile in 28.2 seconds for an average speed of 127.7 mph. The Royal Automobile Club (RAC) of England awarded Stanley the 1906 Dewar Trophy, Europe's most prestigious automotive award. This automobile speed record stood until 1911.

Stanley added a condenser in 1915 making it a **closed-circuit system**. Steamers were quiet-running in contrast to the noisy internal combustion engine. Their only sound was a "whoosh" of the steam exhaust in an open-circuit engine. Those with a condenser were essentially silent. Figure 1-7 shows a Stanley Steamer, sometimes called the "coffin nose" for its front-mounted boiler enclosure. Another sobriquet for the Stanley was "The Flying Teapot."

Stanleys had a reputation as fast cars. They were relatively light and the steam engine's maximum torque available from a dead stop made them "quick off the line." That is, they were quick after you got them started and built up a head of steam in the boiler. Starting a steamer involved quite a ritual. The boiler had to be filled and several valves had to be opened and closed in the correct sequence. A pilot light had to be lit, sometimes resulting in singed eyebrows when the flame occasionally "blew back." Then you had to wait for the burner to heat 15 gallons of water in the boiler to make steam.

Your teapot probably takes several minutes to boil a quart of water on the stove. It took a while to boil 60 quarts in the Stanley. It typically took 15-20 minutes for a Stanley to be ready to roll. Once under way, it used the steam faster than the boiler could replenish it, which resulted in running out of steam after some miles, and a wait by the side of the road while it built up steam again. It used about 2 to 4 quarts of water per mile. If the water tank ran dry, which it did in the early open-circuit cars, the driver hoped for a nearby stream to supply the water to recharge it. Steam power was not for the faint of heart. It helped to have a spirit of adventure.

Locomobile Steamer (1899–1903)

John B. Walker of Locomobile bought the plans for Stanley's automobile in 1899 for $250,000—a large fortune then—about $7 million in 2015 dollars. The Stanley twins became his general managers until they left to form a new company in 1902 and compete with their former employer. Locomobile built over 4000 steamers between 1899 and 1902, but ceased production of steam cars in 1904, switching to internal combustion engines. The Stanleys then bought the rights to their technology back for $20,000, or about eight cents on the dollar compared to what Walker had paid them five years before.

During its production run, the Locomobile offered several body styles, some of which were small and affordable. Figure 1-8 shows a 1900 Locomobile steamer. Prices ranged from $600 to $1400. They used a two-cylinder engine similar to the Stanley. Their most significant improvement was to add a flash boiler (after the

FIGURE 1-7 A "Coffin Nose" Stanley Steamer in the Heritage Museum, Sandwich, MA, and a Stanley engine in the Seal Cove, ME, Auto Museum
(Photos by the author)

CHAPTER 1   MOTIVE POWER

steam generator
water tank

two-cylinders
connecting rod
crank

FIGURE 1-8   1900 Locomobile Steamer, steam generator, and engine in the Seal Cove Auto Museum (*Photos by the author*)

White design—see below), which heated small quantities of water quickly in tubes within the boiler, now properly called a steam generator. This technology had long been used in steam locomotives. This fixed the biggest complaint of owners, the need to wait many minutes after starting to allow the boiler to come to temperature and build steam pressure.

In the U.S., steam cars and electric cars outsold ICE-powered cars prior to Cadillac's introduction of the ICE self-starter in 1912. Before that, IC engines had to be hand-cranked to start. If not properly done, this could (and often did) result in a broken arm, thumb, or wrist. Despite the ritual of valve manipulation, lighting of pilots, and the wait for steam to build up, a steam car started relatively easily, and an electric car only required a switch closure to be on your way.

A Locomobile steamer became the first automobile used in war. The British used it as a tractor and catering vehicle in the Boer War. The British soldiers appreciated having hot water always on tap from the boiler at tea time.

White Steamer (1901–1911)

The owner of the White Sewing Machine Company bought an early Locomobile steamer and found its boiler unreliable. His son, Rolin, improved the boiler with a water-tube, semi-flash design that generated higher-temperature, superheated (dry) steam (800°F) at 600 psi that had higher internal energy than saturated steam (212°F), which could have condensed water droplets in it. The coiled-tube design of the boiler allowed faster steam generation. He patented his boiler in 1900 and persuaded his father to fund the development of an improved steam car, the White. The first cars were built in the sewing-machine factory, but in 1905, the automobile business moved to its own factory.

White built the first automotive condenser system in the U.S., increasing range to as much as 150 miles on a tank of water, and were the first to use a compound-expansion engine in a steam automobile. They also added a two-speed transmission, the only steam car with one. It did not have a clutch. Unlike the Stanley, the engine could be run with the car stationary by putting the transmission in neutral. White steamers were considered quite reliable and were successful in endurance races. White made "Whistling Billy," an open-wheel racer that broke a world speed record in 1905.[1] In 1906, there were more White automobiles in use in the U.S. than any other make. In an interesting video at http://www.youtube.com/watch?v=Lf8miprLH60, Jay Leno demonstrates one of the White steamers in his collection. It takes him about ten minutes to get it rolling from a cold start.

A White Steamer was one of the first automobiles to be put into service by a U.S. President. Theodore Roosevelt bought a 1907 model for the Secret Service to use. The Secret Service followed TR's horse-drawn carriage in it. (TR's successor, William Howard Taft also bought a White—and rode in it.[2]) White built its last steamer in 1911 and switched to gasoline-powered IC engines. In all, around 10,000 White steamers were made, more than Stanley's total output. They were used by the U.S. Army in WWI. Automobile production ended in 1918 but White continued to produce trucks and buses until 1980 when they went bankrupt. Figure 1-9 shows a White steamer and engine.

Doble Steamer (1909–1931)

The Doble is considered to represent the pinnacle of steam car development. Early models used a two-cylinder, double-acting engine. Later ones used a 4-cylinder compound engine. All had condensers. Abner Doble and his three brothers built their first steam car when Abner was still in high school. Abner was a talented engineer and a bit of a perfectionist. He could not stop tinkering with his design, with the result that each new example had small improvements added. Some said that no two Dobles were exactly the same. They only built about 36 cars over the life of the company—a bit over one a year! It is estimated that each car cost about $55,000 to develop. But they sold only for about $20,000, which would be $250,000 in today's dollars. A Model T Ford in 1927 cost $260. Had the Doble brothers not

FIGURE 1-9 A 1910 White Steamer in the Seal Cove Auto Museum (*Photo by the author*) and a 1902 White engine (*Courtesy of Dr. Robert Dyke of the Steam Car Club of Great Britain*)

[1] A YouTube video of Dr. Robert Dyke driving his replica can be seen at http://www.youtube.com/watch?v=0y5uvLfziSM.
[2] President William Howard Taft's White steamer is on display at the Heritage Museum in Sandwich, MA.

FIGURE 1-10 1923 Doble steam car in the Henry Ford Museum *(Photo by the author)*

been the sons of a millionaire, their company would have lasted a much shorter time.

Doble solved the problem of slow starting by automating the processes of valve manipulation and burner ignition (with a spark plug) and by using a flash-tube steam generator that heated 2 quarts of water at a time. The same concept for heating water rapidly is used today in so-called tankless hot water heaters for home use. Instead of keeping a tank of water always hot, the tank-less type has a coil into which cold tap water is introduced and a very high-output heater that can bring the water to the desired temperature before it leaves the coil on the way to the faucet. This is more energy efficient than keeping a tank of water hot. The Doble's burner could put out 2-million **BTU** of heat, creating temperatures as high as 2500°F. A Doble steam car started in 90 seconds after turning a key from the driver's seat to ignite the burner and was ready to roll in 2 minutes. It could go 300 or more miles on one tank of water.

Doble used steam superheated to 800°F and 1200 psi. No Doble boilers ever exploded. Oil was automatically injected into the steam to lubricate the cylinders. The engine developed 2200 ft-lb of torque at stall and delivered it directly to the rear axle. For comparison, a 2015 *C7 Corvette Sting Ray* automobile generates 465 lb-ft of torque at 4700 rpm and significantly less at idle speed. The Corvette engine turns about 1100 rpm at 60 mph. The Doble turned only 900 rpm at 60 mph and reached 60 mph in 15 seconds. For comparison, a Ford Model T of the same vintage needed 40 seconds to reach 50 mph—its top speed. (The Corvette needs less than 4 seconds to get to 60.)

The Doble was a very expensive car. Like Duesenbergs, Rolls-Royces, and other luxury cars, they were sold as a chassis only. The owner then sent it to his favorite body maker to build a custom body to his specifications. The later Doble cars were considered very reliable. Some were said to have gone several hundred thousand miles before requiring major mechanical service. No IC-engined cars of the era could make that claim. Doble went out of business in 1931. Figure 1-10 shows a Doble steam car, and Jay Leno has an interesting video demonstration of his 1925 Doble at http://www.youtube.com/watch?v=ACO-HXvrRz8.

Besler Steam Machines

The Besler brothers bought the Doble factory and patents in the early 1930s. Abner Doble continued to work with the Beslers, who attempted to apply steam engine technology to other types of vehicles. They modified a Travel Air 2000 biplane, substituting a 90-deg V-twin steam engine for its gasoline engine. It flew successfully in 1933.

Henry J. Kaiser, who was manufacturing Kaiser and Frazer automobiles in the 1950s, had Besler convert a 1953 Kaiser Manhattan to steam. It never went into production with that engine, however. In 1969 GM made two experimental Besler steam engine versions—in a Chevrolet Chevelle and a Pontiac Grand Prix.

Steam car production did not survive the Great Depression and had plenty of company in that regard. Many automotive manufacturers went bankrupt during that time. To this author's knowledge, there are no steam cars in production in 2015.

Stirling Engine

Robert Stirling invented the engine that now bears his name in Scotland in 1816. It is an external combustion engine that uses air or other gas such as helium as the working fluid. It is a closed-circuit engine, meaning that the working fluid is sealed within the engine system and is not exchanged with the atmosphere (analogous to a steam engine with a condenser). There are several mechanical implementations of the Stirling engine, called *alpha*, *beta*, and *gamma*. We will discuss only the *alpha* type, which was the original arrangement of the invention. It is of a class called hot-air engines, but is unique among the class in that it contains a **regenerator** that increases its efficiency over that of other hot-air engines.

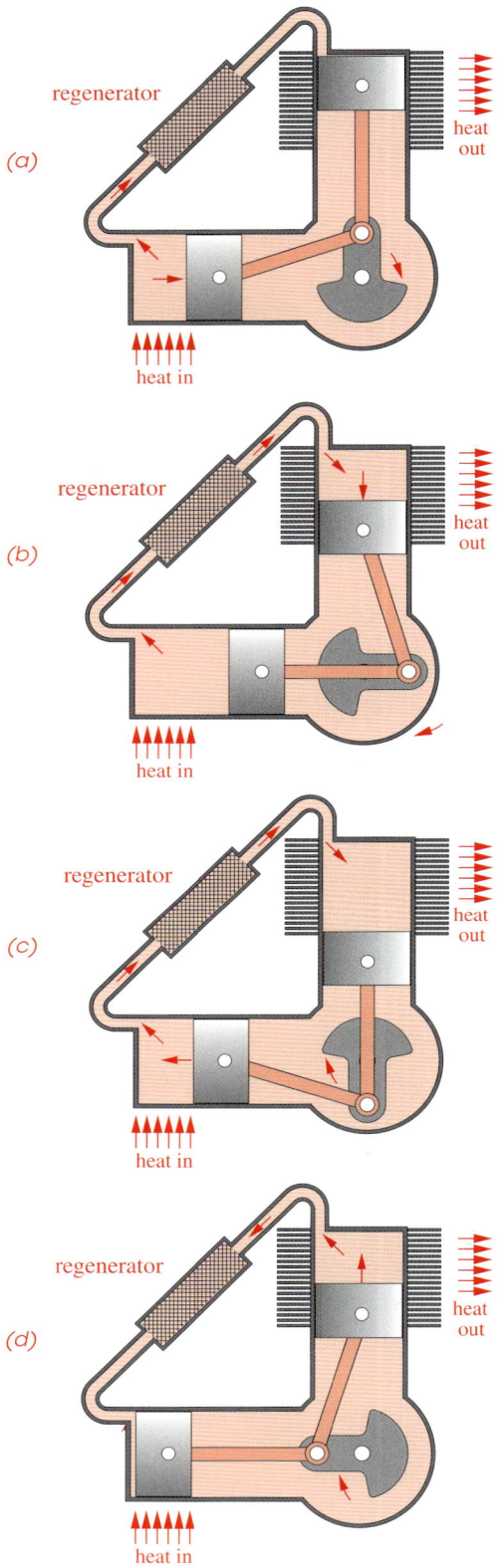

FIGURE 1-11 The Stirling Cycle Engine

CHAPTER 1 MOTIVE POWER

Its working principle is quite simple, as shown in Figure 1-11. There are two pistons connected to a common crankshaft arranged so the piston motions are 90° out of phase. One piston is in the "hot cylinder" and the other runs in the "cold cylinder." The heat source is applied to the hot end of the engine. It can use any fuel as a heat source and can even operate without fuel if solar energy is directed at its hot end. The cold cylinder usually has fins or other means to dissipate heat to the atmosphere or surroundings.

Its operation is based on the so-called ideal gas laws that express the relationship between pressure, temperature, and volume of a gas. When a gas is heated, its pressure and/or volume increase and vice versa. At a constant temperature, volume increases with a decrease in pressure and vice versa.

The four panels of Figure 1-11 depict Stirling engine operation. The hot cylinder is horizontal and the cold cylinder is vertical in the diagrams. The crankshaft turns clockwise. At (a), the hot cylinder is about halfway through its descent and the cold cylinder is at TDC, having emptied its cooled air through the regenerator between the cylinders and into the hot cylinder. The incoming air is heated as the hot piston descends; increased pressure from the heat pushes the hot piston down and also makes the gas flow back through the regenerator to the cold cylinder, driving both pistons downward. At (b), the hot piston is at BDC about to ascend due to flywheel **momentum**, pushing the hot air back through the regenerator and into the cold cylinder, whose piston is descending. At (c), most of the gas is in the cold cylinder where it cools, its volume decreases and pulls both pistons outward toward TDC. At (d) the hot piston is about to descend as flywheel momentum carries the crank through 90 degrees, transferring the gas back to the hot cylinder to complete the cycle.

The regenerator contains material that absorbs and conducts heat readily and has large surface area. An aluminum mesh is sometimes used. When the heated gas passes through, some of its heat energy is stored in the regenerator, cooling the air before it reaches the cold piston. When the cooled gas come back through the regenerator, it picks up some of the stored heat. In effect the regenerator preheats and precools the gas, thus increasing engine efficiency.

Some advantages of this engine are that it is essentially silent running, reliable, and low maintenance. There are no valves. It can run on anything that will burn. Its main disadvantage is that it has a poor power to weight ratio. To get a similar level of power to that of an IC engine will require a Stirling engine of much larger size and weight than an ICE.

Stirling's original intent was to create a safer substitute to the steam engines of his day, whose boilers often exploded. Its first use was to pump water out of a quarry in 1818. A Stirling engine was also used to power a factory in Dundee in 1843. But that engine failed due to problems with seals on the hot piston. The materials of the time were not suitable for the high temperatures required. The Stirling's efficiency is proportional to the temperature differential between the cylinders, requiring very high temperatures on the hot side if the cold side is dumping its heat to atmosphere. Using cold water on the cool side can improve this, making it practical for some marine applications. The Swedish *Gotland Class* submarines use a Stirling engine for their power. Some U.S. submarines use small Stirling engines to drive pumps. Their main attraction in that application is their silence. Another commercial application is in generator sets used on yachts where silent running and cold water available for cooling are advantages.

U.S. automobile manufacturers have experimented with Stirling engines. American Motors put one into a Rambler many years ago, but no one has yet put one into production in an automobile. It has a similar problem to the steam engine in that it needs warmup time to start running. Unlike a steam engine, it cannot be stopped abruptly. Also, it is difficult to control its speed.

The Stirling cycle is reversible, meaning that it can be used for cooling as well. It functions as a **heat pump**. In this case, the crankshaft is driven by an **electric motor** or other means and the cold cylinder is used to reduce the temperature in a closed volume, while heating the separate space occupied by the hot cylinder. It essentially pumps the heat out of a container. It is a very effective cooler that is capable of reducing the temperature of a fluid to cryogenic levels (extremely low temperatures approaching absolute zero). It can create temperatures as low as −200°C (73°K), sufficiently low to liquefy air, oxygen, nitrogen, and argon. This is considered to be the only commercially successful application of Stirling engine technology and is economically competitive with alternative technologies for cryogenic applications.

An intriguing application of a Stirling engine was recently developed by Dean Kamen of DEKA Corp. to address a problem common in third-world countries—the lack of clean water. His company developed a small generator set powered by a Stirling engine with enough output to run a small, separate system that can produce distilled water from dirty water using the electricity generated by the Stirling engine. Each of these systems is about the size of a washing machine and can be operated anywhere. Kamen likes to point out that the system needs only two inputs: dung and dirty water—and outputs electricity and clean water.

Internal Combustion Engines (ICE)

Internal combustion engines (ICE) burn their fuel inside the cylinders. The fuel can be anything that is convertible to a flammable gas to be ignited by some means inside the cylinder after being mixed with air and compressed. Oxygen from the air as well as fuel is required to create the correct mixture for a controlled explosion in the cylinder. The expanding gases from the burning fuel drive the piston downward to turn the crank. Most ICEs use a slider-crank mechanism similar to ECE piston engines as shown schematically with its terminology in Figure 1-12.

Spark-ignition engines use a spark to ignite the compressed mixture of fuel and air at the right point in the cycle when the piston is ready to start its descent. Compression-ignition engines compress the mixture to such a degree that the heat of compression ignites the explosion. Spark-ignition engines are commonly called gasoline engines, though they can run on many fuels such as *natural gas, coal gas, propane, hydrogen, naphtha, benzene, ethanol, methanol*, and *gasoline*. These fuels all have high **volatility**, meaning that they readily mix with or vaporize in air. Compression ignition engines are commonly called diesel engines but

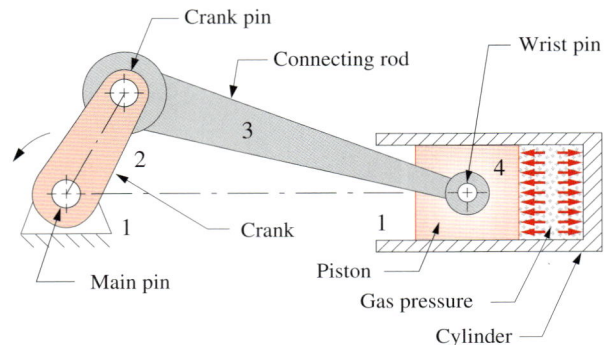

FIGURE 1-12 A slider-crank piston IC engine (schematic)

can also run on multiple fuels such as *vegetable oil*, *kerosene*, and *diesel fuel* (which is the same as home heating oil). These fuels are all much less volatile than those used in spark ignition engines and cannot reliably be fired with a spark. But their low volatility makes them suitable for the compression ignition cycle as they will not pre-ignite at lower pressures than are desired for efficient combustion.

The **compression ratio** (the ratio of the volume of air in the cylinder at BDC to its volume at TDC) is typically between about 8:1 and 10:1 in a spark ignition engine. A compression ignition engine can have a compression ratio of 14:1 to 23:1. If a volatile fuel is used in a compression ignition engine, there is a real danger that it will explode while the piston is ascending on the compression stroke destroying the piston and possibly the cylinder. Don't put gasoline into your diesel engine's fuel tank!

If the fuel for an IC engine is liquid at ambient temperatures and pressures, then some means is required to vaporize it to a gaseous form and mix it with air to ignite it. (See Sidebar *Fuel-Air Ratio*.) The original method for vaporizing volatile liquid fuels such as gasoline was a carburetor that meters the fuel into an air stream where it evaporates rapidly to gaseous form. Modern engines use fuel injection systems to squirt the right amount of liquid fuel directly into the cylinder after the piston has compressed the air in the cylinder to the desired pressure. A spark or the heat of compression ignites the mixture. Carburetors and fuel injection systems will be discussed in a later chapter. Of course if a gaseous fuel such as propane is used then it must simply be mixed with air in the proper proportion.

Stroke Cycles

The term stroke cycle defines the number of strokes of the piston required to complete all events in the cycle. A stroke is the movement of the piston from BDC to TDC or TDC to BDC through 180°. So there are two strokes per revolution of the crank, one up and one down. Several stroke cycles have been invented and each has its advantages and disadvantages. Most have taken the names of their inventors, and that is how we will refer to them here.

Otto Four-Stroke Cycle

Nikolaus Otto is commonly credited with being the inventor of the first practical internal combustion engine. Several inventors had worked on the concept earlier than Otto with some success. But Otto was the first of them to recognize that engine efficiency is a function of compression ratio. Before that discovery all, including Otto, were working on so-called atmospheric engines, which burned the fuel without compressing the fuel/air mixture. These large, stationary atmospheric IC engines could only make 1/2 to 3 HP from 18 liters of displacement and were as tall as 13 feet. Their efficiency was only a few percent. Otto's major contribution was to compress the mixture in the cylinder before igniting it and this significantly improved engine efficiency.

Fuel-air ratio	*There is an optimal ratio of fuel to air for most efficient operation. For gasoline, it is about 14.7:1 (14.7 grams of air to 1 gram of fuel) and this is called the "stoichiometric ratio." At this ratio, all the fuel is consumed in the reaction with none left over. With a carbureted engine, it is difficult to achieve this ratio accurately. But fuel injection systems provide a more accurate ratio, making them more efficient and less polluting. The stoichiometric ratios for other fuels are: natural gas: 17.2, propane: 15.5, ethanol: 9, methanol: 6.4, hydrogen: 34, and diesel: 14.6.*

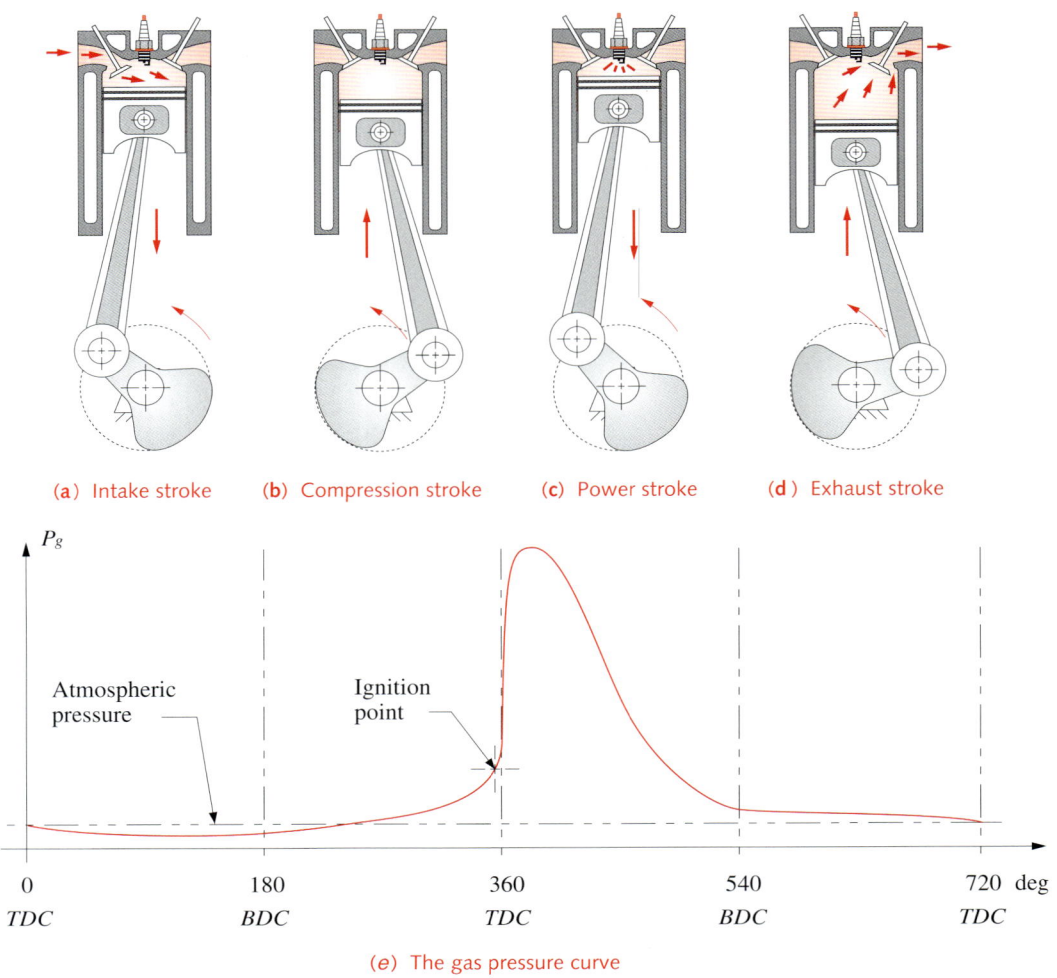

FIGURE 1-13 The Otto four-stroke cycle

Otto built his first engine in 1864, patented it in 1877, and formed a company to manufacture stationary engines. He was the first to define and use the four-stroke cycle, which is now the most common cycle used in modern IC piston engines. As its name implies, the Otto cycle requires four piston strokes or two crank revolutions to complete. The strokes are called *intake, compression, power,* and *exhaust*. In addition to the crank, connecting rod, and piston of Figure 1-12, this engine also requires at least two valves (one for intake and one for exhaust) and a means to open and close them at the right points in the cycle. Figure 1-13 shows a schematic of the four-stroke cycle.

For discussion, we can start the cycle at any point as it repeats every two crank revolutions. Figure 1-13a shows the **intake stroke** which starts with the piston at TDC. A mixture of fuel and air is drawn into the cylinder from the induction system (the carburetor and/or intake manifold) as the piston descends to BDC, increasing the volume of the cylinder and creating a slight negative pressure. In a non-carbureted, direct-injection engine, the fuel is injected into the cylinder shortly before ignition.

During the **compression stroke** in Figure 1-13b, all valves are closed and the gas is compressed as the piston travels from BDC to TDC. Slightly before TDC, a spark is ignited to explode the compressed gas. The pressure from this explosion builds very quickly and pushes the piston down from TDC to BDC during the **power stroke** shown in Figure 1-13c.

The exhaust valve is then opened during the piston's **exhaust stroke** from BDC to TDC (Figure 1-13d) and the piston pushes the spent gases out of the cylinder into the exhaust manifold and thence to the catalytic converter for cleaning and muffler for quieting before being dumped out the tailpipe. The cycle is then ready to repeat with another intake stroke.

Otto ran his early engines on illuminating gas, not gasoline, thus not needing a carburetor. The intake and exhaust valves are opened and closed at the right times in the cycle by a **camshaft** that is driven in synchrony with the crankshaft at half crank speed by gears, chain, or toothed-belt drive.

Figure 1-13e shows the gas pressure curve for one cycle. With a one-cylinder Otto cycle engine, power is delivered to the crankshaft only 25% of the time as there is only one power stroke per two revolutions.

Note in Figure 1-13e that the ignition point occurs slightly before TDC when the piston is still acending. The reason is that it takes a finite (though short) time for the explosion to build to peak pressure after ignition. If the spark were fired at TDC, by the time the pressure peaked, the piston would have already used some of its downward motion. Thus the spark is "advanced" by up to about 15° of crank angle at high rpm and less at low rpm.

The time it takes for the piston to travel through any number of degrees is less at higher rpm. But the time required for the explosion to peak is essentially constant regardless of engine speed. Thus the engine must increase the spark advance at higher rpm to synchronize the arrival of peak explosion pressure with TDC. In a modern engine this is accomplished with the engine control computer. Earlier engines used mechanical means to do this. If the spark is too advanced, pre-ignition will occur, raising peak pressure while the piston is still traveling upward. This is heard as "pinging." If more severe, it is called "knocking" and can destroy pistons.

Otto founded his company in Deutz, Germany with a partner, Eugen Langen in 1864. They reorganized in 1869 and called it Gasmotoren-Fabrik Deutz (The Gas Engine Manufacturing Company of Deutz). Gottlieb Daimler was brought in as technical director and Wilhelm Maybach was made head of engine design. It took until 1876 for Otto to develop the "compressed charge" engine as he called it. He figured out how to feed the fuel into the cylinder such that it burned in progressive fashion instead of in an uncontrolled explosion that had caused earlier engines to suffer from exploding cylinders. He called this approach a "stratified charge." Interestingly, that concept and term reappeared in the 1970s when fuel economy became a concern and Honda developed the Compound Vortex Controlled Combustion (CVCC) stratified charge engine. Deutz also developed a carburetor in 1884, allowing the engine to run on liquid fuel.

Otto's main interest was in producing stationary engines to power factories. Daimler wanted to produce small, light engines suitable for mobile applications such as automobiles and motorcycles. Unable to agree with Otto's approach, Daimler left the company, taking Maybach with him and founded the Daimler company, later called Daimler-Benz, and now better known as Mercedes-Benz. Daimler went on to develop the first practical IC-engine-powered motorcycle and automobile and is now the oldest company continuously producing automobiles. The Deutz company is still in business as the world's oldest engine producer, and now also makes heavy duty vehicles.

Clerk Two-Stroke Cycle

The two-stroke cycle was invented in 1878 by Sir Dugald Clerk, a Scottish engineer, shortly after Otto developed the four-stroke engine. A "pure" two-stroke engine needs no valves and, as its name suggests, has a power stroke once per revolution of the crankshaft as opposed to once per two revolutions for the four stroke. So, theoretically, a two-stroke engine should be twice as powerful as a four-stroke engine of the same displacement. Unfortunately, this is not the case, as the two-stroke cycle is much less efficient than the four-stroke cycle. At best, a two-stroke is only about 1.5 times as powerful as a four stroke for the same displacement.

Figure 1-14 shows a schematic of the **Clerk two-stroke cycle**. This engine does not need any valves, though to increase its efficiency it is sometimes provided with a passive (pressure differential operated) valve at the intake port. It does not have a camshaft or valve train or **cam** drive gears to add weight and bulk to the engine. It requires only two strokes, or 360°, to complete its cycle. There is a passageway, called a transfer port, between the combustion chamber above the piston and the crankcase below. There is also an exhaust port in the side of the cylinder. The piston acts to sequentially block or expose these ports as it moves up and down. The crankcase is sealed, serving also as the intake manifold. There is no oil sump in the crankcase.

Starting at TDC (Figure 1-14a), the two-stroke cycle proceeds as follows: The spark plug ignites the fuel-air charge, compressed on the previous revolution. Expansion of the burning gases drives the piston down, delivering torque to the crankshaft. Partway down, the piston uncovers the exhaust port, allowing the burned (and also some unburned) gases to begin to escape to the exhaust system.

As the piston continues to descend (Figure 1-14b), it compresses the charge of fuel-air mixture in the sealed crankcase. The piston eventually blocks the intake port, preventing blow back through the carburetor. As the piston clears the **transfer port** in the cylinder wall, its downward motion pushes the new fuel-air charge up from the crankcase through transfer port to the combustion chamber. The momentum of the exhaust gases leaving the chamber on the other side helps pull in the new charge as well.

The piston passes BDC (Figure 1-14c) and starts up, pushing out the remaining exhaust gases. The exhaust port is closed by the piston as it ascends, allowing compression of the new charge. As the piston approaches TDC, it exposes the intake port (Figure 1-14d), sucking a new charge of air and fuel into the expanded crankcase from the carburetor. Slightly before TDC, the spark is ignited and the cycle repeats as the piston passes TDC.

Clearly, this Clerk cycle is not as efficient as the Otto cycle in which each event is more cleanly separated from the others. Here there is much mixing of the various phases of the cycle. Unburned hydrocarbons are exhausted in larger quantities. This accounts for the poor fuel economy and dirty emissions of the Clerk engine. It is nevertheless popular in applications where low weight is paramount as it can have a superior power-to-weight ratio due to the absence of a valve train.

Lubrication is also more difficult in the two-stroke engine than in the four-stroke as the crankcase is not available as an oil sump. Thus lubricating oil must be mixed with the fuel. This further increases the emissions problem com-

CHAPTER 1 MOTIVE POWER

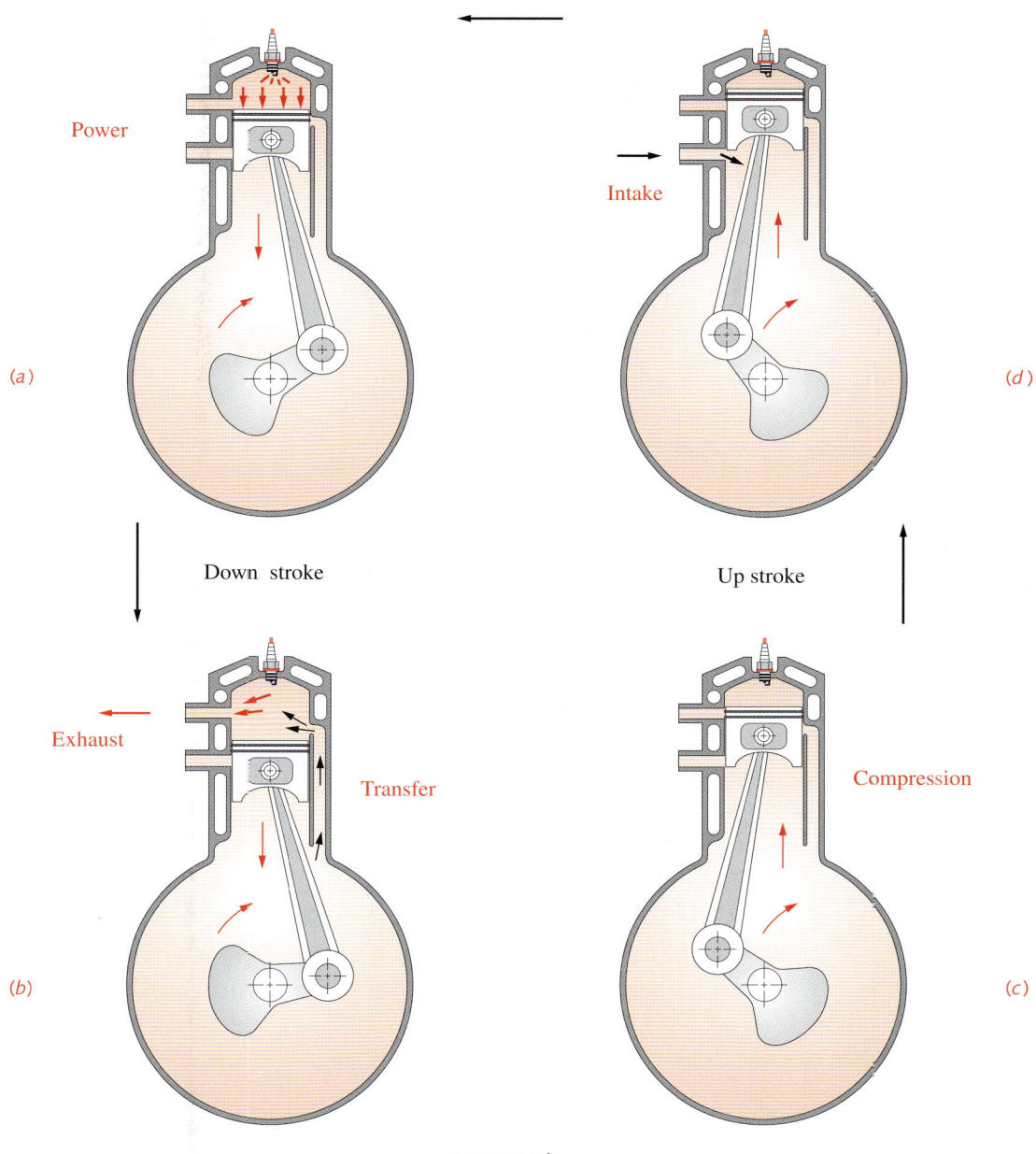

FIGURE 1-14 The Clerk two-stroke cycle

pared to the Otto cycle engine, which burns raw gasoline and pumps its lubricating oil separately throughout the engine. It is difficult to make a two-stroke engine that complies with current air pollution regulations in the U.S. and the European Union. As a result, these engines are increasingly being replaced by four-stroke engines in such applications as outboard motors, off-road motorcycles, and even hand-held "weed-whackers" where the better power-to-weight ratio of the two-stroke long made it the engine of choice.

Diesel Cycle (Compression Ignition)

There are a number of advantages of compression-ignition over spark-ignition in IC engines. Diesel fuel has 11% more energy content per liter than gasoline, which along with the higher compression ratio makes them 30% to 60% more thermally efficient than a gasoline engine. Diesels develop greater torque at lower rpm than a gasoline engine of similar displacement, making them more suitable for pulling heavy loads such as boats and trailers, for example. The U.S. military is also trying to convert all its vehicles to run on diesel or jet fuel (kerosene) primarily because of their lower volatility. A bullet into a gasoline tank results in a conflagration, but a bullet into a diesel or jet fuel tank may merely cause a leak.

Rudolf Diesel invented the engine that bears his name in 1893, a few years after the first Benz automobile appeared with Otto's engine in it. Its working principle is called the Diesel cycle and is very similar to the 4-stroke Otto cycle. Just as with spark-ignition engines, diesel engines can be run on a two-stroke cycle as well. The Detroit Diesel engine company makes two-stroke diesels as well as four-stroke ones. All diesels have no throttles to restrict air flow into the engine in contrast to gasoline engines, which means there is no intake vacuum in a diesel. A two-stroke diesel requires a blower to supply air to the engine because there is no distinct intake stroke as in a 4-stroke engine.

Diesels have no spark plugs because they are compression-ignition. Four-stroke diesels do have intake and exhaust valves like their gasoline counterparts. The speed of a diesel engine is controlled by the amount of fuel injected into the cylinders. All diesels use direct injection, meaning that the fuel is injected directly into the cylinders at very high pressure just before TDC, at about the same point that the spark would be fired in a spark engine. Fuel continues to be injected as the piston descends due to the expanding fuel mixture. Increasingly, gasoline engines are using direct injection also. However, earlier fuel-injected gasoline engines injected fuel into the throttle body to be sucked into the cylinders, just as if from a carburetor. Direct injection provides more precise control of fuel resulting in improved efficiency and lower emissions.

Some advantages of diesel engines over gasoline engines are their higher efficiency, greater torque capacity at lower rpm, and better lubrication of cylinder walls as the fuel has more lubricity than gasoline. Disadvantages include higher weight per HP because the high compression and explosion pressures (about 1.5X a gasoline engine) need thicker cylinder walls and stronger components throughout. Thus, they are most suitable for large vehicles, where engine weight is a smaller percentage of total vehicle weight. Most large trucks use diesel engines as do locomotives, ships, and earth moving equipment as well as stationary power plants and large generator sets. Many automobiles also use diesel engines because of their better efficiency. They typically return more miles per gallon than an equivalent-sized gasoline powered car. Until recently, many states in the U.S. prohibited the sale of diesel-powered passenger automobiles due to their higher nitrogen oxide (NOx) and sulfur emissions that contribute to smog. Fuel suppliers have since developed low-sulfur diesel fuel, and diesel engine manufacturers have introduced systems that inject urea and water to the fuel, which reduces NOx in combination with a catalytic converter. Most states in the U.S. have resumed sales of diesel passenger vehicles as of this writing.

The high cost of fossil fuels has promoted the development of biodiesel fuels that mix a relatively small percentage of vegetable oil with petroleum-based diesel oil. Some enterprising owners of diesel vehicles have converted them to run on pure vegetable oil. A popular (and inexpensive) source of such oil is fast-food restaurants that must dispose of the oil in their fryers when it becomes too dirty. Diesel cars that use discarded cooking oil are said to leave the smell of French fries in their wake.

Atkinson Cycle

James Atkinson invented his eponymous engine in 1892 to get around Otto's patent (Figure 1-15). Atkinson devised a clever linkage arrangement that allows the four strokes of Otto's cycle to complete in one rather than two revolutions of the crank. That makes it theoretically twice as powerful as an Otto engine of the same displacement. In addition, it offers greater flexibility with respect to control of the four events in the cycle. Whereas, in Otto's engine, the four strokes all have the same length, the Atkinson linkage allows the intake and compression stroke to be shorter than the power and exhaust strokes. The **compression ratio** of an engine is defined as *the ratio of the volume of the cylinder at BDC divided by the volume at TDC on the compression stroke*. There is also an **expansion ratio**, defined in the same way but for the power stroke. These two ratios are essentially the same in an Otto engine if we discount the volume of fuel added at the start of the power stroke. The Atkinson engine can have a significantly larger expansion ratio than compression ratio by adjustment of its linkage geometry.

The Otto or Clerk cycle engines both use a very simple mechanism to convert the reciprocating motion of the piston to rotary motion of the crank. It is a four-**link**, or fourbar slider-crank mechanism as shown in Figure 1-12. One of the four links is always fixed and three move. In Figure 1-12, the cylinder and crank pivot comprise the fixed link, while the crank, connecting rod, and piston comprise the three moving links. The kinematic convention is to number the links with the fixed link always 1 and the moving links in an increasing sequence, here 2, 3, and 4, in the order listed above. These numbers are shown on the figure. This is the simplest kinematic arrangement

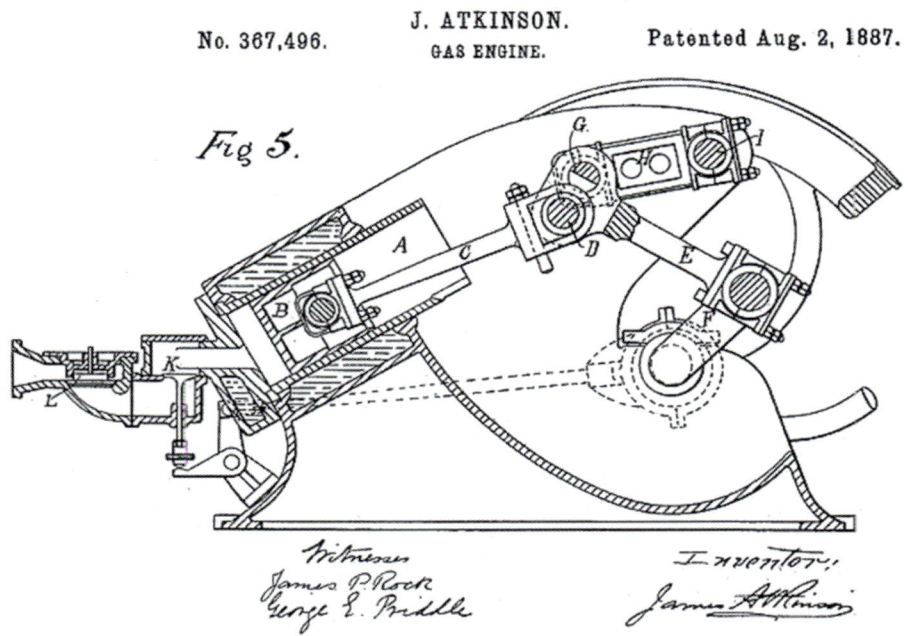

FIGURE 1-15   The Atkinson cycle engine *(Public Domain)*

that will give controlled motion. Four links or bars are the minimum required, but there can be more.

The original Atkinson engine uses a sixbar mechanism as shown in his patent drawing in Figure 1-15. The crank (F) drives the first connecting rod (E), which causes a rocker link (H) pivoted to the cylinder case (A) at (I) to rock back and forth. A second connecting rod (C) connects the first connecting rod (E) to the piston (B). This arrangement is what causes the piston to stroke four times per crank revolution instead of the two of the Otto engine. To understand how this works, look again at Figure 1-12 and picture the motion of its connecting rod as the crank makes one revolution. A point on the connecting rod goes up and down twice for each revolution of the crank. By using this motion to drive the piston, Atkinson gets two up and two down motions of the piston every crank **rotation**.

This linkage can be arranged such that all strokes are the same length as in the Otto engine, or their geometry can be arranged to make two of the strokes shorter than the other two. Why, one might ask, would anyone want to do that? By making the intake and compression strokes shorter than the power and exhaust strokes, the expansion ratio is made larger than the compression ratio. The advantage of this is, by expanding the exploded gases more, they can be brought closer to atmospheric pressure at the end of the power stroke, thus extracting more energy. It also reduces the pumping losses, which refers to the energy required from the crankshaft to compress the gas on the compression stroke.

However, following the truism that there is "no free lunch" in engineering, we know that making something better has to result in making something else worse. What is now worse is that the shortened intake stroke means less fuel and air is taken into the engine and there is thus less energy available per cycle. So, the Atkinson cycle, if made asymmetric in this way, is more efficient than the Otto cycle but is also less powerful. The end result is an engine that gets better mileage but cannot provide as much power as an equivalent-sized Otto engine.

There is little evidence that Atkinson promoted the advantages of using his cycle asymmetrically as described above. It appears that his primary motivation was to get around Otto's patent. But the Atkinson engine was not used very much in the years after its invention, probably due to the complexity and cost of the sixbar mechanism as compared to Otto's simpler design. The Otto engine and its Diesel derivative came to dominate the automotive scene throughout the 20th century.

However, a funny thing happened on the way to the 21st century. Fuel costs went up significantly, prompting customers and government regulators in the Western world to impose increasing demands for higher economy in vehicles. One result has been increased interest in electric and hybrid vehicles, which will be discussed later in this chapter in some detail. But, it is appropriate to bring them up in this discussion of the Atkinson cycle, because the renewed interest in hybrids has given a new lease on life to the Atkinson cycle engine.

The asymmetric Atkinson cycle, with its high efficiency and low power, is particularly well suited for use in combination with another means to make up for its lack of power. One such approach is to pair it with an electric motor, or motors, and use it to drive a generator to replenish the batteries. A generator presents a light load to an engine; the Atkinson cycle can handle it easily while also being very fuel efficient. Table 1-1 shows a list of current hybrid vehicles that use a modified Atkinson-cycle engine in combination with electric motors.

Table 1-1 Hybrid electric vehicles that use a modified Atkinson cycle engine.
Ford C-Max hybrid & plug-in hybrid models
Ford Escape/Mercury Mariner/Mazda Tribute electric
Ford Fusion Hybrid/Mercury Milan Hybrid/Lincoln MKZ Hybrid electric
Honda Accord Hybrid
Honda Accord Plug-in Hybrid
Hyundai Sonata Hybrid
Infiniti M35h Hybrid
Kia Optima Hybrid
Lexus CT 200h
Lexus ES 300h
Lexus IS 300h
Lexus GS 450h hybrid electric
Lexus HS 250h
Lexus RX 450h hybrid electric
Mazda 3 SkyActiv
Mazda 6 SkyActiv-G 2.5L
Mazda CX-5
Mercedes ML450 Hybrid electric
Mercedes S400 Blue Hybrid electric
Toyota Camry Hybrid electric
Toyota Highlander Hybrid
Toyota Prius hybrid electric
Toyota Yaris Hybrid
Source: Wikepedia

The Toyota Prius was the first automobile to use this combination (in 1997). To avoid the cost and complexity of the sixbar linkage, Toyota took a different approach by modifying an Otto engine to behave like an asymmetric Atkinson, but without all the added mechanism. Their approach was to alter the valve timing such that the intake valve is held open into part of the compression stroke, effectively pushing some of the air ingested during the downward piston motion back out the still-open intake valve as the piston starts up on the compression stroke.

This does not waste fuel because, even if fuel is in the cylinder, it merely is pushed back into the intake manifold to be used on the next cycle. With direct injection, an amount of fuel appropriate to the reduced air volume is injected only after the intake valve has closed and the air has been compressed. Thus the asymmetric nature of a true Atkinson engine is simulated by this "trick" of dumping some of the intake air back out the valve before compressing and igniting the mixture in an Otto engine. Toyota claims a 12%–14% efficiency improvement over a conventional Otto engine, mainly due to reduced pumping losses. With modern camshaft design and computer control of fuel injection, this approach required no significant new technology be developed. This is now referred to as a modified Atkinson cycle engine.

AUTOMOTIVE MILESTONES

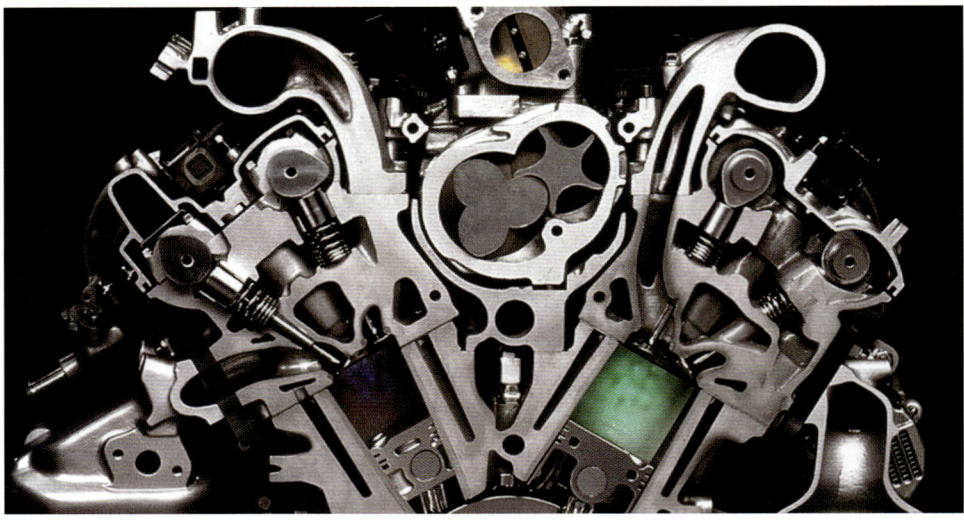

FIGURE 1-16   A Mazda Millenia Miller cycle engine *(Public Domain)*

Miller Cycle

The Miller cycle was patented by Ralph Miller, an American engineer, in 1957. His patent describes "a new and improved method of operating a supercharged, inter-cooled engine." He added a servo-controlled, compression-control valve in the cylinder head with a result similar to that of the modified Atkinson cycle described above.

The intake charge is dumped back into the intake manifold during the first 20 to 30% of the compression stroke. By adding an engine-driven, positive displacement supercharger, he makes up for the deficiency of torque and power due to the short intake charge of the Atkinson engine by forcing more volume of air/fuel into the cylinder. It is essentially a supercharged Atkinson cycle engine.

A Miller cycle engine can be either two- or four-stroke and can run on gasoline or diesel. It has been used as a stationary engine, in ships, and in railway locomotives. It was also used by Mazda in their 1995 Millenia sedan. Figure 1-16 shows a cross-section of a Mazda-Miller engine.

Brayton Cycle

George Brayton, an American engineer, patented his engine in 1872. Though it was a piston engine, his Brayton cycle was later applied to the gas turbine engine and air-breathing jet engines. Instead of compressing the fuel-air mixture with the engine piston's motion, Brayton used an external compressor to compress the air, added fuel to it, and then introduced it to the expansion (engine) cylinder and ignited it to drive the engine piston down. The crank motion drove the vehicle and also drove the compressor.

The modern gas turbine engine has no pistons, but it has an engine-driven compressor, a burner that adds fuel to the compressed air and ignites it, and a turbine The expanding gas drives the turbine, which turns the propeller or other load and also drives the compressor. Fuel is typically kerosene.

Brayton cycle turbines are used to generate electricity and power turboprop aircraft, helicopters, tanks, and high-speed warships such as destroyers and fast frigates. A gas-turbine-driven warship, fitted with two variable-pitch

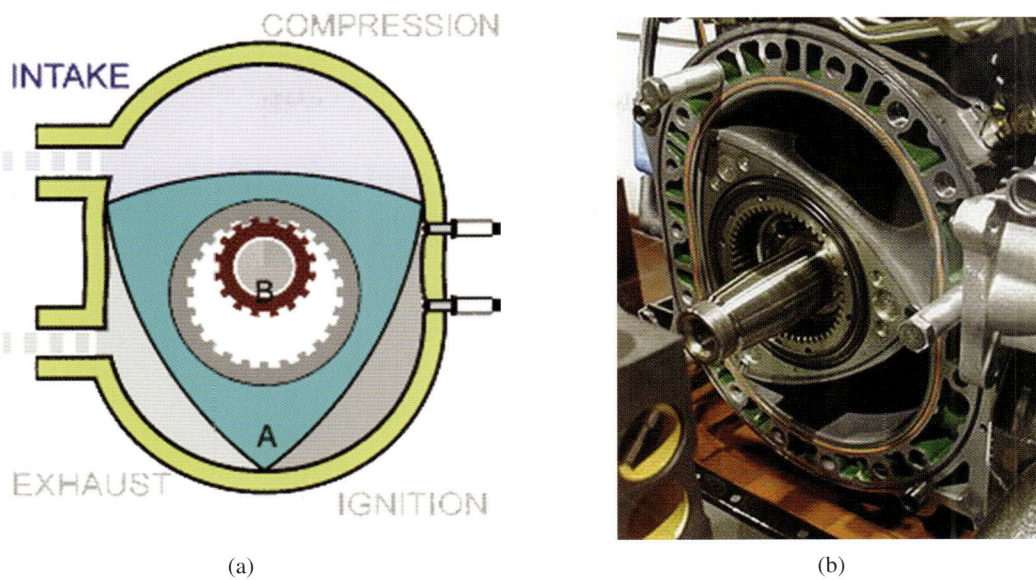

FIGURE 1-17 A Wankel engine *(Courtesy of Y_tambe and Softeis at German Wikipedia)*

propellers, can turn or stop in its own length with both engines at full power (very handy for dodging torpedoes.) A gas turbine turns at very high speed, from 2,000 to 10,000 rpm. A single-speed-reduction gearbox is used to bring the speed down to that needed for the output device.

Wankel Cycle

The Wankel engine, invented by Felix Wankel in 1929 in Germany, is also called a rotary engine because all of its components are in pure rotation. There are no reciprocating parts. This makes it inherently smooth-running compared to a piston engine, which can have high vibration. Figure 1-17 shows a schematic of the rotor and housing and a photo of a cutaway engine. It is a four-cycle engine. The rotor spins on an eccentric on the crankshaft, which is centered in the housing. The housing has an **epitrochoid** shape and the rotor is a triangular shape with bulging sides, sometimes called a Reuleaux triangle. As the crank rotates, a sun gear fixed to the housing causes the rotor to spin because its ring gear is in mesh with the stationary sun gear. The rotor has planetary motion. (Compare this arrangement to Watt's sun and planet drive in Figure 1-1.) The tips of the rotor's three corners are always in contact with the inner surface of the housing and are fitted with both tip and end seals that perform the same sealing function as piston rings.

As the rotor spins at 1/3 the speed of the crankshaft, it forms three moving chambers that each change volume as it turns. These serve as the intake, compression, power, and exhaust chambers in turn as in a 4-stroke piston engine. There are no valves. The chambers expose the intake and exhaust ports in turn as in a two-stroke engine. Figure 1-17a shows the engine with the top chamber exposing the intake port and just about fully charged at maximum volume. As its volume previously increased with rotation, its pressure dropped, sucking in a charge. As it continues to rotate clockwise and shuts off the intake port, it will compress the charge. The chamber at lower right has already compressed its charge, the spark plugs have fired, and the expanding gases are driving the rotor clockwise. The chamber at lower left is exhausting the spent gases through the exhaust port.

The Wankel engine produces three power strokes per rotor revolution. Because that equates to one rotation of the crankshaft, there is one power stroke per revolution of the crank. That is as often as a single-cylinder, two-stroke piston engine and twice as often as a single-cylinder, four-stroke engine. Because of the efficiency of the four-stroke cycle, this makes the Wankel engine more powerful per unit displacement than a piston engine. It is also compact and light weight, due to the absence of a valve train. It has many fewer moving parts than a conventional piston engine and can be quite reliable. Most production Wankels have two rotors, displaced by 180 degrees. Like a two-cylinder piston engine, these produce twice as many power pulses per crankshaft revolution. The two offset rotors also create better static balance than a single-rotor engine.

Early developers were NSU (where Wankel worked and initially developed the engine in the 1950s), Curtiss Wright in the U.S., who licensed it for aircraft applications, and most of the world's major automobile manufacturers, who bought licenses to develop the engine for their applications. The Norton company in England also developed and marketed a Wankel-powered motorcycle. All had problems with the tip seals, which wore excessively. The early engines also used more fuel than desired, and when emission regulations tightened after 1967, the Wankel had problems meeting increasingly stringent emission controls. Eventually, everyone dropped the engine except one company, Mazda. To their credit, they solved the tip seal problems, found a way to clean up the emissions, and continued to produce and sell a few models with Wankel engines until 2012. The Mazda RX-7 and RX-8 were the last models produced with this engine. Still plagued with a relatively high thirst for gasoline in the face of tightening mileage regulations, and more significantly, low sales figures, Wankels were finally retired from Mazda models in 2012. Let's hope they come back. It is a very good engine with lots of advantages over piston engines.

Electric Vehicles

Electric-powered vehicles have been around longer than gasoline-powered ones. The first electric car is credited to Robert Anderson around 1835 in Scotland. By 1900, 28% of cars on the road in the US were electric.

Pure Electric

A pure electric car is one that has batteries for energy storage and one or more electric motors for propulsion, but has no on-board means to recharge the batteries. There were many makes of such cars in the late 19th and early 20th centuries: *Baker Electric*, *Columbia Electric*, *Detroit Electric*, and *Milburn Electric*, among others. The "golden age" of electric cars began around 1910 by which time electricity from power companies had reached a significant fraction of U.S. homes. This allowed owners to recharge the batteries at home.

A 1922 Milburn Electric from the collection of the *Tampa Bay Automobile Museum* is shown in Figure 1-18. This car featured removable racks of fourteen 6V batteries in series (giving 84V) in both the front and rear "trunks." Milburn provided stations where a driver could have his nearly spent batteries quickly removed and replaced with fresh sets. Controls were very simple: the upper lever controlled speed, forward and reverse; the lower lever steered. Pulling the speed lever back while in forward motion caused regenerative braking, using the **kinetic energy** of the car to charge the batteries.

Milburn claimed 100 miles per charge, but President Woodrow Wilson and his secret service agents used Milburns, and they reported 60–70 miles per charge. This was the "Achilles

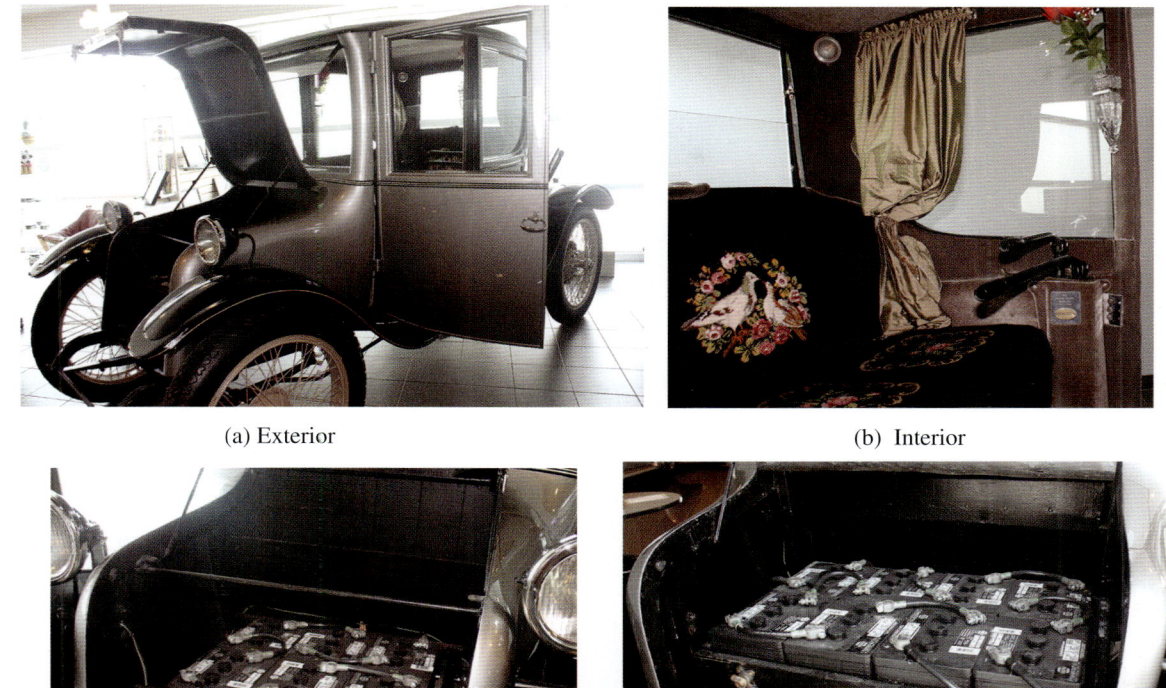

(a) Exterior (b) Interior
(c) Front battery pack (d) Rear battery pack

FIGURE 1-18 A 1922 Milburn Electric *(Photos by the author—taken at the Tampa Bay Automobile Museum)*

heel" of electric cars as most had limited range even down to the present, with a few modern exceptions such as the Tesla. Electric power nevertheless has several advantages over other technologies. Unlike steam cars, they start instantly. Like steam cars, they have maximum torque at stall, so accelerate rapidly from a stop. Unlike ICE cars, they are silent and much less complicated. Electric motors are simple, reliable, and use no power when stopped, unlike an idling IC engine. No clutch or transmission is needed, and regenerative braking reduces wear on the brake system. They are most suited to city use for short trips in traffic, where they are very efficient in terms of energy use.

The high cost of fossil fuels and concerns over carbon emissions' effects on global warming have prompted increased interest in the electric car in the 21st century. Much development has gone into improving battery technology. Modern electric cars no longer use lead-acid batteries. They favor the nickel-metal-hydride or lithium-ion type, which have higher energy density—higher electrical storage capacity per pound. A great deal of effort is now being spent on electric vehicles, but most of them are of the hybrid variety as opposed to pure electrics. Presently several manufacturers offer pure electric cars *(claimed range in miles): BMW i3 (100), Chevrolet Spark (82), Fiat 500e (87), Ford Focus Electric (76), Mercedes B-Class E-Cell (100), Mitsubishi i-MiEV (62), Nissan Leaf (84), Smart ED (68), Tesla Model S (265), Tesla Model X (230), Toyota RAV4 EV (100),* and *VW E-Golf (85).*

Hybrids

One definition of **hybrid** is *composed of elements of different kinds*. Hybrid automobiles typically have two types of motive power, commonly electric motor(s) and a gasoline engine. Other possibilities are electric and diesel, and hydraulic motors paired with either a gasoline or diesel engine. A hybrid automobile can be a **series hybrid**, a **parallel hybrid**, or a **series-parallel (full** or **power-split) hybrid.** The term power-split means that it has two electric motors that can work independently or together to drive the wheels. A series hybrid uses only the electric motor(s) for propulsion with the gas or diesel engine running a generator periodically to provide electricity to recharge the batteries. It is the simplest system as it does not need a transmission or clutches. A parallel system uses both motor and engine to drive the same or different wheels, and a series-parallel arrangement allows both power plants to drive the wheels and the engine to stop while in motion. Most also provide regenerative braking to convert the vehicle's kinetic energy to potential energy when decelerating by using the electric/hydraulic motor as a generator/pump.

Electric/ICE

The concept of a hybrid/electric automobile is far from a new idea. In 1898, the 18-year-old **Ferdinand Porsche** took his first job with the Lohner company and was assigned the task to design a hybrid gas/electric automobile. The **Lohner-Porsche series hybrid** had a 2-cyl gas engine driving the generator and two 3.5 HP electric motors in the front-wheel hubs. An early customer requested four-wheel drive, so another model was designed with electric motors in all four wheel hubs. It carried 44 lead-acid batteries connected to give 80V, and which weighed nearly 2 tons. The electric motors weighed 320 lb apiece and the vehicle weighed about 4 tons. Figure 1-19 shows a picture of a 1900 model. Woods (USA) made a parallel hybrid, gasoline/electric car from 1917 to 1919.

The **Diesel-Electric Locomotive** was mentioned in this book's introduction as an example of an early hybrid as well. In the 1950s, this system replaced steam locomotives in most industrialized countries. It is a series hybrid that uses large diesel engines to drive generators that power electric motors mounted on the "trucks"

FIGURE 1-19   1900 Lohner-Porsche gasoline/electric hybrid *(Public Domain unknown photographer)*

FIGURE 1-20 Toyota Prius engine and transmission *(Public Domain photo by Hatsukari715)*

or wheel assemblies. The high starting torque of electric motors is a significant asset in this application, which has to accelerate the large mass of a multiple-car train. The diesels charge the batteries and also run air compressors to power the air brakes.

The modern surge of interest in hybrid gas/electric automobiles was led by **Toyota** with its **Prius** that debuted in Japan in 1997 and in the U.S. in 2000. The first mass-produced gasoline/electric hybrid car, it is a series-parallel hybrid (power-split) that can run solely on its two AC electric motors (80 HP combined), on its 1.8-liter, 98 HP, 4-cylinder, Atkinson cycle gasoline engine, or all in combination. One electric motor is the main drive motor and the other is a motor-generator connected to the ICE that serves both as a starter motor and generator. These power plants drive the wheels through a planetary transmission that can split the power flow to the wheels to be electric only, a combination, or gasoline only while simultaneously charging the batteries. The ICE is shut off when the car is stopped or when running in full-electric mode with sufficiently charged batteries. Figure 1-20 shows a Toyota Prius engine and transmission.

The transmission continuously adjusts its ratio to keep the engine running in its optimum RPM band for maximum efficiency. This is a significant advantage of hybrid systems that allows the ICE to be sized for average power demands and optimum efficiency; it uses the electric motors to supply the boost needed for **acceleration**. An ICE-only vehicle needs an engine with enough power to accelerate the vehicle from a stop and typically is providing only a small fraction of its peak power at steady-state highway speeds. Another advantage of the hybrid is that its motor/generator can regeneratively brake the vehicle, converting its kinetic energy into charge in the battery. An ICE vehicle throws that kinetic energy away as heat in the braking system every time it slows down.

Many other manufacturers are providing electric/ICE hybrids. The **Chevrolet Volt** has two electric motors, a gasoline engine, and a planetary transmission like the Prius, but its operation is a bit different. Neither the smaller starter motor/generator or the ICE are powerful enough to drive the car on its own, and unlike the Prius, there is no direct mechanical connection between the ICE and the wheels.

The ICE can contribute to vehicle motion by driving through the starter motor, but cannot do so unless the electric motor is also contributing. So it cannot run on gasoline power alone. At least one electric motor must be driving at all times. It has about a 40-mile range on pure electric with a full charge. After that is nearly exhausted, the ICE engine starts and charges the battery, allowing continuous operation. Unlike the early Prius, it is a plug-in hybrid, meaning its batteries can be charged from **mains power**. It takes eight or more hours to fully charge at 110V, but much less time at 240V. Toyota has recently begun to offer a plug-in Prius as have other makes.

The Volt's technology has spawned some controversy and argument in the automotive press over whether it is a true hybrid. The Society of Automotive Engineers (SAE) defines a **hybrid vehicle** as one *that has two or more energy storage systems both of which must provide propulsion power, either together or independently*. General Motors has never called the Volt a hybrid in any of their literature. Rather they have described it as an *electric vehicle with a range-extending gasoline engine* and dubbed it *Extended Range Electric Vehicle* or *E-REV*. The SAE calls it a *plug-in hybrid*. The controversy stems from whether the ICE is "directly connected" to the wheels or not. One can argue it both ways as there are two clutches in the system, one that connects the two electric motors together and one that connects the ICE to the smaller electric motor. All three of these elements are in series, so the path from ICE to wheels goes through the two electric motors, but it is a mechanical path, though not in parallel with the motors. This author considers it a series hybrid, but others may disagree. No question that it plugs in, however.

Initially, hybrid vehicles were offered for their improved economy of operation, but recently makers of high performance cars such as Porsche, BMW, Mercedes, Ferrari, Jaguar, McLaren, Lexus, and others have begun to offer or announce high-performance hybrids (some of them still concept vehicles at this writing) that seem to be bred to improve performance as much or more than economy. The added power available from electric motors coupled with an already powerful ICE promises outstanding acceleration along with a modest increase in fuel economy over their ICE-only counterparts. The results should be interesting. At the 2010 Geneva show, Porsche showed a 918 model concept vehicle that has a 608 HP V8 plus two electric motors making a combined 281 HP for a total of 889 HP. They claim it's capable of 71 mpg in electric mode. It will go into production in September 2014 and 918 units will be produced.

Hydraulic/ICE

Since the 1990s, the Environmental Protection Agency (EPA) has formed R&D partnerships with companies such as Ford Motor Co., Eaton Corp. (a major supplier of automotive technology and equipment), Parker-Hannifin (a supplier of hydraulic equipment), and seven universities to develop **hydraulic-hybrid** vehicles. The focus of this research has been on medium-to-large commercial vehicles such as delivery vans, buses, trash trucks, and other vehicles that have a significant start-stop cycle in their use. Such vehicles have traditionally been diesel powered, but by combining diesel power with a hydraulic motor/pump, hydraulic reservoir, and high pressure accumulator, significant efficiencies of operation can be achieved. Hydraulic hybrids can be less expensive and lighter than electric hybrids because they lack expensive, heavy batteries. Also, accumulators don't wear out like batteries. The hydraulic equipment takes up some room, making them better suited to larger vehicles.

The principle is the same as an electric hybrid vehicle and they can be of the same types, serial, parallel, or series-parallel. Both arrangements use the same components as listed above. A **hydraulic motor** is merely a **hydraulic pump** run backward. Hydraulic pumps can be of several types: centrifugal, vane-type, gear, or piston. A swash-plate, axial-piston, positive displacement pump is often used. A parallel arrangement has one hydraulic motor/pump combination (that is, one device that can be run forward as a motor or backward as a pump) and a series arrangement has two: a separate pump and a motor/pump. Both arrangements have a low-pressure and a high-pressure accumulator, the latter of which is analogous to the battery in a hybrid-electric vehicle. We will call the low pressure accumulator a **reservoir** and the high pressure accumulator just an **accumulator** for simplicity.

An **accumulator** stores **potential energy** in the form of fluid under pressure. It is a simple device, usually shaped as a cylinder with spherical ends and a single opening in one end to which a pipe is connected. The cylinder is initially charged with a gas such as dry nitrogen at moderate pressure. As liquid is pumped into the cylinder, it compresses the gas and the pressure increases, typically to thousands of psi. The nitrogen gas is actually separated from the fluid by a flexible membrane to avoid gas dissolving into the fluid. The fluid is captured in the accumulator by a valve until it is needed. It is also analogous to an electrical capacitor that stores electrical charge. The **reservoir** serves to receive fluid that has passed from the accumulator through the hydraulic motor to drive the vehicle and has dropped in pressure, having converted its energy to vehicle motion.

A series hydraulic hybrid works as follows: The engine is connected only to a pump and not to the wheels. The engine runs initially to pump fluid from the reservoir to the accumulator until it is at full pressure. The engine then shuts off. The vehicle initially accelerates using only the hydraulic motor. As the accumulator pressure drops, the engine turns on to pump fluid from the reservoir into the accumulator and shuts off when it is full. When the vehicle slows or stops, its deceleration turns the drive motor backwards and pumps fluid from the reservoir into the accumulator. This is regenerative braking and captures 70% of the vehicle's kinetic energy as potential energy in the accumulator. In comparison, an electric hybrid only recaptures about 25% of the kinetic energy as charge in the battery. Thus a hydraulic hybrid is much more efficient in start-stop driving than an electric hybrid.

A parallel hydraulic hybrid works as follows: The engine is connected in conventional fashion through a transmission to the drive wheels. The combination hydraulic motor/pump is in the drive shaft or rear end after the transmission and has fluid connections to the reservoir and the accumulator. If the accumulator is charged, under light acceleration the hydraulic motor alone drives the vehicle allowing the engine to disconnect from the transmission with a clutch and stop. Under heavy or prolonged acceleration, the engine and hydraulic motor both drive the wheels and share the load. At constant highway speeds, the engine alone drives the vehicle. When slowing or stopping, its deceleration turns the drive motor backwards and pumps fluid from the reservoir into the accumulator, recapturing 70% of the **kinetic energy** in the accumulator to assist in the next startup. An excellent video showing how a series system works is on the EPA website at: http://www.epa.gov/otaq/technology/research/how-it-works.htm and diagrams showing the function of a parallel hydraulic hybrid are at: http://www.epa.gov/otaq/technology/research/how-it-works-parallel.htm.

United Parcel Service (UPS) was the first company to put hydraulic hybrid trucks into service in 2009. From the EPA website http://www.epa.gov/Region9/air/hydraulic-hybrid/index.html:

EPA and the United Parcel Service (UPS) have developed a hydraulic hybrid delivery vehicle to explore and demonstrate the environmental benefits of the hydraulic hybrid for urban pick-up and delivery fleets. The demonstration vehicle is a 24,000 pound UPS package car, fitted with an EPA-patented full-series hydraulic hybrid drive integrated into the rear axle. . . . In laboratory tests, the city fuel economy of the hydraulic hybrid UPS vehicle is 60% to 70% increased miles per gallon compared to a conventional UPS truck. The CO_2 emissions of the demonstration UPS vehicle are more than 40% lower than a comparable conventional UPS vehicle. The hydraulic hybrid vehicle also achieves approximately 50% lower hydrocarbon and 60% lower particulate matter in laboratory tests. This prototype vehicle has also demonstrated modest reductions in NOx emissions. Optimized production vehicles are expected to have larger NOx reductions. Hydraulic hybrids are able to capture and reuse 70–80% of the otherwise wasted braking energy.

These numbers are impressive and show that for this kind of application, namely short runs between stops with a heavy vehicle in an urban environment, the hydraulic hybrid looks like a winner. Chrysler is also working with the EPA on a hydraulic-hybrid version of a minivan.

Flywheel Hybrids

A flywheel is simply a disk which, when spun at speed, stores kinetic energy. Energy must be supplied to it to increase its speed. When slowed down, it returns some of the stored energy. Kinetic energy is proportional to the spinning **mass** times its **radius** times its **angular velocity** (in **radians per second**) squared. If you double the speed of a flywheel of given mass and radius, you will increase its stored energy by 2 squared, or 4 times. Flywheels designed for these applications are quite exotic, made of strong, light materials such as titanium for the center and carbon fiber composites for the spokes. Heavier material can be used for the rim to provide mass at a large radius. To get maximum energy storage, they are spun at very high speeds up to 60,000 **rpm**. To minimize friction from air at that speed, they are spun in a vacuum.

In the 1950s, flywheel hybrid buses were developed and used in Switzerland. An IC engine provided primary power and drove the vehicle. Switzerland is, of course, mountainous, so traveling downhill the bus would store energy in the flywheel by regenerative braking. On the next uphill, the flywheel would assist the engine by returning some of the stored energy to help drive the vehicle.

The 2010 Porsche 911 GT3R race car is a gasoline/electric hybrid that also uses a flywheel, driven by an electric motor, to store kinetic energy. Two 109-HP electric motors power the front wheels and a 500-HP gasoline engine powers the rear wheels for a total continuous power of 718 HP. The flywheel can release its stored energy rapidly to provide a boost of instantaneous power above that. Some refer to this car as a "Flybrid." This car raced at the 2011, 24 Hours of Nurburgring but did not win. It did better at the 2011 American Le Mans Series at Monterey. There, it turned the fastest GT lap and finished ahead of all other GT cars.

More recently (2013), Volvo announced development of a sedan with an IC engine and a 60,000 rpm, carbon-fiber flywheel running in a vacuum to both assist the engine and to fully power the vehicle with the engine off. Regenerative braking spins up the flywheel. Production plans, if any, are unknown at this writing.

Chapter 2
Chassis Layouts and Drivelines

Every possible arrangement of engine and driveline was tried in the early years of the automobile. All of the currently used arrangements originated in the first few decades of the automobile's development. Mid-engined, rear-engined, front-engined, front wheel drive, rear wheel drive, chain drive, friction drive, gear drive—you name it—are all old ideas. Some of these early designs were abandoned, but many are still in production and a few have come to dominate the industry.

It should not be surprising that the earliest automobiles drew upon existing technology and devices. Many auto manufacturers originally made bicycles or horse-drawn carriages and their early examples borrowed from those technologies. Some took a horse-drawn buggy as a starting point, removed the horse and added an engine and a steering mechanism. Eventually most designs converged on a few arrangements that were considered superior and those are still found in modern automobiles.

Two-, three- and four-wheeled vehicles were all among the earliest examples, but the four-wheel arrangement soon dominated. Nevertheless, two- and three-wheel vehicles (trikes or cycle cars) are still in production today. Many trikes are modified motorcycles and a few are purpose-built three-wheeled motorcycles or automobiles. This chapter will discuss chassis layouts and drivelines.

The First Production Automobiles

Benz

Carl Benz patented the first automobile in 1886 in Germany (see Figure 2-1). It was a three-wheeled cycle with a single-cylinder, 3/4 HP, 4-stroke, horizontal, gasoline engine driving the two rear wheels through a differential. It had a tubular steel frame and wire-spoked wheels. His wife, Bertha took one on a "long distance" drive—180 km (112 mi) round trip—with her two teenage sons, and the attendant publicity generated much demand for the car. Later models featured another Benz invention, patented in 1873,

The opening photograph is of a 1912 Mercer raceabout in The Heritage Museum, Sandwich, MA, taken by the author.

FIGURE 2-1 1886 Benz Tricycle and a closeup of its engine; note exposed mechanism *(Photos by the author)*

double-pivot steering, the predecessor to modern steering mechanisms.

Daimler

About the same time, Daimler and Maybach (after leaving Otto) were developing their own lightweight 4-stroke engine. The first vehicle they put it in was two-wheeled (with "training wheels" for stability) and was the first motorcycle (1885). The frame was wood and the 1/2 HP engine drove the rear wheel via a belt (see Figure 2-2).

FIGURE 2-2 1885 Daimler Motorcycle
(Courtesy of Mercedes-Benz Classic Archives)

Daimler (Germany) introduced the first four-wheel automobile in 1886 (see Figure 2-3). He added a single-cylinder engine to a horse-drawn carriage. The engine was located in front of the rear bench seat, making it the first mid-engined car as well. The rear wheels were belt driven, and instead of using a differential, it used **over-running clutches** at each rear wheel to accommodate the different speeds of the wheels in turns.

Flocken Elektrowagen

The first electric car is believed to be the German Flocken Electrowagen, designed by Anreas Flocken in 1888 in Germany (see Figure 2-4).

FIGURE 2-3 1886 Daimler four-wheeler
(Courtesy of Mercedes-Benz Classic Archives)

FIGURE 2-4 1888 Flocken Elektrowagen *(Courtesy of Franz Haag)*

FIGURE 2-5 1905 Systeme Panhard automobile *(Photo by the author)*

But Robert Anderson is also credited with building an electric carriage in Scotland circa 1837. Others experimented with these vehicles in the interim.

Panhard-Levassor

Panhard-Levassor (France) bought a license for the Otto engine and sold its first car in 1890. In 1891, they designed an automobile based on what they called the *Systeme Panhard* with a front-mounted radiator and front engine driving the rear wheels through a crude transmission mounted behind the engine. In 1895 they developed the first "modern" sliding-gear transmission, which became a standard design until 1928 when Cadillac introduced the **synchromesh transmission**. Transmissions are described in Chapter 5. Panhard's chassis arrangement of 1891 was eventually licensed and adopted by most manufacturers and became the standard for nearly a century. Front-wheel drive was developed a few years later in 1898. From about 1910 to quite recently, the majority of automobile chassis were based on the *Systeme Panhard*. Modern automobiles now favor front-wheel drive (see page 41). An example of a *Systeme Panhard* automobile is shown in Figure 2-5.

Other Early European Makes

Peugeot (France) made its first car, an unsuccessful steam tricycle, in 1889. They succeeded in 1890 with a four-wheeled vehicle using a Panhard-Daimler IC engine. Tatra (Czech Republic) produced the first car in central Europe in 1897.

United States

Work was ongoing in the U.S. at the same time as in Europe. The Duryea brothers founded their company, Duryea Motor Wagon Co., in 1893 and were the first American auto company.

Oldsmobile

Ransom E. Olds had the first automotive production line running by 1902, producing the popular *Curved Dash Olds* from 1901–1907 (Figure 2-6). It was the first mass-produced automobile with over 19,000 built. It had a water-cooled, 4-HP, single-cylinder Otto engine behind and below the seat of the car (see Figure 2-6). The two-speed-forward transmission with reverse was a semi-automatic, planetary arrangement that predicted the modern automatic transmission, as will be discussed in Chapter 5.

FIGURE 2-6   1902 Curved Dash Oldsmobile and its engine *(Photos by the author)*

Ford

Henry Ford built many cars before the famous Model T. His first was the Quadricycle in 1896. It was just a frame with bicycle wheels, a single-cylinder engine, a seat, and tiller steering. After forming the Ford Motor Company in 1903, he began producing an "alphabet soup" of models starting with the Model A in 1903 (not to be confused with the much later Model A of 1928) with a two-cylinder engine under the front seat driving the rear wheels by chain. The much more expensive Model B of 1904 had a 4-cylinder engine in the front, driving the rear wheels. A Model C was a variant on the A and was produced alongside it.

The alphabet then jumped to F and continued to N, R and S, finally arriving at his most successful product, the Model T (see Figure 2-7). This car followed the *Systeme Panhard* with a four-cylinder engine driving the rear wheels through a two-speed plus reverse planetary transmission similar to the Curved Dash Olds. Henry Ford did not invent many new automotive technologies but was a master at perfecting the best he could find, such as Olds' production line and Henry Leland's interchangeable parts. His main contribution was to apply new manufacturing techniques to continually improve the design and reduce production cost.

FIGURE 2-7   1913 Ford Model T and a 1909 Model T engine *(Photos by the author)*

CHAPTER 2 CHASSIS LAYOUTS AND DRIVELINES

Henry Ford said: *I will build a car for the great multitude. It will be large enough for the family, but small enough for the individual to run and care for. It will be constructed of the best materials, by the best men to be hired, after the simplest designs that modern engineering can devise. But it will be so low in price that no man making a good salary will be unable to own one – and enjoy with his family the blessing of hours of pleasure in God's great open spaces.*

The first Model T cost $850 in 1908 when other cars sold for $2000–$3000. By 1927 the price had dropped to $260. Henry put the nation on wheels. He used the best available materials including vanadium steel, a strong alloy, in automotive parts. The high strength of this alloy allowed him to reduce the weight and cost of parts. The Model T went on to sell 15 million cars between 1907 and 1927. This record was not broken until the late 20th century when Volkswagen sold more than 21 million original Beetles over 65 years.

Rear-Wheel Drive (RWD)

Many schemes were devised to deliver the engine's power to the rear drive wheels. Belt and chain drives were common in the early years, especially when the engine was mounted amidships and close to the rear axle. Simplex (1908–1916) made very large, powerful, front-engined cars that used chain drives to each rear wheel from a **jackshaft** that ran across the chassis behind the transmission. Most of the earliest cars were RWD and variously placed the engine in the rear, amidships, or in the front. All of these arrangments can be found in current production autos. BMW, Cadillac, Lexus, Mercedes, and others make front-engined RWD models. Porsche, Ferrari, and others make mid-engined RWD models and Porsche makes both rear- and mid-engined RWD models. Rear-wheel drive is considered a better arrangement than front-wheel drive for superior handling, and the mid-engined variants are the best handling of the breed because they have a low *polar moment of inertia* (**see Sidebar**) due to the engine being closer to the center of gravity (CG) of the vehicle.

A mid-engined or rear-engined RWD arrangment has no driveshaft. The engine is bolted directly to a **transaxle** (a combined transmission and differential) at the rear axle as shown in Figures 2-8 and 2-9. A rear-engined, RWD car such as the Porsche 911 and the original VW Beetle has the engine mounted just behind the rear axles as shown in Figure 2-8. A mid-engined, RWD car such as the *Porsche Boxster/Cayman* has the engine mounted just ahead of the rear axles and behind the driver as shown in Figure 2-9, giving it a lower polar moment

Polar Moment of Inertia

Polar moment of inertia is a property of a mass that is rotated about a center. It is proportional to mass times the distance from the mass to its center of rotation squared. So what does that mean? Take a hammer. Hold it normally, near the end of the handle, and swing it. Now move your hand close to the head and swing it again. It is much easier to rotate the head around a short radius than a long one because the moment of inertia is much less at the short radius. The same principle applies when turning a car. If the heavy mass of the engine is out front or back, a long way from the car's center of gravity (CG), which is near the center of the car (and is the point it wants to rotate about), it is more difficult to make the car turn. Move the mass of the engine close to the CG and it is easier to rotate the car in a turn, giving a better handling, more responsive vehicle.

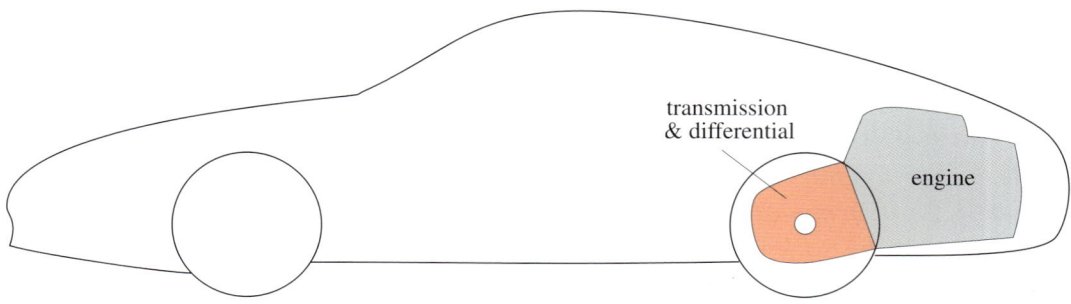

FIGURE 2-8 Rear-engined, rear-wheel drive arrangement (Porsche 911)

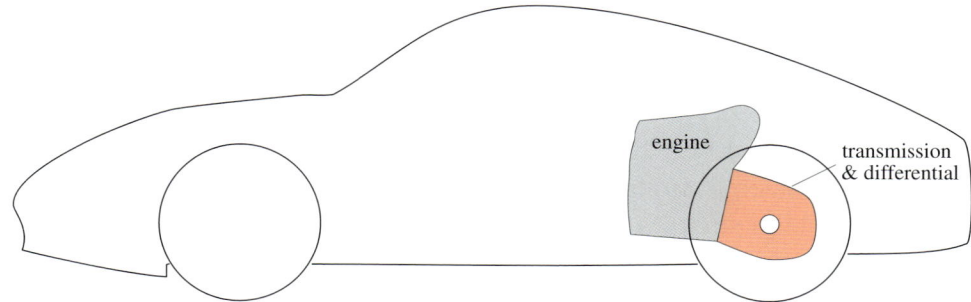

FIGURE 2-9 Mid-engined, rear-wheel drive arrangement (Porsche Cayman)

of inertia. Some front-engined, RWD cars such as the *Maserati Quattroporte* place the engine entirely behind the front wheels to reduce the polar moment of inertia. Maserati refers to it as a *front-mid-engined* vehicle. It has lower polar moment of inertia than a conventional front-engined car with its engine over the front axle and has superior handling.

The original VW Beetle was designed by Ferdinand Porsche on the order of the Nazi tyrant Adolph Hitler, who decreed that there should an inexpensive "people's car" (*Volkswagen* in German) for the masses. He was something of a car enthusiast and admired the Tatra cars produced in Czechoslovakia. Some of the Tatras were rear-engined, rear-drive models with air-cooled, horizontally opposed (boxer) engines. Hitler essentially ordered Porsche to design a car similar to the 1933 Tatra V570 designed by Hans Ledwinka and laid down the specifications to Porsche. He wanted a car that was simple, could carry a family of five at 62 mph (100 kph), get 32 mpg, and be air cooled. He was already planning to build the Autobahn, a high-speed highway system, and wanted a car that could use it. He also named it the people's car. Porsche and Ledwinka worked on the design together and by 1937 had built some prototypes and tested them. Just as it was ready for production in 1938, Tatra sued for patent infringement. Hitler stopped the lawsuit by invading Czechoslovakia in 1938 and taking over Tatra in the process.[1] During WWII, production of VWs was limited to military vehicles. Many VW Beetles and jeep-like variants were produced in Army brown.

Allied bombing had destroyed much of the VW factory and it was handed to the British by the American Army during the allied occupation of post-war Germany. The British auto manufacturers did not think the Beetle was worth producing. It was too different from their idea of what a car should be. An official British report at the time declared: *the vehicle does not meet the fun-*

---
[1] After the war, Germany had to pay Tatra a 3,000,000 DM settlement for stealing their patented inventions.

damental technical requirement of a motor-car ... it is quite unattractive to the average buyer ... To build the car commercially would be a completely uneconomic enterprise.[1] One can only wonder what the fate of the post-war British auto industry might have been had they taken the VW Beetle factory back to England and produced it there. As it turned out, by the 1970s, Britain's auto industry was on life support and by the 21st century had been sold off to former competitors from other countries. Volkswagen ended up with Bentley, BMW with Rolls-Royce and Mini, and Ford with Jaguar, Aston Martin, and Land Rover. Ford subsequently was forced by the 2008 recession to sell Jaguar and Land Rover to Tata of India.

Though the British auto companies did not want the Beetle, the British Army of Occupation found it to be a useful car, and the officer in charge of the factory, Major Ivan Hirst, had it rebuilt and resumed producing Beetles in relatively small quantities and then started to export them. The British turned the factory over to the Germans at the end of the occupation, and the rest is history. When the last one rolled out of the last operating VW Beetle factory in Mexico in 2003, over 21 million cars had been produced to that design, far more than the Model T's 15 million. The original VW Beetle was in every sense an Automotive Milestone. It was first imported to the U.S. in 1949 and that model is shown in Figure 2-10. It was said to have inspired Ed Cole's Chevrolet Corvair design, which used a six-cylinder horizontally opposed, air cooled, rear engine and rear-wheel drive.

As described above, Panhard-Levassor defined the "standard" RWD chassis arrangement in 1891 and most everyone adopted it for decades. The engine is mounted longitudinally over the front axle with the transmission bolted to its rear and mounted on the frame beneath the driver's floorboards. A driveshaft connects the back of the transmission to the differential in the rear axle to drive the wheels, as shown

FIGURE 2-10 1949 Volkswagen Beetle *(Courtesy of Pfan70)*

in Figure 2-11, which depicts a typical RWD sedan. This arrangement does **not** have a low polar moment of inertia due to the heavy engine being far forward of the car's CG.

A **clutch** is placed between the engine and transmission. A clutch allows a mechanical connection between two rotating elements so that torque can be transmitted from one to the other. It also allows that connection to be interrupted such that one part (the engine) can rotate while the other (the transmission) is stopped.[2] An IC engine needs to be running to develop any torque, but the transmission stops turning when the vehicle stops because it is mechanically connected to the drive wheels when in gear. The transmission has a neutral position, which also breaks the connection between its input

---

[1] Quoted in an obituary for Major Ivan Hirst in the The Guardian (London, UK), March 17, 2000.

[2] Clutches and their operation are described in Chapter 5.

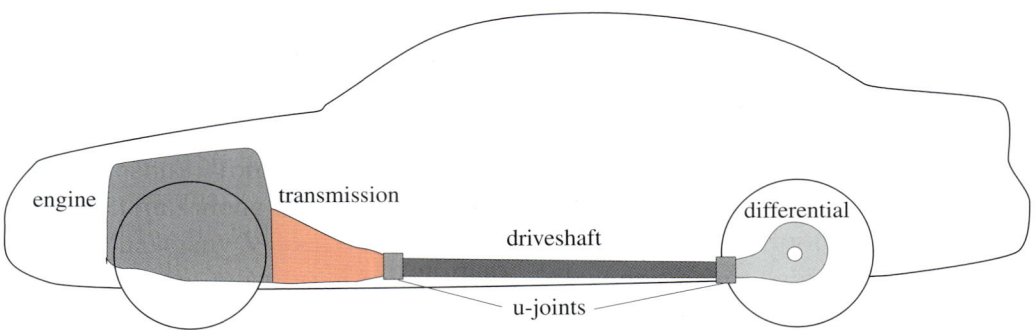

FIGURE 2-11 Front-engined, rear-wheel drive arrangement (BMW 5-Series)

and output, thus allowing the engine to run with the clutch engaged and the car stopped if the transmission is in neutral.

But, to start the vehicle moving, the engine must be running at sufficient RPM to deliver the needed torque and, initially, the transmission must be in a low gear to connect it to the wheels. In that condition with the vehicle stopped, the clutch must be disconnected by stepping on a pedal. To start the vehicle moving, the clutch is engaged gently and the throttle opened to bring engine and transmission smoothly to the same speed as the car begins to move. Subsequent shifts to higher-ratio gears are done by disengaging the clutch while the gears are shifted and then reengaging it to again deliver torque to the wheels. Automatic transmissions do not have a clutch between engine and transmission. Rather they have a **fluid coupling** or **torque converter** that allows slippage to let the engine continue running when the transmission is stopped.[1]

The Panhard-Levassor arrangement shown in Figure 2-11 has the transmission bolted to the engine with the combination mounted to the frame. The power passes from the transmission to the rear axle through a driveshaft that drives a bevel gear train in the axle to turn the power 90° to the wheels. The differential is built into the bevel-gear rear end, which is on the rear axle. Some means is needed to compensate for the change in angle of the driveshaft as the rear axle traverses bumps. Universal joints are fitted to each end of the driveshaft and can transmit torque between two shafts through relatively small angles.

A Cardan universal joint (also called a U-joint or a Hooke's coupling) is shown in Figure 2-12. It consists of four pivots arranged in the form of a cross or plus sign called a spider. Two identical yokes connect to opposite pairs of pivots on the spider. These yokes transfer the torque and rotation between two shafts at an angle but have the problem that the speed of the driveshaft will vary within each revolution when a constant speed is applied to the input side of the joint if there is an angle between the yokes. To counter this problem, they must be used in pairs set 90° out of phase and with the same deviation angle for both joints. The second joint subtracts the speed error introduced by the first with the result that the shafts on either side of the driveshaft turn at the same constant speed. An animation of a Cardan joint can be seen at: http://en.wikipedia.org/wiki/Constant-velocity_joint#mediaviewer/File:Universal_joint.gif

FIGURE 2-12 A Cardan universal joint or Hooke's coupling (Photo by the author)

---

[1] The operation of various transmissions is discussed in detail in Chapter 5

CHAPTER 2   CHASSIS LAYOUTS AND DRIVELINES

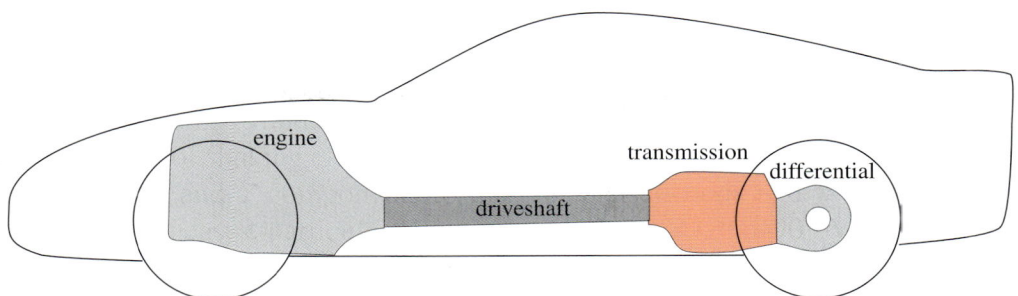

FIGURE 2-13   Front-engined, rear-wheel drive arrangement with transaxle (C6 Corvette)

Some front-engine, RWD arrangements place the transmission at the rear axle, bolted to or incorporated with the differential instead of bolting it to the engine. This is called a **transaxle** and its advantage is to better balance the car's mass evenly between front and rear wheels. The driveshaft then connects the engine crankshaft through a clutch to the transaxle and turns at engine speed rather than at the transmission's output speed. Figure 2-13 shows a C6 Corvette which has a transaxle.

A front-rear balance close to 50%–50% is considered to provide superior handling. A nose-heavy car will tend to **understeer** or resist turning and threaten to break the front end loose in a tight turn. A rear-heavy car may want to **oversteer** or turn too readily, threatening to break the rear end loose in a tight turn. It is often said that if you lose control of an understeering car in a turn, you will go off the road looking where you are going. But lose control in a turn with an oversteering car and you may go off the road looking at where you have been.

The arrangement shown in Figure 2-13 has the differential and transaxle mounted on the car's frame or body structure rather than on the rear axle. Each rear wheel assembly is independently sprung. This is a superior arrangement because the heavy mass of the differential and/or transaxle are not part of the unsprung mass of the wheels. This markedly improves the vehicle's ride. The **sprung/unsprung ratio** of a vehicle is discussed in the chapter on suspensions.

Front-Wheel Drive (FWD)

As described in Chapter 1, the *Fardier de Cugno*—the first self-propelled vehicle—was built in 1769, and it was front-wheel drive. But, front-wheel drive (FWD) automobiles really began to be experimented with in the 1890s. A French company patented a FWD vehicle in 1898 and it competed in a race in 1899. They had sold about 400 units by 1901. The German, 1900 Lohner-Porsche, electric-hybrid FWD car was described in Chapter 1. The first U.S. patent for a FWD car was issued around 1900. But FWD did not catch on in those early years and the RWD arrangements came to dominate the market until much later.

It was not until the 1920s that FWD began to appear again, mainly on race cars designed to run in the Indianapolis 500 and similar events. Harry A. Miller designed a FWD race car that ran in the 1925 Indianapolis 500. Alvis in England built some FWD race cars to compete in hill climbs in the same year and made a production version of its race car in 1928 that did not sell well. Tracta (**Tract**ion-**a**vant) in France was formed in 1926 to build a FWD car designed by Jean-Albert Gregoire as a race car, and it was also offered for road use in 1927. A Tracta FWD car won its class at Le Mans in 1929.

No production FWD cars appeared in the U.S. until the Ruxton and the Cord L29 in 1929. In England, the Birmingham Small Arms Co. (BSA) started production of a three-wheeled

41

FWD car that continued to 1936. Other makes were produced in the 1930s in Germany and France. Cord brought out the model 810/812 FWD cars in 1934-1937 and they were moderately successful. But the Cord was an expensive car and the depression made it short-lived. No FWD cars were produced in the U.S. after the Cord 812 until the 1966 Oldsmobile Toronado and its companion, the Cadillac Eldorado. The Eldorado also had the distinction of having the largest displacement V8 engine used in a post-WWII production automobile—8.2L or 500 cu. in. Citroën and other European manufacturers continued to market FWD autos down to the start of WWII and also after the war. Citroën introduced its *Traction Avant* (front drive) auto in 1934, which became the first mass-produced FWD car in the world. Andre Citroën was a fan of Henry Ford, visited Ford's factories several times, and openly adopted Ford's ideas for mass production.

Citroën at that time was the largest automotive producer in Europe and the 4th largest in the world. In 1948, Citroën introduced the *Deux Chevaux* (*2CV* meaning two horsepower or *two horses*[1]) FWD small car for the masses. Despite the name, it had a 12-HP two-cylinder, horizontally opposed, air-cooled engine. Citroën's VP Paul Boulanger's intent was to provide the French people, especially farmers still recovering from the war, with a suitable alternative to the horse and cart. The 2CV was extremely simple. Its interior could be hosed out and was large enough to carry a pig or other small livestock or produce to market. The roof had a nearly full-length canvas sunroof, which when open allowed tall objects to be carried as in a truck. Nearly 4 million 2CVs were produced between 1948 and 1990 and it put France on wheels, much as Ford had done much earlier in the U.S. Figure 2-14 shows a 2CV. A 1953

FIGURE 2-14 A Citroën Deux Chevaux *(Public Domain)*

review in Autocar described the *extraordinary ingenuity of this design, which is undoubtedly the most original since the Model T Ford.*[2] L. J. K. Setright, an automotive journalist, called the 2CV *the most intelligent application of minimalism ever to succeed as a car.*[3]

There are two basic chassis layouts used with FWD, one places the engine longitudinally in the chassis (front to rear). Others place the engine transversely (side to side). All the early examples (until the 1950s) placed the engine in a conventional longitudinal orientation in the chassis behind the front axle, except that most turned the engine backwards, with the transmission and differential in line with the front axle as shown in Figure 2-15a. This is sometimes called a mid-engined FWD. Some of these mid-engined FWD arrangements put the transmission in front of the axle and others put the transmission in the same case as the differential, side by side. Some current FWD cars (VW, Audi, Subaru) use a longitudinal engine arrangement but place the engine ahead of the front axle and the transmission at the front axle as shown in Figure 2-15b. This is convenient for their all-wheel drive variants as the trans-

---

[1] CV, an acronym for Chevaux (horse), refers to "Tax Horsepower" on which the vehicle's tax was calculated. It comes from a simplistic formula and grossly understates the actual HP.

[2] Compiled by R.M. Clarke. Citroën 2CV Ultimate Portfolio. p. 13. ISBN 1-85520-426-6.

[3] L.J.K. Setright. Drive On!: A Social History of the Motor Car. ISBN 978-1-86207-698-3.

CHAPTER 2 CHASSIS LAYOUTS AND DRIVELINES

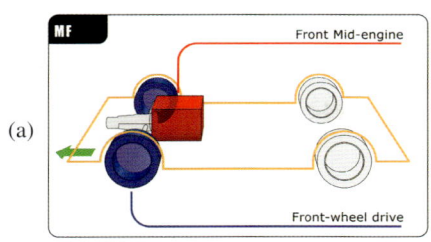

(a)

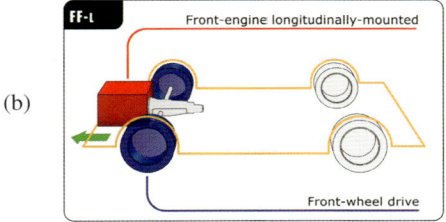

(b)

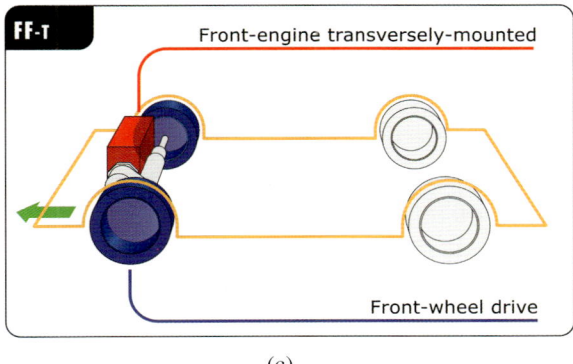

(c)

FIGURE 2-15 Three variations of FWD arrangements *(Courtesy of Moebiusuibeom-en)*

mission output can be taken to the rear wheels with a driveshaft. Most manufacturers, however, have adopted the newer transverse engine arrangement as shown in Figure 2-15c and described below.

Transverse Engine FWD

The 1959 Austin/Morris Mini-Minor is generally credited with starting the "revolution" to transverse-engined FWD cars. However, there were a few earlier examples such as the short-lived English Lloyd in 1946 and the 1955 Suzuki Suzulight in Japan.

Still, the Mini was a ground-breaking, automotive milestone. Designed by Sir Alec Is-sigonis, it pioneered a very efficient layout that maximized usable passenger space (about 80% of total floorpan area) in a very small body as shown in Figure 2-15c. He placed four very small (10-in) wheels at the four corners, suspended them on rubber springs, and placed the engine crossways over the front axle. The transmission and differential were within the engine's oil sump and shared its lubrication. The output was taken by two short "half-shaft" axles to the front wheels. The result was a light, economical car that would accommodate a family of four and also handled extremely well due to the wheels being out at the corners. It also rode well due to the rubber suspension. It was enormously popular in England and also had some success in the U.S. and around the world. The original design, shown in Figure 2-16, remained in production until 2000. BMW bought the company in 1994 and brought out a new, redesigned Mini in 2001 that is still in production.

Most of the FWD cars made today have a transverse-engine layout similar in concept to the Mini. However, they differ in many significant details. The incorporation of the transmission and differential within the engine posed some problems of serviceability and lubrication. Gears generally like thicker oil than an en-

FIGURE 2-16 1959 Morris Mini-Minor *(Courtesy of Mark Brown, Hampton, New Brunswick, Ca.)*

43

FIGURE 2-17   1969 Fiat 128 driveline

gine uses. The Mini engine had to be removed to change a clutch. Also its radiator had to be at the side of the engine, not optimum for good cooling air flow.

In 1969 Dante Giacosa, an engineer at Fiat, developed a slightly different transverse-engine arrangement for the Fiat 128. It placed the engine transversly but mounted a separate transmission on the back of the engine in its normal location and with its own lubrication, as shown in Figure 2-17. One tradeoff was that the two half-shaft axles were now different lengths; this difference can cause some problems with steering. But the radiator was in the normal location in the front of the car. Fiat claimed that 80% of its space was passenger room and used the advertising slogan, *It's smaller than a VW bug but has more room inside than a Cadillac Eldorado.* Although some manufacturers placed the transmission beside the engine (which gave more nearly equal-length half-shafts), most followed the design of the Fiat 128 for their transverse-engined FWD vehicles. It became the new standard and was widely copied.

In Europe, FWD cars have been quite common from WWII to the present. Simca and Citroën concentrated on FWD in France, and Citroën developed some quite luxurious models such as the DS19. VW replaced the Beetle with the FWD Golf/Rabbit in 1974 and Ford brought out the Fiesta in Europe in 1976. Japan converted most of its small car models to FWD. The 1973 Honda Civic and 1976 Honda Accord, and later the 1983 Toyota Corolla were FWD. Most of the cars listed were based largely on the Fiat 128 model.

With the exception of the FWD Oldsmobile Toronado and Cadillac Eldorado, all postwar U.S. cars were rear-wheel drive until the late seventies. The first "gas crisis" occurred in 1973. The Organization of Petroleum Exporting Countries (OPEC) in the Middle East reduced production to drive up the cost of oil. The result was supply shortages, long lines at U.S. gas stations and a sudden realization that the West was somewhat at the mercy, economically, of those countries from which we imported the majority of our oil. This spurred public interest in more economical vehicles. Detroit, at the time, produced only larger cars with large and thirsty engines. The small, light, and economical Japanese and European imports such as the Honda Civic/Accord and VW Rabbit became very popular and sales of domestic autos dropped. Fiat also sold the 128 and its mid-engined, sports-car version, the X1/9, in the U.S.

Detroit automakers scrambled to bring out smaller and more economical cars. Many of these eventually adopted the front-drive, transverse engine design pioneered by the Fiat 128 and Simca Horizon. Chrysler at the time was partnered with Simca and in 1978 quickly re-badged the Simca Horizon as the Dodge Omni/Plymouth Horizon. Ford began to import its German-Ford Fiesta to the U.S. GM had no FWD options available for import until 1979 when its German subsidiary, Opel, brought out the Kadett, which Opel's English subsidiary, Vauxhall, also sold as the Astra. GM designed a new FWD car, the Citation, in 1980, followed it with the Celebrity in 1982, and Ford debuted the FWD Taurus/Sable in 1986. Eventually, by the mid 1990s, U.S. automakers sold mostly FWD models. The Corvette, Camaro, Mustang and Ford/Mercury full-size sedans were about the only RWD U.S. made automobiles left by 2000. Now RWD is having a small renaissance in the U.S. with the Chrysler 300, several Cadillac models, the Mustang, Camaro, and Corvette. Imports such as BMW, Mercedes, Porsche, Lexus, and Infiniti stuck with RWD throughout the FWD revolution. Most light trucks and larger SUVs that are designed on truck chassis such as the Ford Explorer and Expedition, and the Chevrolet Tahoe and Suburban, also remained RWD throughout.

Constant Velocity Joints

Universal joints were discussed briefly in the section on rear-wheel drive and it was noted that the Cardan or Hooke's joints (Figure 2-12) must be used in pairs to cancel the velocity oscillations introduced by each. Gerolamo Cardano, an Italian mathematician and physician, invented the Cardan joint in the 16th century, but Hooke, a 17th century English scientist, recognized its velocity variation problem. He discovered that using two

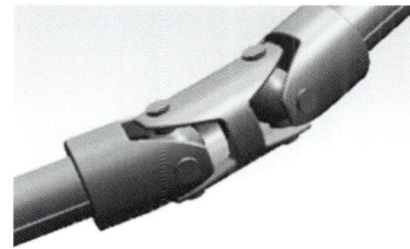

FIGURE 2-18 Double Cardan joint *(Public Domain)*

of them arranged 90° out of phase and having equal angular deviation would cancel the error. Figure 2-18 shows a double Cardan joint as is commonly used in steering columns.

Cardan joints can accommodate only a limited range of angles between shafts (< about 25°) without problems. This range is not enough to comfortably accommodate the desired angle for steered wheels. So their application to front-wheel drive had the two disadvantages of velocity variation and limited angular capacity. Nevertheless, Citroën used them on their Traction Avant in 1934 and they are still used on the front axles of Land Rovers and other off-road vehicles because they are simple and very strong.

The first constant velocity (CV) universal joints were invented about the same time that FWD race cars were being developed. Tracta invented its version for its FWD cars and filed for a French patent in 1926. Figure 2-19 shows

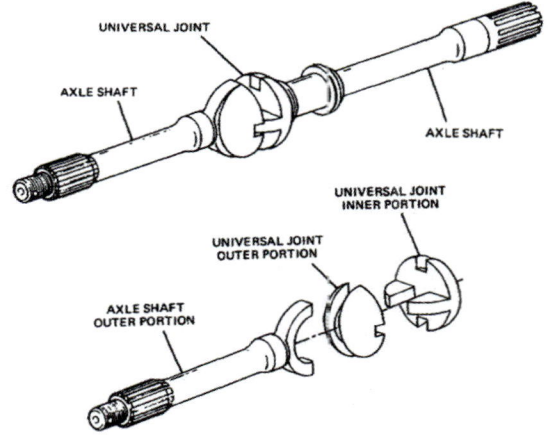

FIGURE 2-19 Tracta CV joint *(Public Domain)*

a Tracta joint. It uses two matching tongue and groove joints set at 90° to each other. There is a fair bit of friction in the joint as the surfaces slide on one another. The intermediate grooved piece in which the two shafts' tongues fit has variable velocity as it turns. However, the 90° phase between the two axles cancels the error and the two shafts turn always at the same velocity.

At about the same time that Tracta was inventing its CV joint, Rzeppa in the U.S. was also developing one that turned out to be superior to Tracta's. Rzeppa used six ball bearings captured in six curved grooves in the two halves. The requirement for constant velocity transmission is that the plane of the balls must always intersect the apex of the two shaft axes. This is called the *homokinetic plane*. Rzeppa's patent drawing and a photo of his joint are shown in Figure 2-20. The balls provide less friction than the sliding joints of the Tracta. The Rzeppa can accommodate 48° or more of angular deviation. Many present FWD cars use Rzeppa CV joints. Other CV designs such as the Weiss, the Tripod and the Thompson are also sometimes used. An animation of a Rzeppa joint can be found at: http://en.wikipedia.org/wiki/Constant-velocity_joint#mediaviewer/File:Simple_CV_Joint_animated.gif

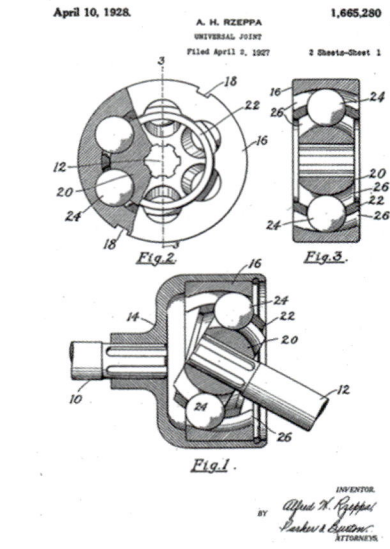

FIGURE 2-20   A Rzeppa CV joint *(Courtesy of Nutzdatenbegleiter)*

Four-Wheel Drive (4WD) and All Wheel Drive (AWD)

Since most automobiles have four wheels, it is not surprising that one might wonder why both *All Wheel Drive* (AWD) and *Four Wheel Drive* (4WD) are used to describe such a vehicle. Don't they mean the same thing? Well, as it happens, there is a difference, and 4WD and AWD are not the same. Historically, the first four-wheel drive vehicles (mostly light trucks) were properly called *part-time four-wheel drive*. Their owner-manuals caution that four-wheel drive should not be selected unless one is driving in slippery conditions (on snow/ice) or off-road on dirt where wheels can slip. If you drive a part-time 4WD vehicle on dry pavement with the transfer case in 4WD, the vehicle will fight you in turns with the front wheels periodically losing contact and hopping as you turn. This is a potentially dangerous condition that could lead to an accident. It also can damage the driveline. Thus the owner-manual's caution to avoid this situation in a part-time 4WD vehicle.

These part-time 4WD vehicles are basically RWD and drive both axles via a **transfer case** that takes power from the transmission tailshaft and transfers it to the front axle and/or rear axle.

CHAPTER 2   CHASSIS LAYOUTS AND DRIVELINES

The transfer case is engaged by a shift lever in the cockpit when the vehicle is stationary (Figure 2-21) and the driver can select neutral, 2WD, or either of two speeds in 4WD—low or high range. In 2WD only the rear wheels are driven. Once engaged in 4WD, there is a direct mechanical connection between front and rear axles that splits the torque equally between them. Also, in 4WD the front-wheel hubs have to be locked to their driveshafts, requiring, in early vehicles, that the driver get out and turn a locking hub on each front wheel. Some newer versions of these vehicles (typically pickup trucks and the SUVs based on them) shift the transfer case electronically by a selector on the dash and lock/unlock the hubs automatically as well. The hubs are unlocked when in two-wheel drive to allow the front wheels to freewheel and reduce wear on the front driveline.

Just as vehicles need a differential in the axle to allow left and right wheels to rotate at different speeds when turning, a full-time four-wheel drive (AWD) vehicle needs an additional differential between the rear and front axles to allow the front wheels to turn at different speeds than the rear wheels when turning, especially on dry pavement. Figure 2-22 shows a center differential on an AWD Jeep Grand Cherokee.

FIGURE 2-21   Transfer case shift knob
(Courtesy of Gnangarra via Wikipedia)

A transfer case looks a bit like the center differential in Figure 2-22 but it does not have a differential in it.

The need for a third differential for full-time four-wheel drive was recognized as early as 1893 when an English engineer, Bramah Diplock, patented a four-wheel drive system that included three differentials (one in the front axle, one in the rear axle, and one between the axles) as well as four-wheel steering. The 1900 Lohner-Porsche four-wheel-drive vehicle was described in Chapter 1, and it would now be called an Individual Wheel Drive (IWD) vehicle, a term recently coined to describe hybrid or pure-electric vehicles that have electric motors

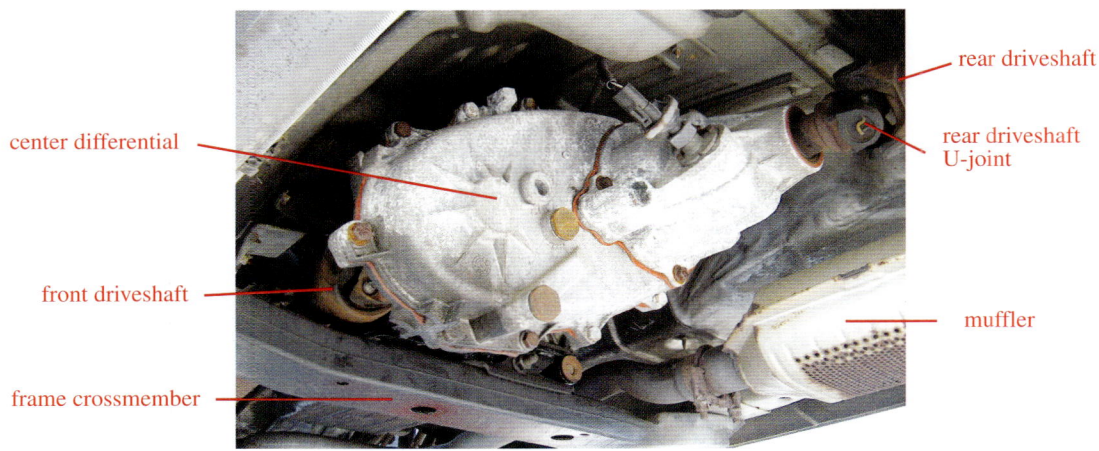

FIGURE 2-22   Center differential of an AWD Jeep Grand Cherokee *(Public Domain)*

47

at each wheel. An IWD vehicle does not need any differentials as the individual wheel motors can operate at different speeds.

The first four-wheel drive automobile with an internal-combustion engine was the Spyker 60 HP, built by the Dutch Spijker brothers in 1903, shown in Figure 2-23. It also had the first six-cylinder engine in any automobile and four wheel brakes. Presumably, it had a center differential as it was used in racing. Other 4WD examples were built in the early 20th century in the U.S., Britain, and Russia. During WWI, these were aimed primarily at military applications. The Four-Wheel Drive Corporation in Wisconsin delivered about 15,000 4WD trucks to the U.S. Army during WWI.

Production of 4WD vehicles took off during WWII with the Jeep leading the way. The Jeep was designed by Bantam, but most were produced by Willys and Ford. (The official Army name for it was *General Purpose Vehicle* (*GP*)—the GIs dubbed it the Jeep.) An updated civilian derivative of the WWII Jeep is still produced as the Wrangler by Chrysler Corp. After WWII, Kaiser, who had bought Willys, introduced the Jeep Wagoneer in 1963, a station-wagon-like 4WD vehicle that became quite popular. It offered *Quadra-Trac* full-time 4WD with center differential for the first time in 1973. Its descendents, now called Jeep Cherokee and Grand Cherokee, are still produced by Chrysler.

The first full-time 4WD system in a production sports GT was the Jensen FF, sold from 1966 to 1971, shown negotiating snow in Figure 2-24. Since 1980, many manufacturers have offered full-time 4WD variants in passenger cars, SUVs, and sports cars. Somewhere along the line, the marketing departments must have decided that "Full-Time Four-Wheel-Drive" was too large a mouthful for advertising purposes and coined the term *All-Wheel Drive* (AWD) to refer to systems with a third differential. The term 4WD is still

FIGURE 2-23 1903 Spyker Four-Wheel Drive Automobile
(Public Domain)

used for vehicles that are only part-time 4WD, and that is the difference between the two terms.

Figure 2-25 shows four arrangements for 4WD or AWD. Figure 2-25a shows a vehicle that starts as front-engined RWD and has either a transfer case added to create part-time 4WD or adds a center differential for an AWD arrangement. Most 4WD trucks and large SUVs use the transfer-case version of Figure 2-25a. Examples are the Ford F-series, Explorer and Expedition, the Chevrolet and GMC K-series, Tahoe/Sierra and Suburban/Yukon, and the Dodge Ram. The first mass-production automobile made in America that used the AWD arrangement of Figure 2-25a was the American Motors Eagle in 1980 as shown in Figure 2-26. It was offered in sedan, station wagon,

FIGURE 2-24 1966 Jensen FF four-wheel drive GT
(Courtesy of Francis Pullen)

CHAPTER 2 CHASSIS LAYOUTS AND DRIVELINES

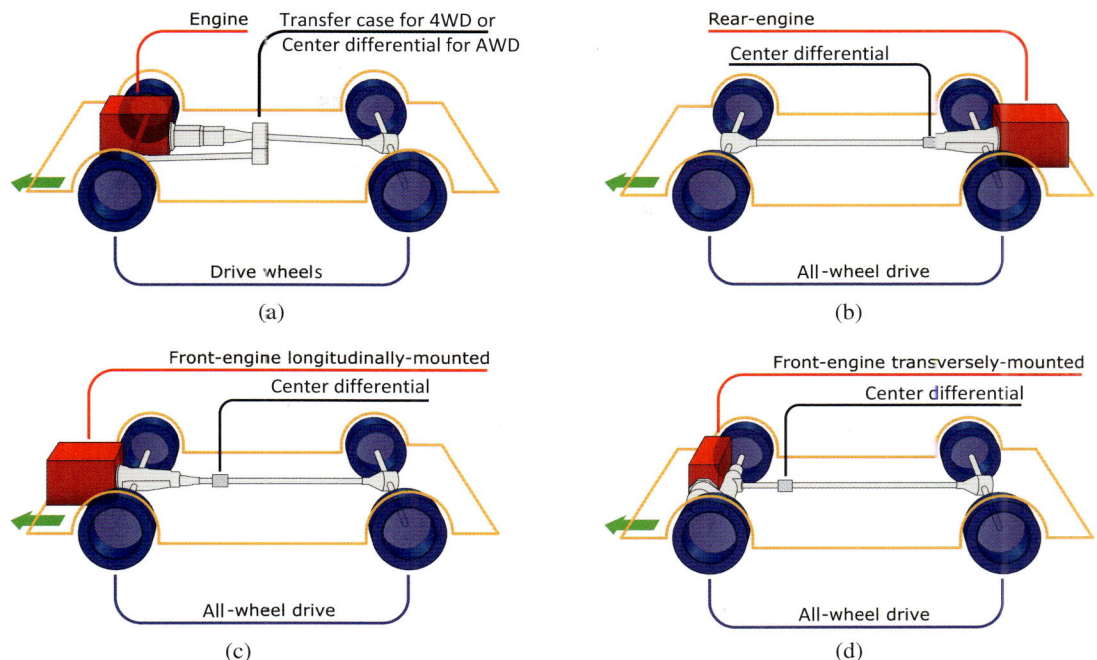

FIGURE 2-25 Four variations of AWD/4WD *(Adapted from Moebiusuibeom-en)*

and convertible body styles. This car proved very popular and started a trend that continues today with additional marques adding AWD in all of the variants of Figure 2-25 to their lineup. Examples of a Figure 2-25a AWD arrangement are found in some current Jeep, Dodge, BMW, Mercedes, and Maserati models.

Figure 2-25b shows a vehicle that starts as a rear-engined RWD model and adds a center differential and driveshaft to drive the front wheels in AWD. An example is the Porsche 911 AWD model. Figure 2-25c shows a vehicle that starts with longitudinal, front-engined FWD and adds a center differential and driveshaft to power the rear wheels in AWD. Audi and Subaru use this arrangement on most of their models. Audi introduced its Quattro AWD series in 1980 and dominated rally racing for a decade. Subaru be-

FIGURE 2-26 1980 AMC Eagle All-Wheel Drive Automobile *(Courtesy of Christopher Ziemnowicz)*

gan offering AWD in some models in 1972 and made all their exports to the U.S. AWD in 1996. Their recent BRZ sports car is the only exception.

Figure 2-25d shows a vehicle that starts with transverse-engined FWD and adds a center differential and driveshaft to power the rear wheels in AWD. An example is the Toyota Rav4. Another Toyota offering, the 4Runner, has driver-selectable 2WD, 4WD, and AWD. For economy on dry pavement, 2WD is selected. For slippery conditions, or for improved handling on dry pavement, AWD can be selected, which uses a center differential. Selecting 4WD locks the center differential, making it useful for snow-plowing and the like. Other arrangements, not shown, start with a mid-engined, RWD car and add a center differential and driveshaft to the front wheels. Examples are the Audi R8 and some Lamborgini models.

One of the most unusual arrangements was the Citroën 2CV Sahara model, shown in Figure 2-27. To make a vehicle capable of traversing the sands of the Sahara Desert, Citroën took their front-engined FWD 2CV and added a second powertrain with another identical engine turned backwards to drive the rear wheels. The car could be run on either engine independently or both simultaneously. Not only did it give superior traction in the desert, but it did not leave you stranded if one engine quit. They built 694 examples that were used by French oil companies, the military, and police. About 27 are still in existence, one being in the Revs Institute Museum in Naples, FL and one in the Tampa Bay Automobile Museum (Florida).

An example in the collection of the Revs Institute, Naples, FL

An example in the collection of the Tampa Bay Auto Museum, Pinellas Park, FL

FIGURE 2-27 Citroën 2CV Sahara, two-engined, all wheel drive vehicle *(Photos by the author)*

Chapter 3
Engine Configurations

A large variety of engine configurations have been tried and most of them are still in production. The simplest inline configurations have their multiple **cylinders** arranged in a single plane. V-arrangements put half the cylinders in one plane and the others in a second plane arranged at an angle to the first. Boxer or opposed engines have opposite cylinders at 180°. W-configurations have cylinders arranged in either three or four mutually intersecting planes. Radial and rotary engines have their cylinders arranged like the spokes of a wheel. The number of cylinders can range from one to as many as thirty. This chapter will discuss these configurations in some detail and explain their advantages and disadvantages. We will also discuss the art and science of engine balancing to reduce vibration. The discussion will focus primarily on Otto four-stroke engine designs as these dominate the automotive industry. Where significant differences exist for two-stroke engines, they will also be discussed. Figure 3-1 shows some of these configurations.

Engine Construction

The earliest engines were mostly single-cylinder or multi-cylinder inline and were typically made from several cast-iron castings. The crankcase, as its name implies, was a case to house the **crankshaft** and was only tall enough to contain the swing of the **crank** throws. It provided internal webs to surround and fully support the crankshaft's main pins in bearings. The bottom surface of the crankcase was machined flat to allow the lower half of the crankcase, which was typically another casting and later a steel stamping, to be bolted in place. The top surface of the crankcase was machined flat to allow the cast-iron cylinders to be bolted to it. Some early engines had a separate casting for each **cylinder** that bolted to the crankcase. The casting housed one **piston** and **connecting rod**, and formed a combustion chamber surrounded by water passages for cooling.

In early engines, there was no separate head as in modern engines. The cylinder had to be lowered over the piston and connecting rod at assembly. It was more common to have the cylinders cast in pairs with two pistons and two combustion chambers in a single casting. A four-cylinder engine of this type would have two cylinder castings, a six-cylinder three. This approach was probably

The opening photograph is of a Jaguar, Series 1, E-Type DOHC, six-cylinder inline engine taken by the author at the Larz Anderson Auto Museum's British Car Day, 2005, Brookline, MA.

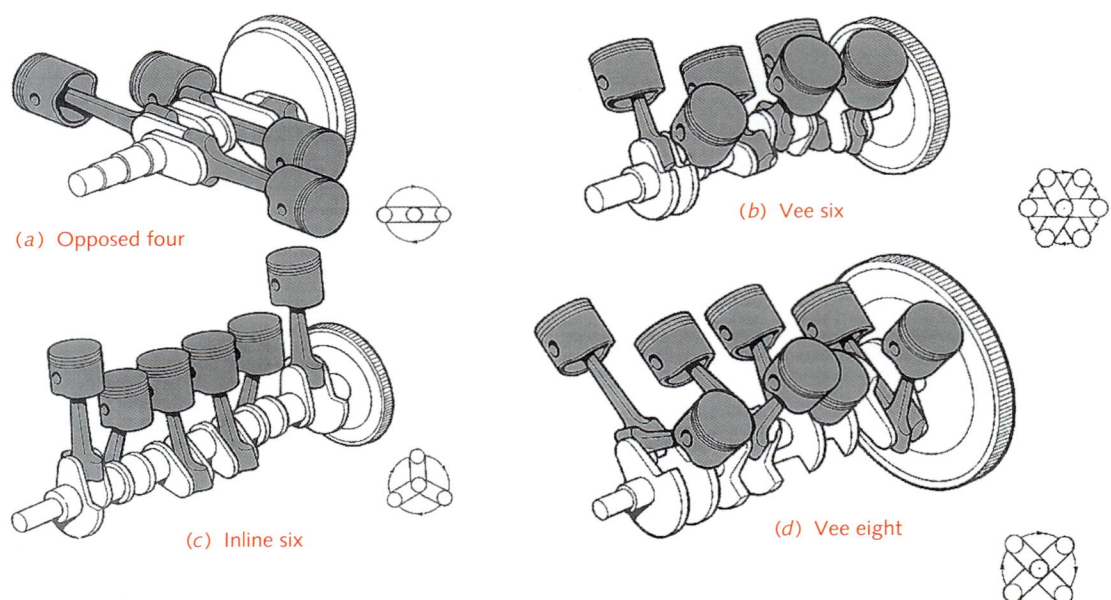

(a) Opposed four (b) Vee six (c) Inline six (d) Vee eight

FIGURE 3-1 Various multicylinder engine configurations

a result of the limited casting technology of the time. Figure 3-2 shows a 1905 De Dion Bouton (French) two-cylinder engine with a crankcase cast in two halves and its two cylinders cast together. Figure 3-3 shows a 1905 Wolseley (English) six-cylinder engine with cylinders cast in pairs, a cast crankcase extending well below the crankshaft centerline to support the bearings, and a stamped-steel oil pan.

Later engines were cast "en-bloc," a French term meaning *all together* or *in a block*, with a single casting containing both the crankcase and all the cylinder **bores**. The bores were open at top and bottom. A separate casting, called the head, was bolted to the top of the block to form the combustion chambers. The en bloc configurations were a more difficult casting to make, but they simplified the engine assembly process. The crankshaft was installed first with the block upside down. The pistons and attached **connecting rods** or **conrods** were assembled from the top with the block right-side up and the conrods fastened around the crank pins from below. This arrangement is the standard for engines today. Figure 3-4 shows a four-cylinder engine cast en bloc, circa 1919, and Figure 3-5 shows a modern BMW six-cylinder block casting. Note the extensive ribbing to provide strength with minimum weight.

FIGURE 3-2 1905 DeDion Bouton, two-cylinder engine with separate crankcase and cylinders
(Public Domain)

CHAPTER 3 ENGINE CONFIGURATIONS

FIGURE 3-3 1905 Wolesly, six-cylinder engine with separate crankcase and cylinders cast in pairs
(Scanned by Andy Dingley from Public Domain source.)

FIGURE 3-4 1919 four-cylinder engine cast en bloc
(Scanned by Andy Dingley from Public Domain source.)

FIGURE 3-5 Modern BMW six-cylinder engine cast en bloc
(Courtesy of Carter)

Cylinder Arrangements

The simplest arrangement is a single-cylinder engine. These are quite common, being used on lawnmowers, snowblowers, outboard motors, chain saws, and other small machines. Their main advantage is low cost and light weight. They have some disadvantages, however. They fire only once every two revolutions if a 4-stroke or once per revolution if a 2-stroke. The force from the piston's acceleration up and down on every rotation cannot be fully balanced with only one cylinder, and this causes significant vibration. This force can be quite large, especially if the engine is run at high speed. This **shaking force**, also called an **inertial force**, is due to the piston's mass being accelerated and decelerated twice per crankshaft revolution. The magnitude of shaking force is equal to the **mass of the piston** times the **crank radius** times the **square of the crank speed**. Thus, **doubling the crank speed** will **quadruple the shaking force** (two squared equals four). Even in a small engine, this force can amount to thousands of pounds

53

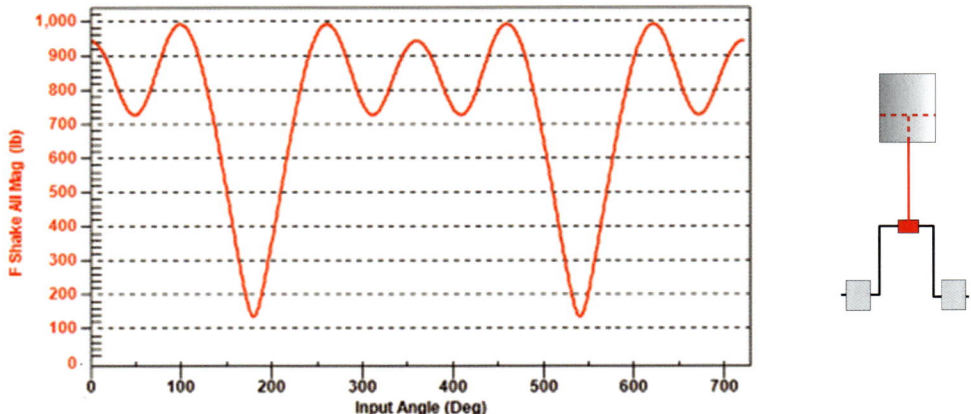

FIGURE 3-6 Shaking force in a 410 cc motorcycle engine at 3400 rpm over two crankshaft revolutions

at high crankshaft speeds. It will be felt at the motor mounts and cause vibration. Figure 3-6 shows a plot of the shaking force of a small, properly balanced, single-cylinder motorcycle engine of 410 cc displacement, 76.2 mm (3 in) bore by 89.8 mm (3.5 in) stroke at 3400 rpm. Its peak value is nearly 1000 lb. It is shown over two cycles (720°) and repeats each cycle.

Multi-Cylinder Inline Engines

When multiple cylinders are placed in any configuration, the engine designer is faced with some very basic design decisions. The first is how to arrange the multiple cranks on the crankshaft. They could be at any angles with respect to one another and this decision will profoundly affect the engine's behavior with respect to its sequence of power pulses and its shaking forces. The angle that each crank makes with respect to a reference or zero angle is called its **phase angle**. The phase angle of the crank for the first cylinder (typically at the front of the engine) is always defined as zero degrees and is the reference for all the other cranks' phase angles.

There are two factors that are affected by the choice of phase angles: the spacing of power pulses within the cycle and the potential cancellation of piston shaking effects. Unfortunately these two factors are often in conflict, meaning that making one better can make the other worse. The engine designer needs to decide which of these factors is more important if they cannot make both be good.

Two-Cylinder (Twin) Inline Engines

Gottlieb Daimler designed an inline twin engine in 1895 that was used in the Panhard auto of the same year. De Dion presented the inline twin shown in Figure 3-2 in 1905. Automobiles soon moved to four and more cylinders to get more power. Most modern use of inline twins was in motorcycles such as Triumph, BSA, Norton, and others in England. Honda and Yamaha also used them in some models. Some very small "micro-cars" from NSU, Mitsubishi, Subaru, Fiat, and others used them including the current Fiat 500, Panda, and Punto models. Few of these cars were imported to the U.S.

The ideal spacing of power pulses from multi-cylinders is considered to be that which spaces them evenly over the cycle. This is called *even firing*. Remember that the cycle here is two revolutions of the crank in a 4-stroke engine, or 720°. That is how far the crank must turn to complete one combustion

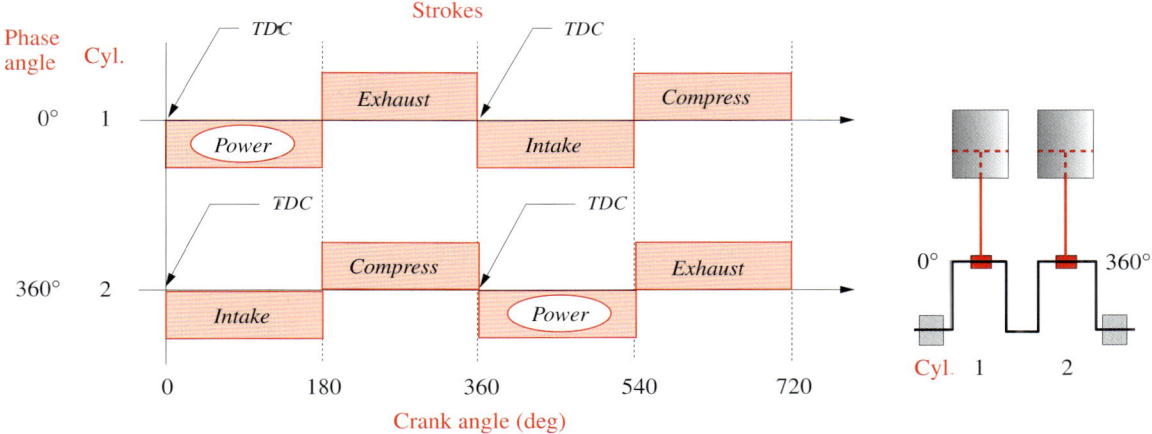

FIGURE 3-7 Even-firing, four-stroke, inline two-cylinder crank phase diagram with $\phi_i = 0, 360°$

cycle. As a simple example, consider a two-cylinder, 4-stroke, inline engine. The ideal power pulse spacing in this case would be to have cylinder one fire at zero degrees and cylinder two fire at 360 degrees, as shown schematically in Figure 3-7. Here the up and down motions of the pistons are shown schematically as positive and negative rectangles. Each up or down motion consumes 180° of crank rotation. One up motion plus one down motion is one full crank rotation or 360°. A power pulse can only occur on a down motion. The other down motion is the intake stroke. The up motion after the intake stroke compresses the fuel-air mixture in preparation for the power stroke, and the second up motion exhausts the spent charge from the cylinder.

Note that the two pistons in Figure 3-7 go up and down together, which doubles the shaking force compared to a single cylinder with the same bore and stroke. This engine is even firing as shown by the evenly spaced gas force pulses in Figure 3-8. But, unfortunately, this engine has no cancellation of shaking forces. Compare the shaking force of Figure 3-8 to that of the single cylinder of Figure 3-6, which has the same bore, stroke, and masses per cylinder. Shaking force is doubled. We have chosen even firing over better balance in this example. The balance can be improved, but only at the expense of even

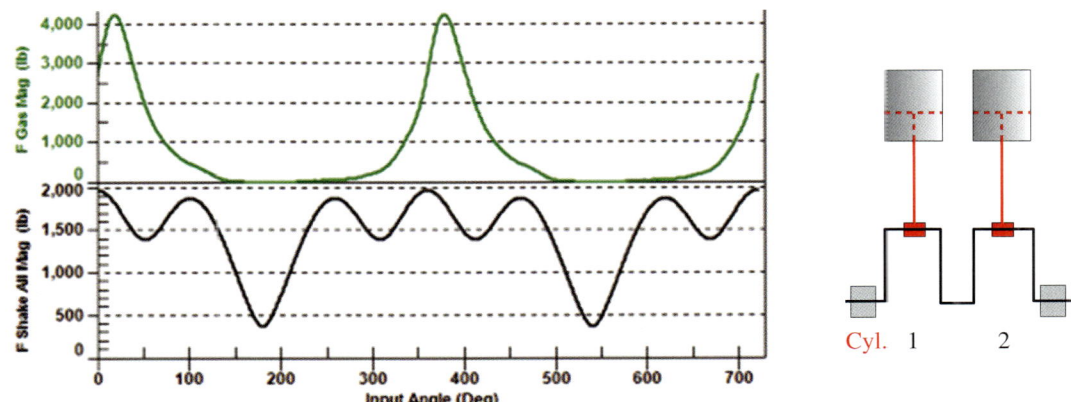

FIGURE 3-8 Gas and shaking force in a two-cylinder, inline engine at 3400 rpm over two crankshaft revolutions

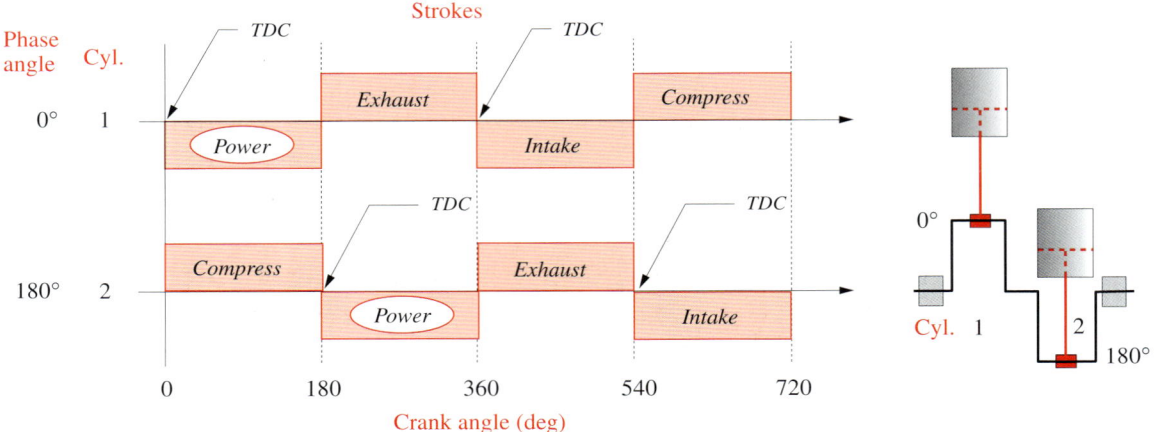

FIGURE 3-9 Uneven-firing, four-stroke, inline two-cylinder crank phase diagram with $\phi_i = 0, 180°$

firing with only two cylinders. It will require at least six cylinders inline to get both even firing and good balance in the same engine.

To balance inertial forces, we need to arrange the crankshaft so that one piston is always moving in the opposite direction to another. Figure 3-9 shows such an arrangement for this two-cylinder inline engine. The only change versus Figure 3-7 is to place the second crank at 180° with respect to the first. Then one piston will generate an upward force while the other generates a downward force and vice versa. Figure 3-10 shows that the balance condition is improved over that of Figure 3-8 in that the peak shaking force is reduced from 2000 lb to 800 lb.

The reason the shaking force is not zero in Figure 3-10 is that piston motion is not the only source of shaking force in an engine. The connecting rod oscillates in angle twice per revolution creating a second harmonic of unbalanced force at twice crankshaft speed. The crank's rotation speed is called the first harmonic or fundamental frequency. Notice in the top panel of Figure 3-10 that the first harmonic (labeled F Shake 1) is now zero because the opposite piston motions are canceling it. In the lower panel, there are four complete oscillations of

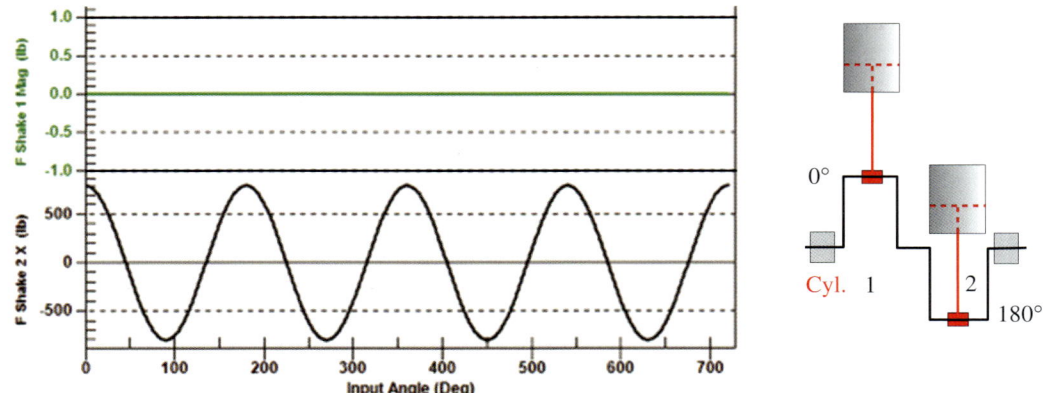

FIGURE 3-10 Shaking force in a 0-180° two-cylinder engine at 3400 rpm over two crankshaft revolutions

CHAPTER 3 ENGINE CONFIGURATIONS

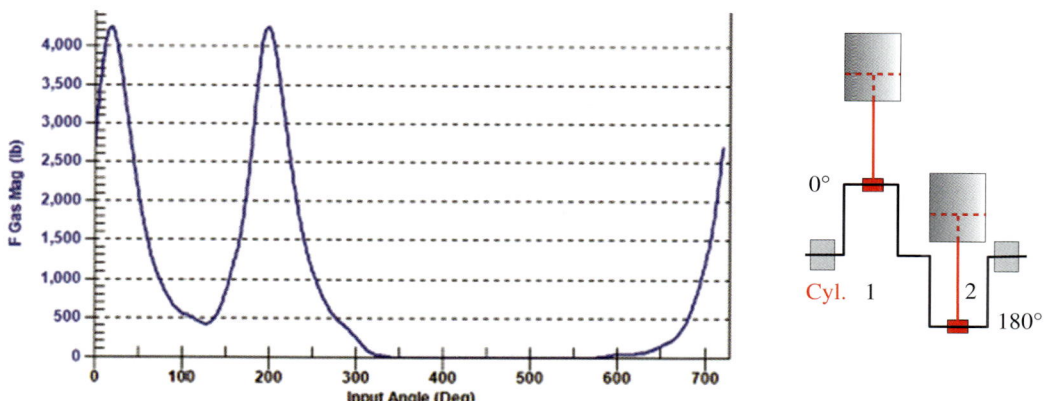

FIGURE 3-11 Gas force in an uneven-firing, two-cylinder engine at 3400 rpm over two crankshaft revolutions

shaking force over two revolutions of the crankshaft. This is the second harmonic of shaking force (labeled F shake 2), oscillating twice per revolution. There are also higher harmonics at fourth, sixth, and all even harmonics, but the magnitudes of harmonics above the second are so small that they can be ignored. In Figures 3-6 and 3-8, the non-zero first and second harmonics are combined (summed) to create the odd-looking curves.

However, as shown in Figure 3-11, placing the crank throws of an inline twin at 0 and 180° results in the firing pulses being unevenly spaced over the cycle. So there is a clear trade-off between even firing and improved dynamic balance in this and in many engines. It is up to the engine designer to choose between these less than optimal solutions. Uneven firing will sound and run roughly at low speeds. Unbalance will create vibrations, and these become worse at high speeds. A slow-running engine can withstand more unbalance than a fast-running one.

Most British inline twin motorcycles, and some smaller displacement Hondas, used a 0, 360° crankshaft to get even firing. Some larger Honda and Yamaha inline twins over 500cc displacement used a 0, 360° crankshaft but added a balance shaft to cancel the primary vibration.

Other Honda twins use a 0, 180° crankshaft and accept uneven firing in exchange for better balance. Yamaha used a 0, 270° crankshaft in its TRX850 inline twin which, though uneven-firing, provided about 2/3 the vibration of the even-firing 0, 360 crank and also had zero secondary force. They added a balance shaft to cancel its primary force. Balance shafts are discussed in a later section of this chapter.

The 0, 270° crank in an inline twin has 315° between firing pulses, which sounds a bit like a Harley-Davidson V-twin with its 405° between pulses. This may have been some of its appeal. The distinctive, "potato-potato" sound of a Harley-Davidson motorcycle is due to it being an uneven-firing engine. The Harley also suffers from significant vibration. Its engine is a V-twin, which are as difficult to make even-firing with reasonable vibration as is the inline two-cylinder. V-twins are discussed in a later section of this chapter. The cure for this problem is to add more cylinders, and the early automobile-engine designers recognized this. They moved to four-cylinder engines early and not much later to six and more cylinder combinations.

Four-Cylinder Inline Engines

Until 1901, all automobile engines were one or two cylinders. John Wilkinson, a Cornell engineering graduate, had been working on the design of a four-cylinder, air cooled engine since 1897 and by 1901 had it working well. He mounted the engine cross-ways in the front of the chassis to promote cooling and drove the rear wheels via chain. Wilkinson brought his prototype to the H. H. Franklin Co. in Syracuse, NY. Mr. Franklin had developed a process for die-casting metals and manufactured them. Franklin was interested in automotive technology and, after Wilkinson took him for a ride in his machine, Franklin agreed to partner with him to manufacture it. Franklin was the businessman and Wilkinson was the engineer. This began a long and profitable business partnership that manufactured the only air-cooled cars until the depression ended it in 1934. In 1902, Locomobile produced the first water-cooled, four-cylinder, inline engine. After that many other companies jumped on the four-cylinder bandwagon. Henry Ford adopted it for the Model T; ultimately it became, and still is, the most used automobile engine configuration in the world.

One crankshaft arrangement has been used for the 4-stroke, four-cylinder inline engine as it gives the best compromise between even firing and reasonably low vibration. Its phase angles are 0, 180, 180, 0° from front to back as shown in Figure 3-12. This arrangement gives even firing and cancels the first **harmonic** of **shaking force**, and **shaking moment** because two pistons are always moving opposite to the other two and the crank throws are mirror-symmetric around the middle of the crankshaft. However, the second harmonic at twice crankshaft speed, which is an indirect result of connecting-rod motion, is out of balance in this engine as well.

Shaking moment refers to the turning force (force times distance) or **moment** that is created by pairs of oppositely directed forces (called a **couple**) that are generated by pairs of oppositely moving pistons. A moment has the units lb-in or N-m. Figure 3-13 shows these couples as pairs of straight arrows and their moments as circular arrows Their effect is to try to rock the engine forward and backward in its mounts like a bucking bronco. The 0, 180, 180, 0° arrangement of crank throws serves to generate two oppositely directed shaking moments that act to cancel each other. One is trying to rock the en-

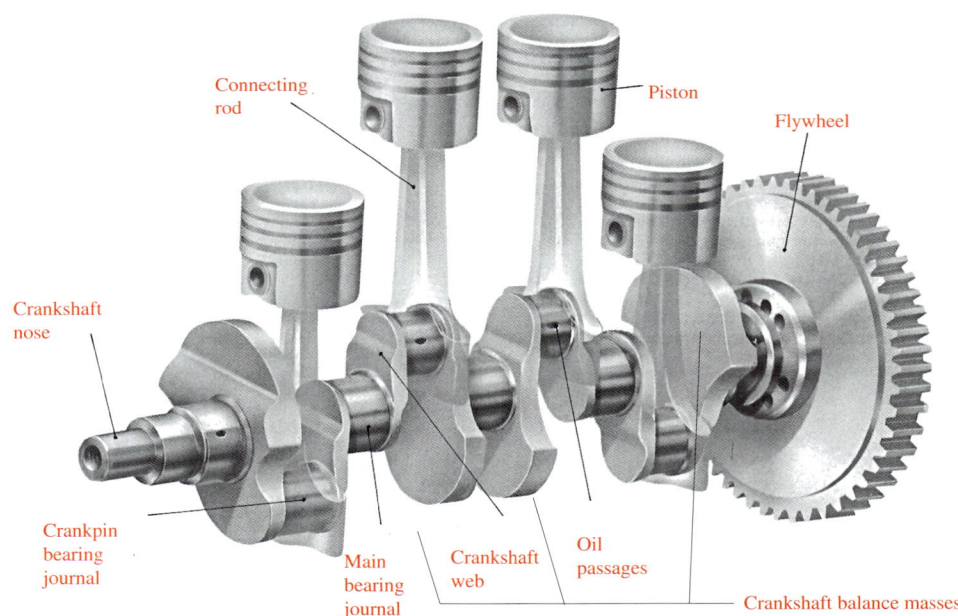

FIGURE 3-12 Crankshaft from an inline, four-cylinder engine with pistons, connecting rods, and flywheel

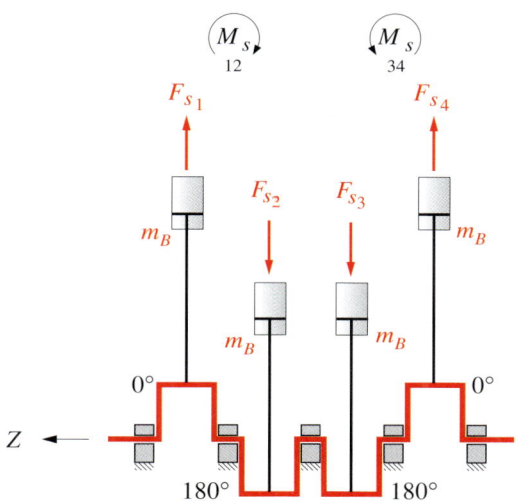

FIGURE 3-13 Four-cylinder 0-180-180-0 crankshafts cancel primary moments as well as primary forces.

two-cylinder engine of Figure 3-10 has both unbalanced primary and secondary shaking moments.

The magnitudes of the secondary components of force and moment are about 1/3 to 1/4 of the magnitudes of the primary components. Engine designers consider it important to balance or cancel the primary components because they are large, but are often willing to live with secondary forces and moments being out of balance. Proper motor-mount design can serve to hide the smaller secondary vibrations from the driver and passengers. Unless some steps are taken to cancel it, all four-cylinder inline engines suffer from secondary vibration at twice crankshaft speed. If it becomes too large to hide, balance shafts running at twice crankshaft speed can be added to cancel the secondaries at some additional weight and cost. Balance shafts are discussed in a later section of this chapter. The upper limit of displacement for four cylinders is considered to be about 2.7 liters, though some manufacturers have made larger examples. Four-cylinder engines larger than about 2-liter displacement often add sec-

gine clockwise and the other counterclockwise simultaneously. However, cancellation affects only the first harmonic, also called the **primary moment**. The **secondary moment** from the second harmonic in the **conrod** motion does not cancel. So this engine is said to have primary balance (primary forces and moments cancel), but it has secondary imbalance. Note that the

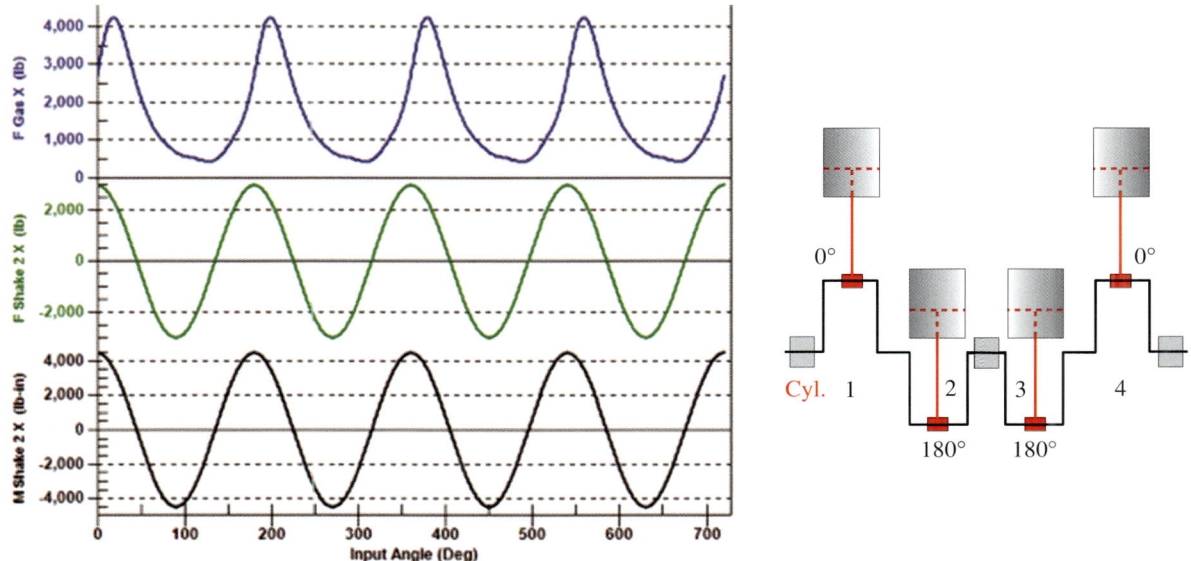

FIGURE 3-14 Gas force and secondary imbalance in a four-cylinder inline engine at 3400 rpm

AUTOMOTIVE MILESTONES

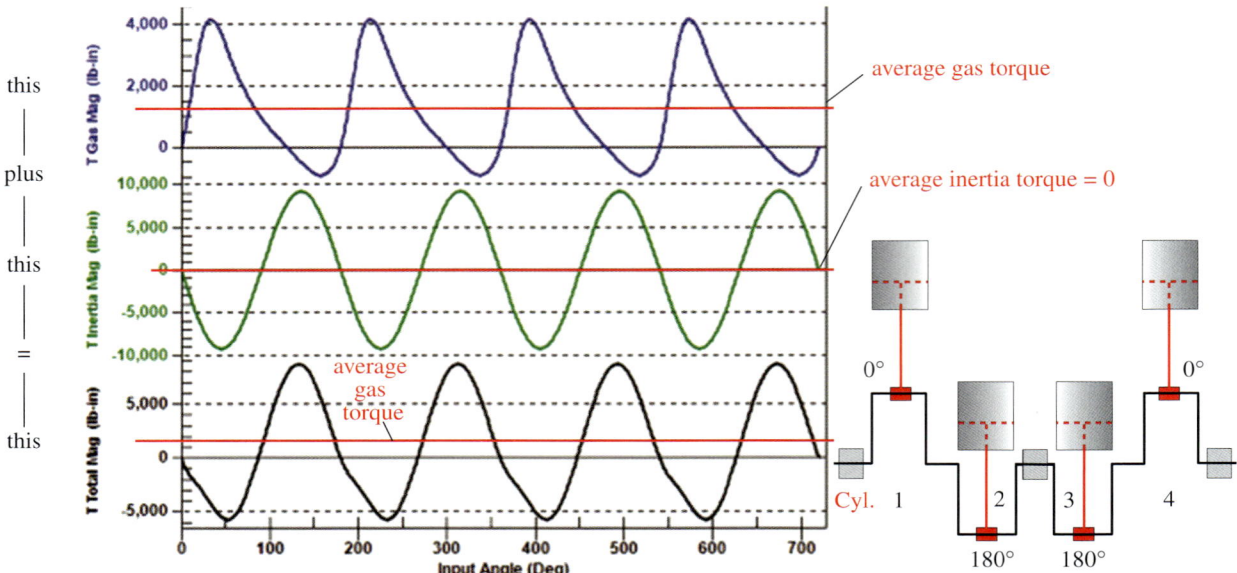

FIGURE 3-15 Gas torque, inertia torque, and total torque in a four-cylinder inline engine at 3400 rpm

ondary balance shafts to cancel their larger secondary forces due to the heavier pistons. Figure 3-14 shows the gas force, secondary shaking force and secondary shaking moment for a standard inline four-cylinder, four-stroke engine, respectively, from top to bottom.

What drives the vehicle, however, is the torque developed by the piston forces acting to turn the crankshaft. There are two components of engine torque: gas torque due to the gas force, and an inertia torque due to the inertia force associated with piston acceleration. Total torque is the sum of these two components. Gas torque, which drives the vehicle, has a positive average value over the cycle. The average value of inertia torque is zero because it has equal positive and negative pulses over the cycle; therefore, the inertia torque does nothing to drive the vehicle. It is an unwanted and potentially detrimental component that serves only to increase stresses in the crankshaft.

Ideally we would like to cancel the inertia torque with a suitable arrangement of the crankshaft phase angles, just as we seek to cancel the inertial/shaking forces and moments by the same technique. With a sufficient number of cylinders and proper choice of crankshaft phase angles, it is possible to cancel all three of these unwanted inertial components: force, moment, and torque. Unfortunately, the minimum number of inline cylinders to achieve this goal is eight. Thus, all the arrangements so far discussed have some or all of these factors present in one or more harmonics. The inline four-cylinder engine, as described above, has zero primary inertia torque, but its secondary inertia torque is nonzero. Figure 3-15 shows the components of torque for this engine, gas torque, inertia torque, and their sum—total torque—from top to bottom, respectively. Note that the gas torque is the only one that drives the vehicle.

Five-Cylinder Inline Engines

In the 1930s, Henry Ford developed an experimental five-cylinder engine but never produced it. The five-cylinder inline engine has not been used as much as other configurations, but it has appeared in some Acura, Audi, GM, Land Rov-

CHAPTER 3 ENGINE CONFIGURATIONS

er, and Volvo models in recent years. It first appeared as a Mercedes diesel in 1974 and shortly after as a gasoline engine in the Audi 100. It was adopted mainly because it is a little shorter than a six-cylinder inline of the same displacement and thus packages in a smaller space.

However, it is not the best balanced of engines. Its standard crankshaft phase angles are 0, 144, 72, 216, 288°. It is even-firing and has complete force balance through the second harmonic, but both its primary and secondary shaking moments are unbalanced. The primary moment can be counterbalanced by adding a balance shaft turning at crank speed and adding corresponding counterweights to the crankshaft, but the secondary moment requires two balance shafts running at twice crankshaft speed. These unbalanced moments are sometimes masked by motor mount design. Its gas force, shaking force, and shaking moment are shown in Figure 3-16. This engine has the nice feature of having zero inertia torque through the third harmonic. No five-cylinder engines are currently in production.

Six-Cylinder Inline Engines

The first six-cylinder engine appeared in the 1903 Spyker shown in Figure 2-23. Napier made the first series-produced inline six in 1903. By 1906, Pierce-Arrow, Stevens-Duryea, and Franklin were offering six-cylinder engines. Oldsmobile followed in 1908, Buick in 1910, and Packard in 1912. By 1909, about 62 manufacturers were offering this engine worldwide. The inline six has been in continuous production since its beginnings and some very famous engines have used this layout. Jaguar won Le Mans in '51, '53, '55, '56, and '57 with their dual-overhead camshaft, hemi-head six in their C- and D-type racing cars. The D-Type morphed into the production E-Type, produced from 1961 through 1986 with an updated D-Type engine. A Jaguar, E-Type engine is shown in Figure 3-17. The same engine was used in their XJ sedans. The Chevrolet "Stovebolt Six," overhead-valve engine was produced from 1929 until 1990. Six-cylinder engines are the second most popular in the world after the inline four.

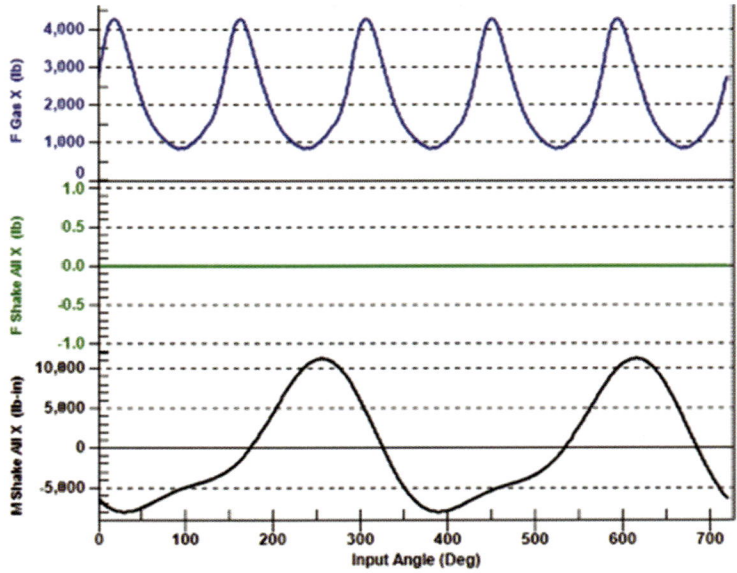

FIGURE 3-16 Gas force and balance conditions in a five-cylinder inline engine at 3400 rpm

FIGURE 3-17 Jaguar E-Type six-cylinder inline engine
(Courtesy of Morven)

The standard crankshaft arrangement for a six-cylinder inline engine has phase angles of 0, 240, 120, 120, 240, 0° from front to back. This gives even firing and complete balance of shaking force and shaking moment through the second harmonic. Thus, the inline six is considered to be one of the best configurations for inertia balance and smooth operation. It has six evenly spaced power pulses, zero shaking force, and zero shaking moment as shown in Figure 3-18. Six is the fewest number of cylinders for which this is true, absent additional balancing mechanisms being added to an engine with fewer cylinders. However, it has one flaw in its balance condition, making it less than perfect in that regard. Its third harmonic of inertia torque is nonzero as shown in Figure 3-19. This is not as bad a disadvantage as having the inertial forces or moments out of balance because oscillations in inertia torque can be absorbed and hidden by the flywheel, thus masking them from driver and passengers. Note in the bottom panel of Figure 3-19 the effect that a flywheel has to reduce the oscillations in torque in the total torque shown in the panel immediately above it. Nevertheless, the oscillations in inertia torque still affect stress and deflection within the rather long crankshaft and this must be addressed by adding a torsional damper to the crankshaft as will be discussed in a later section of this chapter.

The average values of the functions in all panels except the inertia torque (second from top) in Figure 3-19 are identical, and all derive from the gas torque function. The inline six was a very popular configuration prior to WWII and

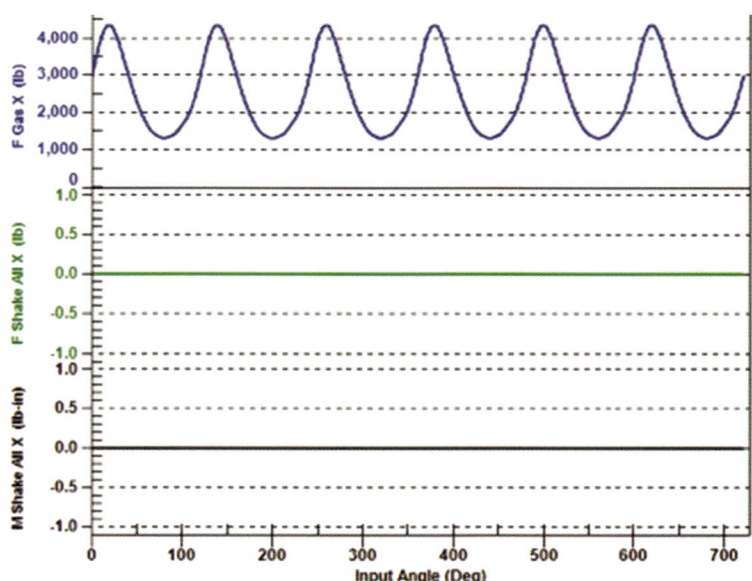

FIGURE 3-18 Gas force and balance conditions in a six-cylinder inline engine at 3400 rpm

CHAPTER 3 ENGINE CONFIGURATIONS

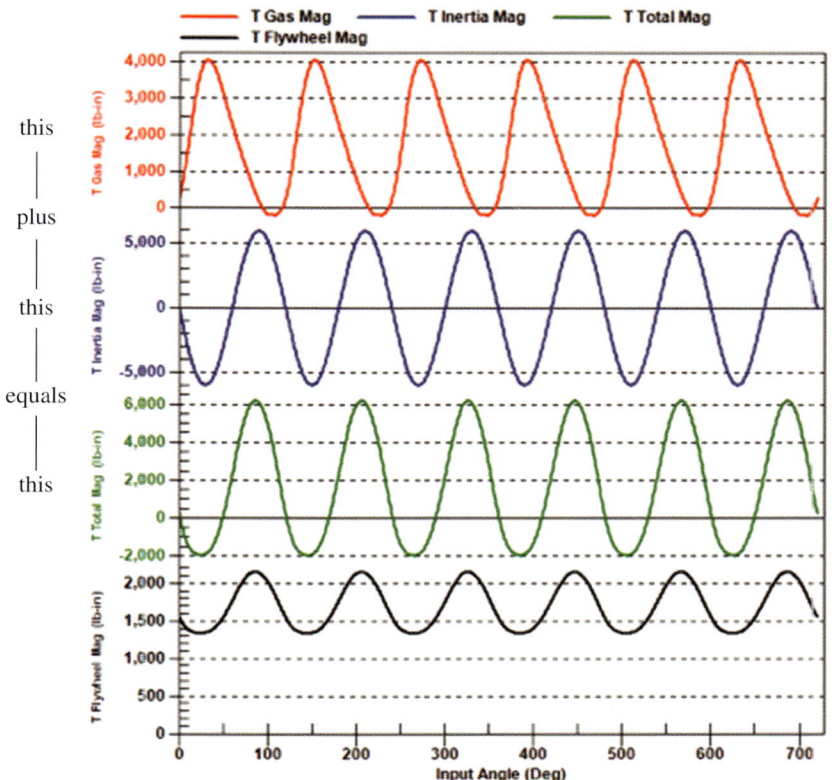

FIGURE 3-19 Gas, inertia, total, and flywheel torque in a six-cylinder inline engine at 3400 rpm

is still made by major manufacturers such as BMW. Its only disadvantage was that it's rather long compared to V-configurations. It has been supplanted by the shorter V6 in many applications, but a V6 is not as well balanced as the inline six and requires balance shafts to correct its flaws. Mercedes recently announced that it is planning to phase out its V6 line and go back to the inline six, which is an engine that they have been producing for over 100 years.

Three-Cylinder Inline Engines

A three-cylinder inline engine, also called an inline triple, is essentially an inline six cut in half. Vauxhall made a triple in 1904. Many were used in later motorcycles such as the Triumph Trident, Laverda, Benelli, and Yamaha SX750. They were also used as boat and tractor engines. Modern automotive uses have been limited to small economy cars, and many of these European and Japanese models were not imported to the U.S. A 1979 Suzuki Alto/Fronte automobile had a 543cc triple, smaller than many motorcycle engines. Other automobile marques that have used them in the past are the Subaru Justy, the Saab in the 1970s (a 2-stroke), the Geo Metro, Chevrolet (Suzuki) Sprint, and more recently, the Smart ForTwo. The last four have been imported to the U.S.

With the new U.S. Corporate Average Fuel Economy (CAFE) standard requiring the average fuel economy for cars and trucks sold in the U.S. be 54.5 mpg by 2025, manufacturers are bringing out many new inline triples to achieve that economy. Ford is selling an EcoBoost triple in the 2014 Fiesta that makes 120 HP. BMW is offering a turbo three. VW is currently selling gas and diesel triples, and GM has also announced one. Many of these are paired with

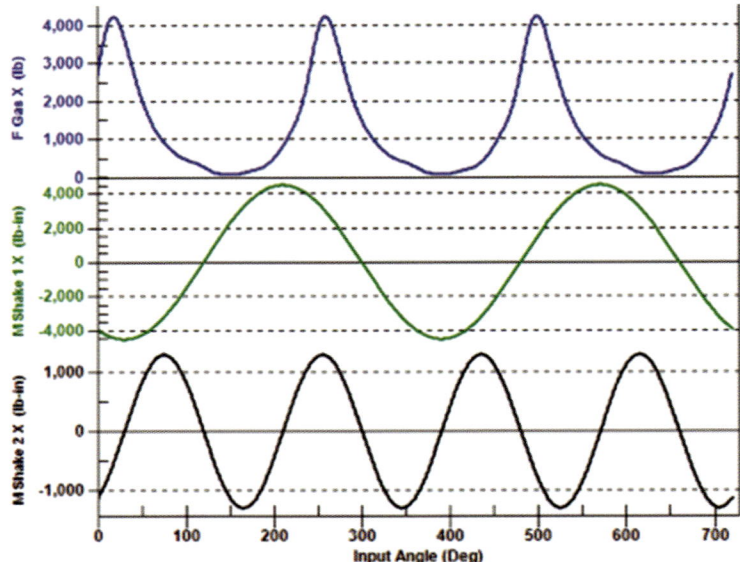

FIGURE 3-20 Gas force and unbalanced moment conditions in a three-cylinder inline engine at 3400 rpm

electric motors in hybrid arrangements. Audi, Citroën, and Peugeot are also reported to be working to develop inline triples. Nissan is reported to be developing a 400-HP triple for Le Mans. It may be that the time has come for this less common engine configuration to shine.

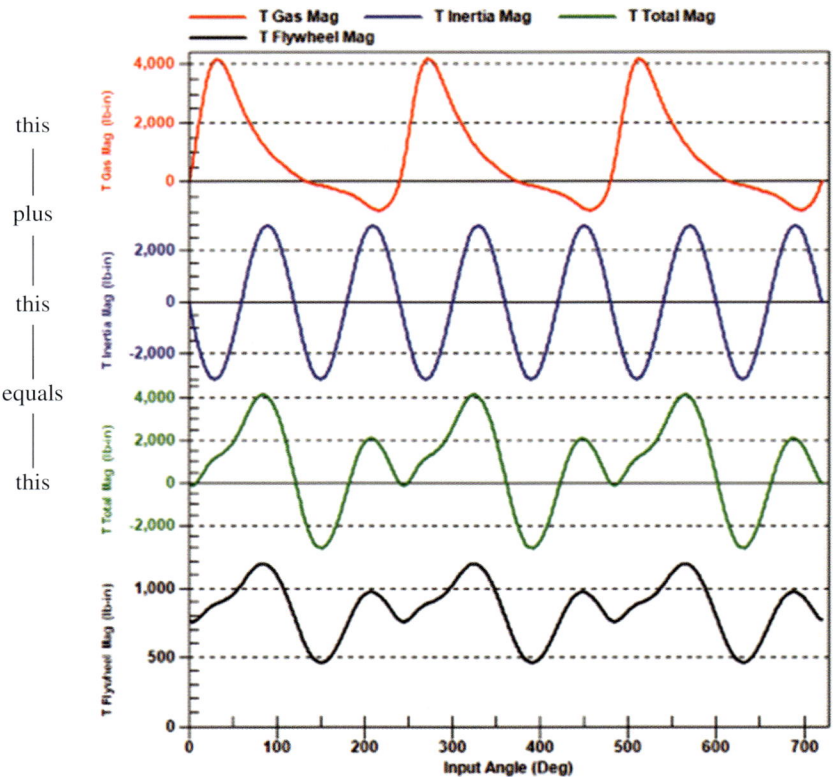

FIGURE 3-21 Gas, inertia, total, and flywheel torque in a three-cylinder inline engine at 3400 rpm

CHAPTER 3   ENGINE CONFIGURATIONS

The inline triple has phase angles of 0, 240, 120° and shares force balance with the six-cylinder—its shaking force is zero. But because it lacks the second, mirrored set of crank throws of the inline six, its primary and secondary moments are both nonzero. It is even-firing, however. Figure 3-20 shows the gas force and unbalanced primary and secondary moments (from top to bottom) of the inline, 4-stroke triple. It is possible to counter the primary moment with a single balance shaft in combination with counterweights on the crankshaft. The secondary moment needs two balance shafts, turning at twice crank speed.

The inline triple has the same inertia torque condition as the inline six—its third harmonic is nonzero. Figure 3-21 shows the gas, inertia, total, and flywheel torque for the inline triple from top to bottom, respectively. The total torque looks distorted compared to that of the inline six because there are only three gas torque pulses rather than six to combine with six pulses of inertia torque over 720°.

Eight-Cylinder Inline Engines

During WWI (1914–1918), the airplane began to be used on a larger scale than it had been since the Wrights first flew in 1903. Aircraft of the time were rather fragile, being made largely of wood and fabric. An engine with significant vibration could shake a delicate aircraft apart. Accordingly, much effort was directed at development of low-vibration engines. One of these was the unique *rotary engine* in which its stationary crankshaft was fastened to the fuselage and the entire engine block, with propeller attached, rotated around the crankshaft. Because there was nothing oscillating in this engine, it did not vibrate. This interesting engine will be discussed in its own section later in this chapter.

More conventional engine designs were developed based on the state of the art at the time. Six-cylinder engines had demonstrated their low vibration, except for the inertial torque issue, and it was logical for designers to pursue configurations having even more cylinders. Note that the V8 had already been invented, and Cadillac produced one starting in 1914. (They produced essentially the same engine until 1948 when they redesigned and updated it.) But the V8 does not have perfect balance either (without added counterweights) as will be discussed in a later section. Thus was born the straight-eight engine with eight cylinders inline. Bugatti, Mercedes, and others developed them for aircraft use during WWI. In 1919, the first automotive straight-eight was offered by Isotta-Fraschini (Italy). Leyland Motors (England) introduced one in 1920 and Duesenberg produced the first American straight-eight in 1921. Duesenberg went on to offer very powerful, straight-eights into the 1930s. One of their later examples from their Model J (1928–1937) is shown in Figure 3-22. Duesenbergs were made virtually by hand in small quantities, won the Indianapolis 500 many times, and were considered by many to be the best cars in the world.

Packard made the first mass-produced straight-eight in 1923 and it became the mainstay of their luxury line until 1955 when they offered a V8. By that time, most makers had switched to V8s in their top lines. By 1930 many American manufacturers of premium and even lesser brands had adopted the straight-eight engine. Besides Duesenberg and Packard, these included Auburn, Buick, Chandler, Chrysler, DeSoto, Dodge, Gardner, Hudson, Locomobile, Marmon, Nash, Oldsmobile, Pierce-Arrow, Pontiac, Stearns-Knight, Studebaker, and Stutz. The straight-eight engine became an icon of luxury automobiles. Many European makers also adopted this engine in their more expensive models. Alfa-Romeo, Bugatti, Daimler, Isotta-Fraschini, and Mercedes Benz all made them. Figure 3-23 shows a jewel-like,

FIGURE 3-22 Duesenberg Model J straight-eight, dual-overhead-cam engine, circa 1930 *(Courtesy of Larry Stevens)*

dual-overhead-camshaft, 1933 Bugatti Grand Prix straight-eight engine. This automobile is in the Ralph Lauren Collection.

Straight-eight engines use a crankshaft with phase angles of 0, 90, 270, 180, 180, 270, 90, 0°. Note that this arrangement is mirror symmetric about the engine's mid-plane, which balances the shaking moment. It has zero shaking force, zero shaking moment, and zero inertial torque through the third harmonic, which is considered to be "perfect balance." In 1942, Packard bragged in advertisements that one could balance a nickel on edge on the head of an idling Packard straight-eight engine and the coin would not topple. A few years later Rolls-Royce copied this claim for their similar engine, presumably using an English coin instead.

One of the reasons the Packard straight-eight was so smooth was that, in addition to its inherent balance, their engineers recognized that the very long crankshaft needed to be well supported. Thus, they provided it with nine main bearings, seven between the eight crank throws plus one

FIGURE 3-23 1933 Bugatti Grand Prix, dual-overhead-cam, straight-eight engine *(Courtesy of Sfoskett)*

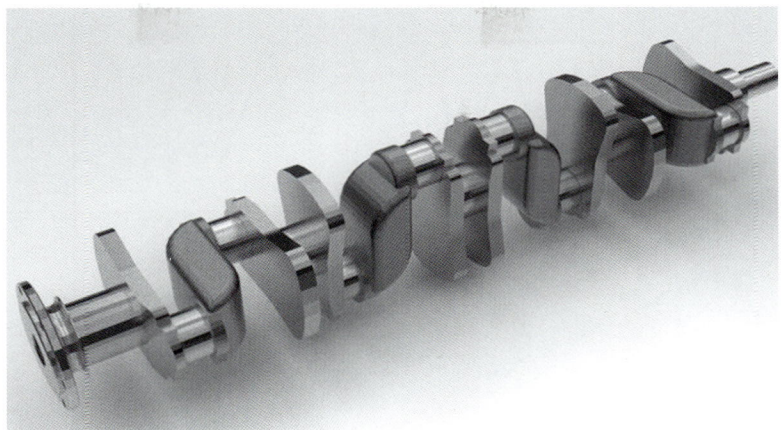

FIGURE 3-24   1937 Buick, straight-eight engine crankshaft *(Courtesy of Dominic Notman)*

at each end. This made the crank quite stiff in bending, which reduced vibrations from the piston forces. Contrast this to the Buick straight-eight crankshaft, shown in Figure 3-24, which has only five main bearings, one between each pair of crank throws plus one at each end—much less stiff in bending.

Another inherent problem with long engines such as the inline six, and even worse in the inline eight, is torsional vibration set up in the convoluted crankshaft by the piston forces acting on the crank pins to twist the shaft. These torsional vibrations, if not controlled, could eventually cause the crankshaft to fail from fatigue. These engines need a torsional damper to be fitted to the end of the crankshaft to control this problem. The working of this clever device will be discussed in a later section of this chapter.

Because of the number of cylinders, these engines also had very strong and smooth torque down to low RPM. They could accelerate a 5000 lb car from just above idle speed (600 RPM) in top gear without complaint or stumbling. This virtually eliminated the need to down-shift the transmission once the car had been accelerated through the gears from a stop to even a low road speed. Thus they were very easy to drive in an era when shifting a manual transmission could be a noisy, difficult, and stressful proposition.

Figure 3-25 shows the gas force, shaking force, and shaking moment for a straight-eight engine. The only nonzero function is the one you want to be nonzero, the gas force. Figure 3-26 shows the gas torque, inertia torque, total torque, and flywheel torque for the same engine. The inertia torque is zero, unlike the inline six, and thus the total torque is equal to the gas torque. The flywheel torque has the same average value as the total torque but its oscillations about the average are reduced by the flywheel's ability to store and return energy during the crankshaft's revolution. The flywheel is mounted to the back end of the crankshaft, so the transmission and drive line only feel the smoother, flywheel torque. But the crankshaft internally feels the total torque oscillations as they pass through to the flywheel, and that is what causes torsional vibration.

Twelve-Cylinder Inline Engines

In 1905 Wolseley, in England, built a 360 HP, 12-cylinder inline engine for marine use. The Duesenberg brothers built another in 1910 for

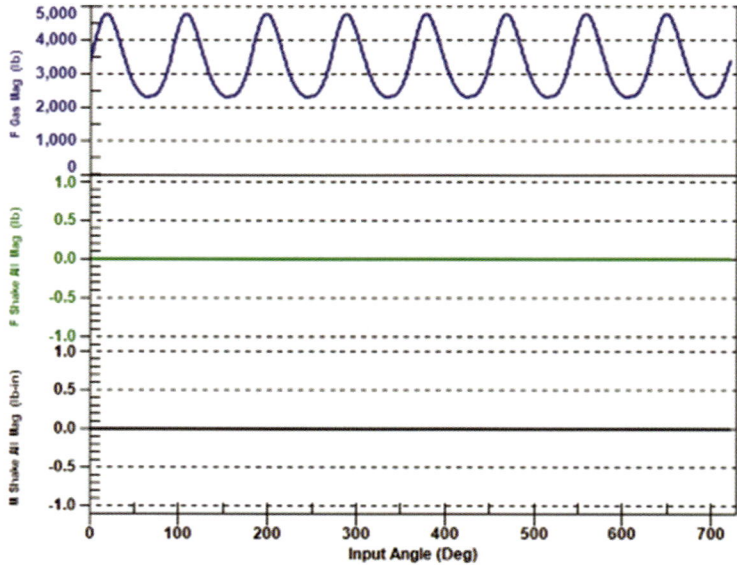

FIGURE 3-25 Gas force and balance conditions in an eight-cylinder inline engine at 3400 rpm

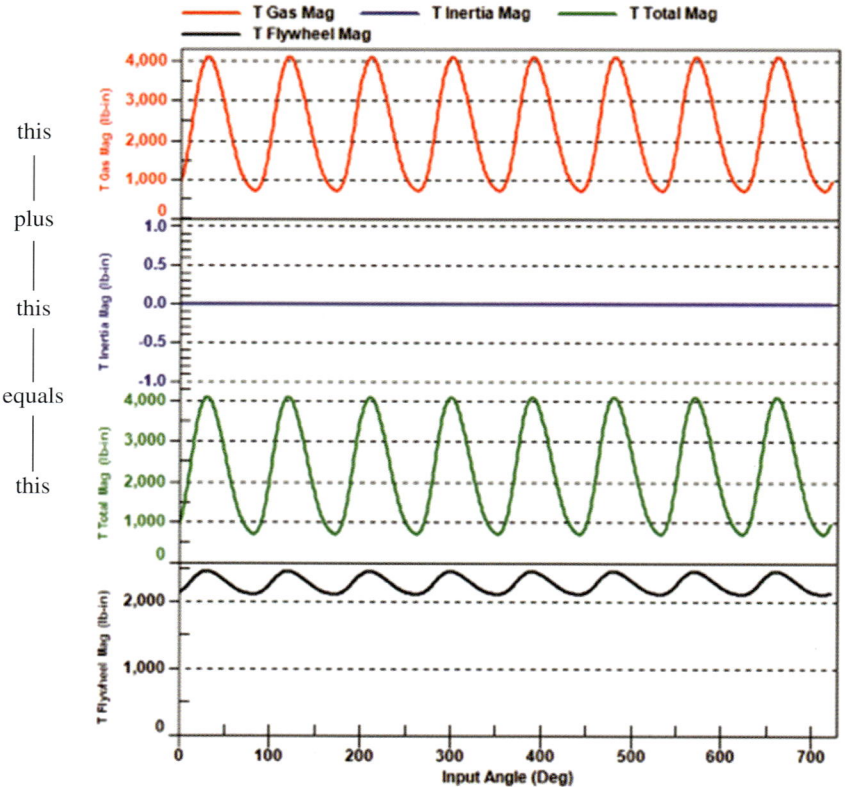

FIGURE 3-26 Gas, inertia, total, and flywheel torque in an eight-cylinder inline engine at 3400 rpm

CHAPTER 3 ENGINE CONFIGURATIONS

FIGURE 3-27 1905 Wolseley twelve-cylinder inline marine engine *(Public Domain)*

Commodore James A. Pugh's racing boat, the *Disturber*. Pugh's goal was to win the Harmsworth Trophy in England, but WWI intervened and the race was canceled. Engines this long were too large for automobiles but were used in ships and large trucks. They had the same balance condition as the two six-cylinder inlines that were joined nose to tail on a common crankshaft (see Figures 3-18 and 3-19), but had twice as many power pulses spaced at 60°. Figure 3-27 shows the 1905 Wolesley marine engine. The very long crankshaft probably survived because marine engines turn at slow, propeller speeds, resulting in smaller torsional vibrations.

Multi-Cylinder V-Engines

V-engines have two banks of cylinders arranged with a V-angle between them. The geometry of a four-cylinder, V-engine is shown in Figure 3-28. The reference X-axis is taken midway between

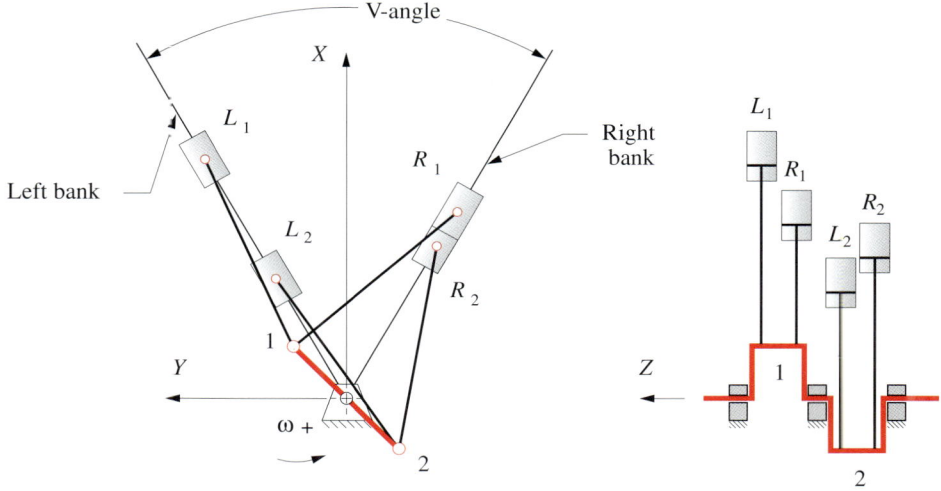

FIGURE 3-28 V-engine geometry

the banks and the V-angle is shown symmetric about the X-axis. Unlike an inline engine in which all forces and moments are within the X-Z plane, a V-engine has forces and moments with X and Y components. Torque is always about the Z-axis, which runs along the crankshaft axis. A V-engine can have any V-angle more than 0° and less than 180°, but most V-engines have an optimum V-angle that will give even firing with good balance. The optimum V-angle varies with the number of cylinders.

V-engines are far from a new idea. Daimler made the first, a V-twin, in 1889. V-engines were in production for use in boats by 1903. The first V12 engine was produced, also for marine use, in 1904. A V-engine is essentially two inline engines joined at some V-angle to a common crankcase. The two, inline banks share the same crankshaft and usually also share the same set of crankpins. V-engines have been produced with 2, 3, 4, 5, 6, 8, 10, 12, 14, 16, 18, 20, and 24 cylinders. We will discuss only the more common ones having 2, 4, 6, 8, 10, 12, and 16 cylinders.

One advantage of the V-configuration, especially for larger numbers of cylinders, is its shorter length, which packages better in an automobile engine compartment and also has a shorter, stiffer crankshaft that is less prone to torsional vibration. The crankshaft is stiffer, not just due to being shorter, but for the fact that most V-engines place two connecting rods (one from each cylinder in opposite banks) on the same crankpin. A V8 engine has only four crank throws rather than the eight of an inline-8 engine. This arrangement is depicted schematically for a V4 in Figure 3-28 and with more detail in Figure 3-29, which shows the more common arrangement with two conrods side by side at (a) and a *fork and blade* arrangement where one conrod fits within the other at (b). The fork and blade eliminates the small offset between the two pistons and gives better moment balance, but is more expensive. Some V6 engines need to use as many crank throws as cylinders to achieve even firing.

In general, a V-engine takes on the same balance characteristics as the inline engines that make up its banks, with the difference that if

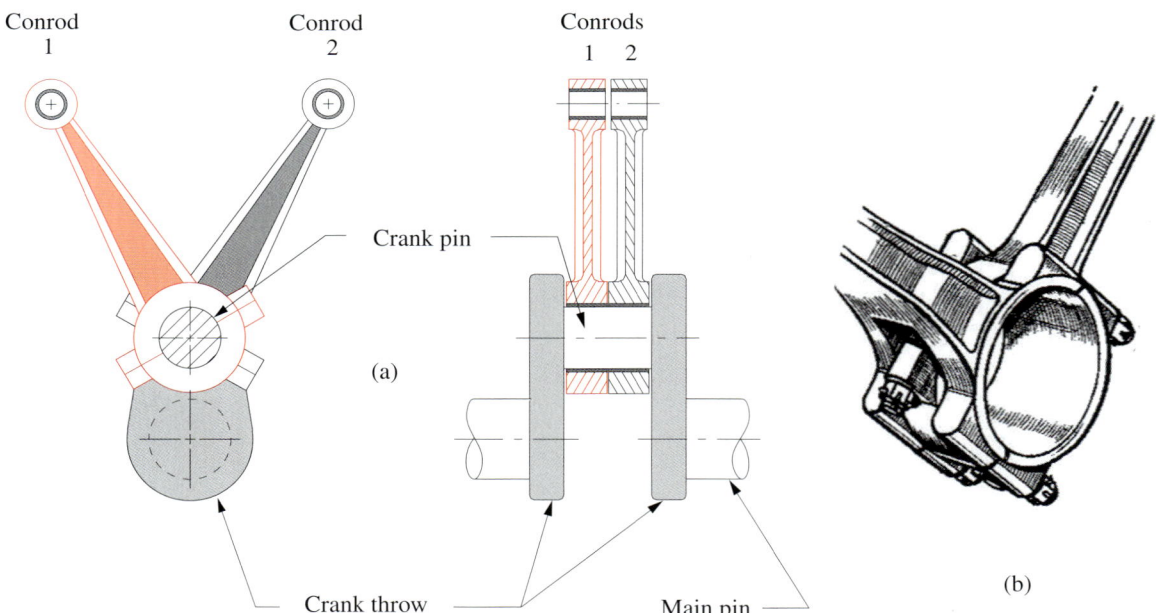

FIGURE 3-29 Two connecting rods on a common crank throw (a) side by side, (b) fork and blade *(Public Domain)*

either the shaking force or shaking moment is nonzero, then it can have both *X* and *Y* components rather than these effects being confined to the plane of one bank of cylinders. Each bank generates its own set of shaking effects and they add vectorially as *X* and *Y* components to the whole. In some rare cases, particular V-angles can cause additional cancellation of shaking effects, but the crank phase angles primarily determine the balance condition of a V-engine.

V-Twin Engines

V-twin engines were used in a few early automobiles such as Daimler's second car in 1899, some early Peugeots, in the Morgan three-wheeler (England) from 1911 to 1939, and in a few other microcars since. But the primary application of the V-twin is in motorcycles. Motorcycle manufacturers have used V-angles ranging from 26° to 170° in V-twins. The choice is driven more by packaging considerations than engine dynamics. Regardless of its V-angle, a V-twin is not well balanced and most configurations are uneven firing. With certain combinations of phase angles and V-angle, such as a 0-270° crankshaft and a 90° V-angle, it will even-fire but be unbalanced. With particular crankshaft phase-angle and V-angle combinations and a counterweighted crankshaft, primary force balance can be achieved. For example, with a 0-180° crankshaft and a 90° V-angle, it fires unevenly but can be counterweighted for primary force balance. The secondary force and all moments and torques are still unbalanced.

Most V-twins use a single crankpin with various V-angles. Harley-Davidson uses a single crankpin and a 45° V-angle, which gives it the distinctive, "potato-potato," uneven-firing ex-

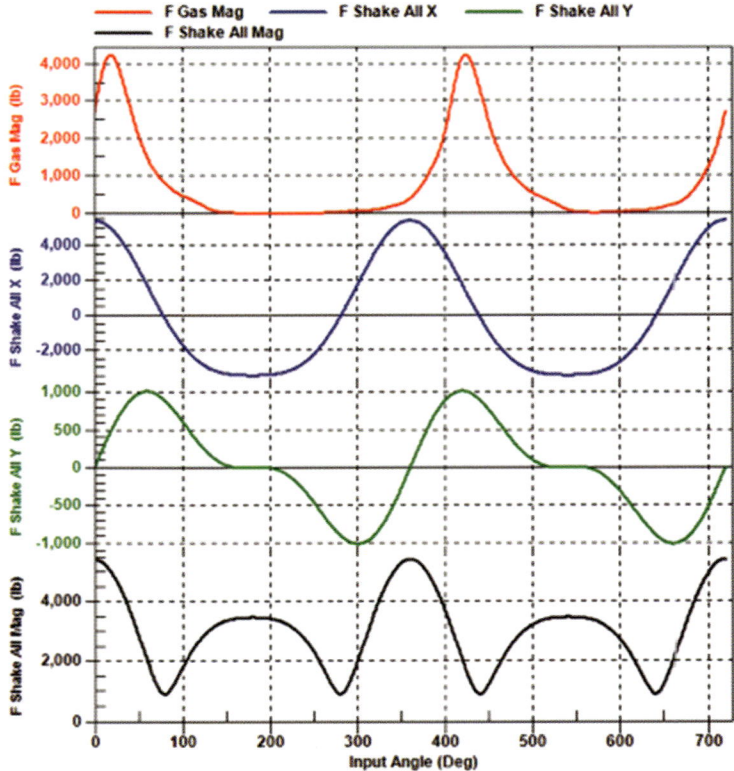

FIGURE 3-30 Gas and shaking force magnitude and X, Y, components in a Harley-Davidson engine at 3400 rpm

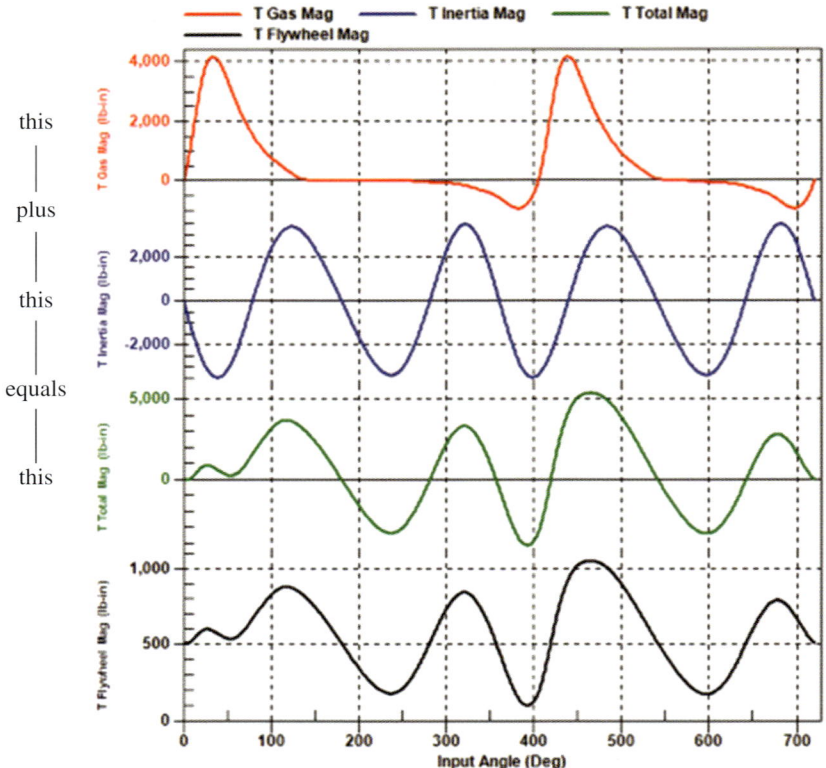

FIGURE 3-31 Gas, inertia, total, and flywheel torque in a Harley-Davidson V-twin engine at 3400 rpm

haust note. The forces in a Harley engine are shown in Figure 3-30 and its torques are shown in Figure 3-31. Honda imitated this arrangement in its VT1100 Shadow which looked and sounded like a Harley. Other Honda V-twins such as the 1983 Shadow 750 V-twin used a two-pin 0-76° crankshaft to get primary force balance in a 52° V-engine with a counterweight on the crankshaft. The Suzuki VX800 used a 0-45° crankshaft with a 45° V-angle. Moto-Guzzi (Italy) used a 0-180° crankshaft with a 120° V-angle in 1934. Modern Moto-Guzzis are 90° V-twins. Some V-twins add counterbalance shafts to reduce vibration.

V4 Engines

V4s have not been used in many automobiles but are more common in motorcycles. Marmon made one in 1902. Lancia (Italy) manufactured one from the 1920s through the 1960s that was used in several of their cars. Ford made one for some of their English and German cars, and it was also used in some Saabs. Ducati, Honda, Suzuki, and Yamaha have offered V4 motorcycles in recent years. Honda still markets a few models with this engine. Some outboard marine engines also use V4s.

The V4 engine is not even-firing if pairs of pistons share crankpins. With four crankpins they can be made to even-fire. With a 0-180° crankshaft and a 90° V-angle, primary force and torque are zero, but the primary moment and all secondaries are nonzero. Some examples add counterbalance shafts to reduce vibration.

V8 Engines

In 1903, a Frenchman, Clement Ader, built the first V8 engine for an automobile and entered it in the Paris-Madrid race. It was not very successful in the race. Another Frenchman, Leon Levavasseur, built a V8 for use in speedboats and aircraft in 1904. It was not until 1910 that DeDion made the first production V8 engine for automobiles. This engine set the pattern for many V8 engines that followed in later years. It used a 90° V-angle, the optimum angle for a V8, placed the camshaft in the center of the V, and was a flathead, but it only had two main bearings! Cadillac brought out the first American V8 in 1914 and a V8 became the only engine offered in their cars until quite recently, other than their V12 and V16 of the 1930s. The V8 was based on the DeDion design, but Cadillac added a third main bearing and made many improvements. It soon surpassed its predecessor in technical development.

All of these early V8 engines used two conventional, inline fours combined on a common crankcase, with the four-cylinder's 0, 180, 180, 0°, single-plane crankshaft (Figure 3-32a) and two, fork and blade connecting rods per crank throw as shown in Figure 3-29b. Though even-firing, these V8s suffered from the same vibration issues as the inline four-cylinder (Figures 3-14 and 3-15), namely nonzero secondary forces and moments in each bank, which combined from the two banks to create a side-to-side shaking and rocking in the engine at twice crank speed. In the early 1920s, Charles Kettering, GM's head of research, challenged his engineers to solve this vibration problem, which they did by designing a so-called cross-plane crankshaft in 1923 that used phase angles of 0, 90, 270, 180° as shown in Figure 3-32b. Still even-firing, this arrangement cancels all inertia and shaking components of torque, force, and couples except for the primary moment. This was a potential problem because the magnitude of the primary moment is large, especially at high crankshaft speeds.

But the 90° V-angle makes the X and Y components of this primary moment equal in magnitude and 90° out of phase. This means that they create a rotating moment of constant magnitude at crankshaft speed that plots like a circle when viewed end-on to the crankshaft as shown in Figure 3-33. In this plot, you are looking down the Z-axis of rotation (the crankshaft axis). The X-axis is between the two cylinder banks and the Y-axis is transverse to the engine as defined in Figure 3-28. (The engine is effectively on its side in the plot of Figure 3-33.) The outer circle shown in Figure 3-33 rings the tips of a set of vectors (the red lines) that rotate about the crank centerline at crankshaft speed. Each red line represents the (constant) magnitude and (changing) direction of the shaking couple (moment) at each 3° of rotation of the crankshaft. To visualize what this rotating, rocking moment is doing to the engine, imagine a magical, flying horse bucking forward and backward and simultaneously doing barrel rolls.

This constant magnitude primary moment can be cancelled by simply adding a pair of counterweights at opposite ends and on opposite sides of the crankshaft (at 180° and 0°) in Figure 3-32b. The pair of counterweights cre-

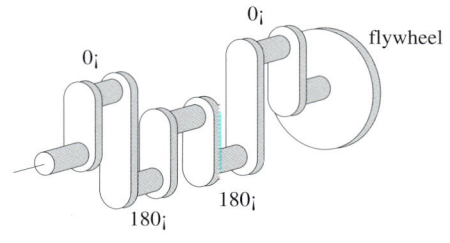

(a) Flat-plane crank

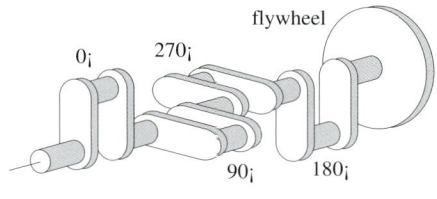

(b) Cross-plane crank

FIGURE 3-32 Flat-plane and cross-plane crankshafts

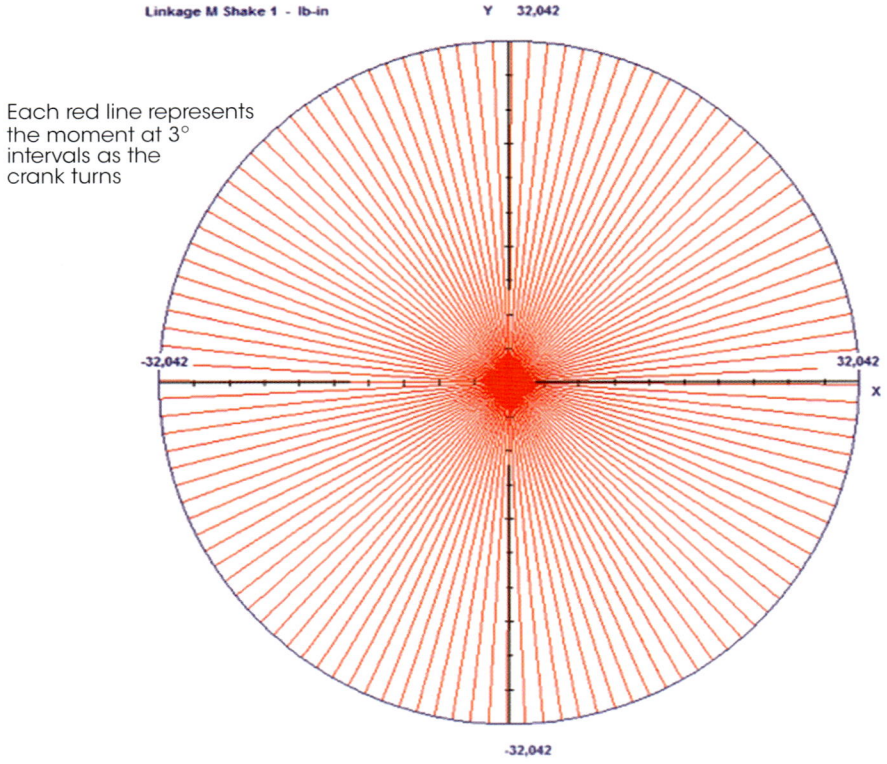

FIGURE 3-33 Polar plot of the rotating, primary shaking moment in a cross-plane V8 engine at 3400 rpm

ate a rotating, rocking couple of opposite sign and constant magnitude, thus cancelling the primary moment's effect. Most modern V8 engines use this cross-plane crankshaft with counterweights to provide as vibration-free an engine as the straight-eight. Some manufacturers, notably Ferrari, Lotus, and Cosworth, still use a flat-plane crankshaft in their V8s because they want faster rotational acceleration response when revving the engine. The inertia of the counterweights on the cross-plane crankshaft makes it more sluggish in acceleration. These manufacturers accept the flat-plane crank's secondary vibration in return for its "race-car" response to engine acceleration.

Peerless introduced its V8 in 1916. It was similar to the Cadillac, but Peerless used two conrods side by side as in Figure 3-29a instead of the fork and blade design of Figure 3-29b. Their use introduces a small moment due to opposite pistons now not being exactly in line; however, it is less expensive to manufacture than the fork-in blade arrangement. Modern V8s use conrods side by side on the crankpin like Figure 3-29a.

Peerless copied Cadillac's cross-plane crankshaft after it came out. Lincoln also brought out a V8 soon after Peerless, and it became that company's mainstay as it was with Cadillac. Their competitors in the luxury market such as Packard, Duesenberg, and others, continued to use the straight-eight engine. Oldsmobile and Pontiac (also parts of GM) brought out their own V8s in 1929 and improved on the Cadillac design by casting the crankcase and both cylinder banks in one piece, making it stiffer and cheaper to manufacture. Up till then, most V8s had been made from three castings—a crankcase and two cylinder banks—bolted together. Cadillac adopted the one-piece cast block shortly thereafter.

Pontiac's V8 did not sell well, in part because it kept the flat-plane crank and so suffered from vibration. Pontiac made its V8 only until 1932 and then brought out a new straight-eight engine in 1933 to replace it. Pontiac continued to sell its straight-eight with minor changes and upgrades until 1954 when they replaced it with a new overhead-valve V8 in 1955.

Ford introduced the first V8 in a low-priced car in 1932. As was typical, Henry Ford learned from his competitors, improved upon their designs and, most importantly, improved their production methods. Most V8-engine producers to that time were low volume, selling expensive cars to the high end of the market. Ford aimed at the lower end of the market and went for high production to reduce cost. Cadillac's annual production at that time amounted to only about 3 to 4 weeks of Ford's production. Oldsmobile then was producing about 4000–5000 engines a year. Ford was producing 3000 **per day** at the same time.

Ford invested $50 million[1] in an automated foundry to cast blocks for his engines. He put the entire casting process on an assembly line. Molds were conveyed to the pouring line at the rate of 100 per hour. A moving, two-ton, pouring furnace followed the molding conveyor and poured the molten metal into the molds on the move. Up until that time, all crankshafts had been made by the forging process. This is essentially semi-automated blacksmithing, in which machine-powered, shaped hammers, called forging dies, pound red-hot metal into the desired shape. This gives an extremely strong part.

In contrast, casting melts the metal and pours it into a mold. Castings then were much weaker than forgings, too weak to withstand the high loads on crankshafts. Ford was, among other things, a very good materials engineer. He set his engineers the task of developing stronger casting materials. The result was a cast-iron crankshaft, made from higher-strength cast alloys than had been available before. Previously, the crankshaft counterweights had to be machined separately, then bolted to the crankshaft. The counterweights could now be cast as an integral part of the crankshaft, saving much labor and cost. Ford's flathead V8 engine became very popular and was kept in production with minor improvements until 1953, when it was replaced by a new overhead-valve V8 in 1954.

During WWII, from early 1942 through 1945, all production of nonmilitary vehicles was suspended, and all automobile factories produced jeeps, trucks, tanks, aircraft, and weapons. When the war ended, there was pent-up demand for new cars, and people were in a better position to buy them because the war had ended the economic depression and people had money to spend.

1946 to 1949 were watershed years for the auto industry in the U.S. It would take Europe and Japan longer to put their industries back into shape because of the damage inflicted on them by bombing. But the U.S. infrastructure was intact and more robust than ever due to the war effort. It takes several years for any automobile company to design, build, test, tool-up, and ready a new car design for production.

The late 40s saw most auto companies offering essentially the same models that they had been selling in 1941. An exception was Studebaker, who presented a radical new design of automobile in 1946, with the front and rear ends looking quite similar. It was also much lower to the ground than prewar models due to adoption of the hypoid rear end (see Chapter 5) that allowed the driveshaft to be below the center of the axle and the floor lower. All makes adopted the **hypoid** rear end and designed "longer, lower, wider" automobiles post-war. Hudson dusted off a design that they had been planning to make in 1942 and presented the "Step-Down"

---

[1] $870 million in 2014 dollars

Hudson in 1948 with its floor dropped inside of a perimeter frame. It was only 60-in-high, 7-in lower than the Buick and 6-in lower than the Chrysler in 1948 and had the lowest CG.

The war had driven technology ahead by a large measure. Gasoline was now better, with higher octane that could withstand higher compression ratios without pre-ignition (knock). The engineering that had gone into development of high-performance aircraft engines during the war had taught engineers how to design a better, more powerful automobile engine.

Overhead valve (OHV) engines had been around since the beginning of the automobile. Buicks made only OHV engines from their first engine in 1904 to the present. The Chevrolet OHV "Stovebolt Six" was produced from 1929 to 1970. But most of the high-volume producers had stuck with the flathead design prior to WWII, especially with their V8 engines. Most production V8 engines were flatheads until the fifties.

Cadillac and Oldsmobile introduced the first high-compression, overhead-valve V8 engines in 1949 and were followed by virtually all of the other manufacturers by the mid-fifties. The '49 Cadillac and Olds V8s were separate, clean-sheet-of-paper, light-weight redesigns with higher compression ratios, improved combustion chamber design, overhead valves on each bank operated from a single camshaft in the valley of the block by pushrods and rocker arms, five main bearings, and **slipper pistons** with their bottoms contoured to fit around the crankshaft counterweights for a lower engine. Ricardo invented the slipper piston in 1920 in England. Figure 4-4 in the next chapter shows a slipper piston.

Valve trains will be explored in detail in their own chapter because they are as complicated as the engine configurations covered in this chapter. For now, suffice it to say that moving the valves from beside the pistons to above the pistons allowed improved breathing and combustion, and that translated to significantly more power than could be extracted from the old flathead-engine designs.

In 1949, Ford brought out a new line of cars with the bodies completely redesigned and much lower. But Ford's new OHV V8 engine would not be ready until 1954. Until then, Ford continued to use its flathead V8 in the new bodies. GM, Chrysler, and other makers also introduced restyled bodies in the late 40s and early 50s and eventually added new OHV V8s as well.

Chrysler introduced its hemi-head OHV engine in 1951. The hemi-head designation refers to the combustion chamber shape, being an approximate hemisphere above the pistons. Alfa Romeos in the 1930s all had dual-overhead camshaft, hemi-head inline-4s. Jaguar used a hemi-head in its 1948 dual-overhead camshaft, inline-6, XJ engine that won Le Mans several years in a row (see the previous section on inline-six engines). This XJ engine later powered all of the Jaguar line (E-Type and XJ sedans) until the late '80s. A hemispherical combustion chamber with the valves arranged at angles on each side of the chamber allows good breathing and high power. But it proved difficult to make hemi engines meet later emission requirements and they are now less common. Chrysler still manufactures what they call "hemi-head" engines that meet current emission standards, but these have a flatter and more complex combustion chamber shape than a true hemisphere. Some consider the 1951 Chrysler hemi with 1-HP/in^3 to be the pinnacle of high-power V8 engine design of that era. Others cite the 1955 Chevrolet, "small block" V8 as an icon, the genes of which spawned a generation of engines down to the current Corvette engines. The last straight-8 engine made in the U.S. was replaced by a V8 in the 1955 Packard, which had merged with Studebaker in 1954. By the late 1980s, 80% of automobiles sold in the U.S. were V8s, all of them OHV.

Until recent years, the V8 engine dominated the high end of automotive performance offerings. Since fuel economy became the mantra of automotive technology, the relatively thirsty V8 has become something of a pariah and may be becoming an endangered species. As of this writing, V6 engines, either naturally aspirated or turbocharged, and, increasingly, turbocharged four-cylinder engines are having a resurgence because of their better fuel economy compared to a V8. With modern, computer-controlled engine management and advanced engine technology, smaller displacement engines with fewer cylinders are proving to be a performance competitor to the venerable V8 engine of the fabulous fifties when gasoline cost only twenty-five cents per gallon. The old slogan of hot-rodders from the fifties (your author included) *"there is no replacement for displacement"* may now be obsolete.

V6 Engines

Marmon built and sold a car with a V6 engine in 1905, but the first series production V6 was from Lancia (Italy) in 1950. The optimum V-angle for a V6 for even-firing with a three-throw crankshaft is 120°, but this engine is too wide to fit well in an engine compartment. A 60° V-angle will give even firing with a six-throw crankshaft, and that is what Lancia used. GM also introduced a 60° V-6 for GMC trucks and SUVs in 1959. Buick wanted a V6 for its Special model in 1962, but also wanted to avoid the large expenditure of tooling up a separate production line for it. So they chose to cut two cylinders off their V8 to make a 90° V6 that could be assembled on the same line as the V8s. They also used a 0, 240, 120°, three-throw crankshaft, which made the engine uneven-fire. It fired at alternate intervals of 90° and 150° rather than every 120° due to the 90° V-angle, as shown in Figure 3-34. As a result it ran roughly at idle and customers complained. In 1977, they fixed it by offsetting the crankpins by ±15° to get even firing. The V6 is force-balanced and its inertia torque has only a third harmonic like the inline six. But both primary and secondary moments are nonzero. A balance shaft is often added in the block between the banks to cancel the primary moment in combination with counterweights on the crankshaft.

The V6 engine has surged in popularity in recent years due to its compact size (shorter than an inline-4 and narrower than a V8) which fits well in a transverse-engine FWD arrangement. The drive to increase engine economy has also motivated manufacturers to replace V8s with V6s even in RWD models. In 2014, the V6 was the second largest selling engine after the inline 4. Most makers offer at least one V6 in their line-

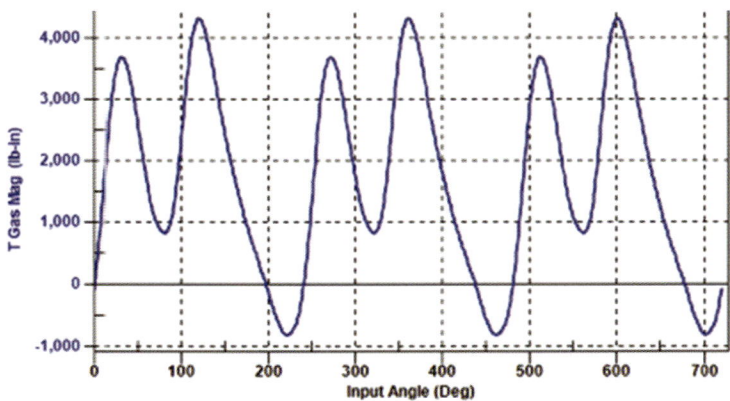

FIGURE 3-34 Uneven firing gas torque in a 90-degree V6 engine with a three-throw crankshaft at 3400 rpm

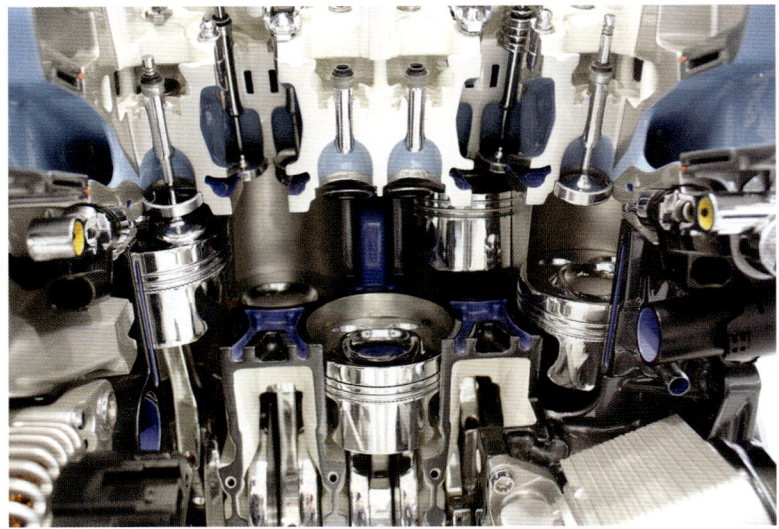

FIGURE 3-35 Volkswagen VR15, 15-degree V6 engine with overlapping, staggered cylinders *(Photo by the author)*

up. Performance car manufacturers are offering turbocharged V6 engines that make as much horsepower and torque as larger displacement V8s. An example is the Maserati Quattroporte that in 2007 was powered by a 395 HP, 4.2 liter V8. In 2013, the engine was changed to a 3.0 liter, turbocharged V6 that makes 404 HP and has equivalent performance to the earlier V8.

Volkswagen offers an unusual V6 that is closer to an inline six in configuration. Called the VR15, it is a 15° V that is essentially an inline six in which every other cylinder is offset by 15° and moved closer to its neighbor to shorten the engine as shown in Figure 3-35. It was designed to get the advantage of six cylinders in a block no longer than an inline five. This allowed it to fit in their models that were designed for a five-cylinder engine. VW/Audi has also found some creative ways to use this engine as will be discussed in the section on W-engines.

V10 Engines

The V10 is two inline-fives on a common, five-throw crankshaft. Its optimum V-angle is 72°, but for the same reasons some manufacturers made 90° V6 engines, most manufacturers use a 90° V-angle to make the V10 engine on their V8 lines. With a 72° V-angle it is even firing. With a V-angle of 90° it is not even-firing, unless the crankpins are splayed ±18°. Its balance condition is the same as the inline five. Forces and torque are in balance but moments are not, unless balance shafts are added.

Dodge made the first V10 for its pickup trucks and then used a version of it in the Dodge Viper in 1991. They engaged Lamborgini to help develop the aluminum block version for the Viper. Lamborgini put a 90° V10 in their Gallardo in 2003 that was later fitted to the Audi R8. (Audi owns Lamborgini). Ford built a 90° V10 for its truck line from 1997 to 2010. The Porsche Carrera GT also used a V10.

V12 Engines

The V12 is two, inline sixes, sharing the same crankshaft. Because the inline six is inherently well-balanced, a V12 at any V-angle will also be as balanced. But for even-firing with cylinders sharing crank throws, even firing is achieved with V-angles of 60, 120, or 180°. The last of these is called a flat-12. Packard made the first V12 in 1916 and called it the "Twin-Six."

CHAPTER 3 ENGINE CONFIGURATIONS

FIGURE 3-36 1916 Packard Twin Six V12 engine *(Photo by the author)*

(Figure 3-36) V12 engines were used in aircraft in both WWI and WWII. When the U.S. entered WWI in 1917, the government asked Packard's and Hall-Scott's top engine designers to come up with a design for a V12 aircraft engine. The result, called the Liberty 12, was produced by Buick, Ford, Cadillac, Lincoln, Marmon, and Packard during the war. By the time it was in production, the war had ended and it had only limited use in that conflict. But some aircraft through WWII continued to be powered by the smooth V12 engine. Rolls-Royce's Merlin V12 powered P51 Mustangs, Spitfires, Hurricanes, and Lancaster bombers.

Between the wars, many luxury car manufacturers produced V12s. In addition to Packard, Daimler, Hispano-Suiza, Cadillac, Auburn, Franklin, Rolls-Royce, and Pierce Arrow offered them. Lincoln continued to produce a flathead V12 until 1948. After WWII, the V8 became the mainstay for most high-end manufacturers. V12s came back in the 1970s with Jaguar in its E-Type and XJ12. BMW added a V12 to its 7-series line in 1988 and that engine is still in production in BMWs and Rolls-Royce cars. (BMW now owns Rolls-Royce.) Ferrari (Figure 3-37) has had V12

FIGURE 3-37 1961 Ferrari V12 engine *(Courtesy of Sfoskett)*

AUTOMOTIVE MILESTONES

FIGURE 3-38 1930 Cadillac V16 engine *(Courtesy of Herranderssvensson)*

engines in its top line models since it resumed producing cars after WWII in 1947. Aston Martin, Lamborgini, and Mercedes-Benz also currently produce V12s.

V16 Engines

A V16 is two straight-eight engines on a common crankshaft. Since the straight-eight is perfectly balanced, any V-angle can be used, but even-firing will only occur at a V-angle of 45° or 135° with shared crank throws. Cadillac produced a 45° V16 with overhead valves from 1930 to 1937. (Figure 3-38.) They redesigned it in 1938 as a 135° flathead V16, which allowed a lower cowl and hood line. These engines were their top-of-line offerings until 1940. They still produced their

FIGURE 3-39 1931 Marmon aluminum V16 engine *(Courtesy of Liftarn)*

80

flathead V8 as the base engine during that time. Marmon introduced its 45° V16 in 1931 made of aluminum. (Figure 3-39.) They stopped production of this engine in 1933 after producing only 400. There are no V16 engines in production automobiles today, but V16 diesels are used in railway locomotives.

Multi-Cylinder, Horizontally Opposed Boxer Engines

Horizontally opposed engines were created to package more efficiently in an engine compartment and to have better balance than an inline engine of the same number of cylinders. They also have the same advantage over inline engines as V-engines—a shorter, stiffer crankshaft that is less subject to torsional vibration and fatigue failure. Opposed, boxer configurations with the cylinders arranged 180° apart were used in some of the earliest engines. The boxer designation means that each piston is on its own crankpin and opposite pistons are arranged to move toward and away from each other to cancel their forces. The piston motions are like a boxer punching his gloves together. Figure 3-40 shows the crankshaft and piston arrangements for two, four, and six-cylinder boxer engines.

The pistons in opposite banks of V-engines usually share the same crankpin. If a 180° V-engine is so arranged, then those piston-pairs move in the same direction and do not cancel each other's inertial forces. This arrangement is not considered to be a boxer engine. Sometimes non-boxer, 180° V-arrangements are referred to as flat-engines. The flat-12 mentioned in the section on V12s is such an engine. Because the inline sixes from which the flat-12 is made are balanced (except for inertia torque), it is not necessary to use 12 crankpins. Even when opposite pistons share crankpins on the 6-throw crank of the flat 12, the engine will still be force balanced.

Opposed Boxer Twins

Some of the earliest engines designed were opposed twins. Their designers apparently recognized the advantages of having pistons moving in opposite directions to cancel inertial forces. Benz showed a boxer twin in 1897. Henry Ford used a boxer twin in his first car, the Model A of 1903. The Jowett automobile used one from 1906 to 1937. After WWII, several manufacturers made small cars with this engine such as the famous Citroën Deux Cheveaux, described in Chapter 2. But these engines have been more popular in motorcycles than in automobiles since 1905. BMW has been making boxer-twin motorcycles continuously since 1923. Figure 3-41 shows a 1967 BMW boxer motorcycle en-

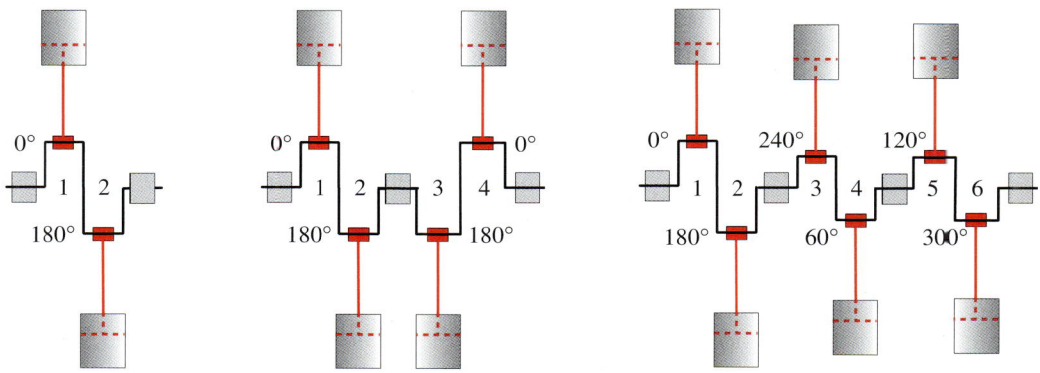

FIGURE 3-40 Boxer engine arrangements for two, four, and six cylinders

FIGURE 3-41   BMW Boxer engine *(Photo by Jeff Dean)*

gine. The boxer twin is the only two-cylinder arrangement that has both even firing and complete shaking force cancellation. But its shaking moment and inertia torque are both nonzero.

Lanchester designed an extremely clever opposed twin in 1897 for use in his first automobile. As shown in Figure 3-42, this engine had not only two opposed pistons but also two crankshafts, geared to rotate in opposite directions. Two sets of connecting rods (six in total) connected the pistons to the two crankshafts. His arrangement cancels all harmonics of shaking force and shaking moment. The scissors action of the conrod pairs serves to cancel the higher harmonics and the oppositely rotating cranks cancel the inertia torque. It is as perfectly balanced as an inline-eight. Pure genius.

Opposed Boxer Four

The four cylinder boxer is even-firing and has a better balance condition than an inline four—its primary and secondary forces and primary moments are balanced. Only secondary moments and inertia torque are nonzero. The first boxer four was built by Wilson-Pilcher in 1900 and was made until 1907 in England. Tatra (Czechoslovakia) used them extensively in the early 1930s. The story of Hitler ordering Porsche to copy the Tatra for the Volkswagen Beetle was told in Chapter 2.

The boxer four has been a popular engine, used in the original VW Beetle, early Porsches, and most current Subarus. Porsche, which has used boxer sixes since the 1970s, recently announced that they will once again make a four-cylinder boxer engine. Honda introduced a 1200 cc boxer four in its first generation Gold Wing motorcycle in 1975. Opposed fours made by Lycoming and Continental have also been used as aircraft engines in small planes such as Cessnas and Piper Cubs.

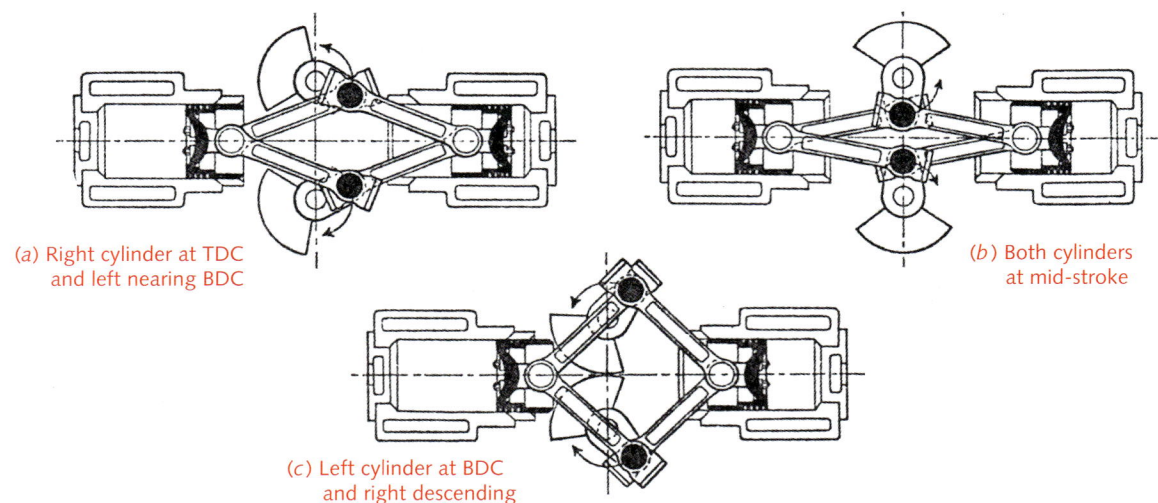

(a) Right cylinder at TDC and left nearing BDC

(b) Both cylinders at mid-stroke

(c) Left cylinder at BDC and right descending

FIGURE 3-42   Perfectly balanced Lanchester two-cylinder, horizontally opposed engine (1897)

Opposed Boxer Six

The boxer six is even-firing and has similar balance to an inline six. Wilson-Pilcher also made the first opposed six engine in 1900. The 1948 Tucker used an opposed six converted by Lycoming from one of its aircraft engines. In 1955, the Citroën DS used an opposed six made from three of its Deux Chevaux's opposed twins.

Chevrolet put an opposed six in its rear-engined, RWD Corvair of 1960. In his book, *Unsafe at Any Speed*, Ralph Nader attacked this car as a dangerous design that could spin out and roll over because of its rear engine and swing-arm rear suspension. The bad press that resulted hurt sales and GM dropped the car in 1969. Ironically, the Corvair was essentially the same chassis/engine design as the VW Beetle, which GM had copied. Had Nader attacked VW instead, the Beetle might not have become the best selling car of all time.

Porsche made the boxer six engine the mainstay of its sports car lines. Since 1965, Porsche has used this engine in all 911s, Boxsters, and Caymans. Subaru also uses an opposed six in some of its larger models. In 1988, Honda replaced the boxer four in its Gold Wing motorcycle with a 1500 cc opposed six and enlarged the engine to 1800 cc in 2001. The boxer six engine in this motorcycle is very impressive (as are all Porsche boxer engines as well). The Honda is completely vibration free and has enough torque to accelerate the 900-lb bike from 1000 rpm in top gear with no hesitation.

W-Engines

The classical W-engine has three banks of cylinders on a common crankcase with the center cylinders vertical and the other two forming a V around it. Its name comes from its resemblance to the letter W in a front view. The British also call it a "broad arrow" configuration, after a

FIGURE 3-43 1909 Anzani, three-cylinder, W-engine used in Bleriot's airplane *(Courtesy of TRSL)*

three-pronged mark used to define their government land boundaries. Three-bank, W-engines have been made in W3, W12, W16, and W18 configurations. Figure 3-43 shows a W3 built by Anzani (Italy) in 1906 (three banks of single cylinders) for use in their motorcycles. This engine also powered Bleriot's airplane on the first flight across the English Channel.

Napier built its W12 Lion aircraft engine in 1917 and Isotta-Fraschini made a W18 for aircraft use in the late 1920s. A Napier Lion W12 is shown in Figure 3-44. A three-bank, W12 with 60° between each bank and a 0, 180, 180, 0 crankshaft with triplets of cylinders sharing crankpins, it is even firing and has primary force and moment balance. Its secondary forces and moments as well as its inertia torque are all nonzero.

FIGURE 3-44 1917 Napier Lion, W12 aircraft engine *(Courtesy of Dinesh Advani)*

An alternative W-arrangement uses four banks of cylinders instead of three. These engines make creative use of VW's novel VR15 engine, (Figure 3-35) which is a narrow-angle V6—really an inline six with alternate cylinders staggered at 15° and squeezed closer together. The VR15 engine has only one head for two "banks" of cylinders, so it looks like an inline engine.

VW first made a W8 by lopping off two staggered cylinders from a VR15, making it a VR4, and then mating two of these on a common crankshaft. From the front, it looks like a V-engine because it only has two heads, but technically, there are two sets of two banks of four staggered cylinders each. This arrangement is sometimes called a V-V to distinguish it from the true, three-bank W-engine. In effect, this W8 is a shortened V8 with poorer balance. This engine was used in the VW Passat from 2001 to 2004 but did not sell well.

They next mated two complete VR15s, putting two of them on a common crankshaft to make a W12, which looks like a V12, but is somewhat shorter for a given bore size. This engine was used in the Audi A8, the VW Phaeton, the VW Toureg, and the Bentley Continental GT. The Audi W12 is shown in Figure 3-45. This engine is still being produced at this writing.

Continuing the exercise, they added two cylinders to the VR15, making it a narrow-angle V8, and then mated two of these on a common crankshaft to make the 1000-HP, 8L Bugatti Veyron W16, shown in Figure 3-46 and still in production as of this writing. (VW owns Bugatti.) Quite a cornucopia of engines from one oddly arranged V6.

Rotary and Radial Engines

Both rotary and radial engines have their cylinders arranged radially around the crankshaft and so look alike when not running. They both have odd numbers of cylinders, typically five, seven, or nine in a given row, with either one or multiple rows. Their difference is that a radial engine has its cylinders stationary and the crankshaft rotates within the crankcase in conventional fashion. But a rotary engine inverts this arrangement with the crankshaft stationary and the entire crankcase, cylinders, and all other accessories rotating around it.

FIGURE 3-45   Volkswagen/Audi W12 engine
(Courtesy of Hasse A on sv.wikipedia)

FIGURE 3-46   Bugatti Veyron W16 engine
(Courtesy of Florian Lindner)

CHAPTER 3 ENGINE CONFIGURATIONS

ROTARY ENGINES The main advantage of the rotary engine is its complete lack of vibration because everything moving is in pure rotation, including the pistons. Another advantage is that the large rotating mass of the cylinders acts as a big flywheel, obviating the need for a physical flywheel, which saves weight. Since rotary engines were primarily used in aircraft, their low weight and absence of vibration were significant advantages. The propeller was bolted to the crankcase.

The earliest applications of rotary engines were in automobiles and motorcycles. Aircraft use soon followed. The first rotary engine was patented in England in 1847, but it used steam for propulsion. Felix Millet (France) built a five-cylinder rotary into a bicycle wheel in 1897 as shown in Figure 3-47. Balzer (U.S.A.) made a three-cylinder rotary and installed it in a buggy in 1894. The Adams-Farwell Co. of Iowa, made a five-cylinder rotary and installed it in their automobile in 1906. At the 1908 Paris auto show, the Seguin brothers of France showed their first Gnome, five-cylinder rotary for aircraft use. They subsequently produced an entire line of rotary engines with one and two rows of seven to nine cylinders each. The Gnome engine was later licensed to German manufacturers (before WWI) and this resulted in French and German pilots facing one another in dogfights during the war flying planes powered by the same engines. Many WWI aircraft were powered by rotary engines.

Rotary engines had some peculiarities that did not make their pilots happy. When a mass is spun at speed, it creates a gyroscopic effect[1] that resists any attempt to turn its spin axis transversly in one direction, and enhances its rotation in the opposite direction. Because of this, a rotary-powered airplane was sluggish and climbed when turning to the left, but in a right turn, it would dive to the right and turn so fast that the pilot could lose control. Experienced pilots could turn this flaw to their advantage in dogfights by breaking to the right faster than an oncoming opponent could turn left to follow, in order to escape his fire.

Another disconcerting problem was that the rotary engine had no stationary crankcase to store lubricant. So, though it was a four-stroke engine, the oil was mixed with the fuel as in a two-stroke, and it was a "loss-lubrication" system. Lubricant was expelled in the exhaust stream after passing through the cylinders. This would have been merely messy if the lubricant used were not castor oil, a vegetable oil that does not burn well or mix with gasoline. Your mother may have forced you to take a spoonful of this foul-tasting liquid when you had a stomach-ache, because it is a powerful laxative.

[1] See the Sidebar on Gyroscopic Forces in Chapter 6 (p. 174).

FIGURE 3-47 1897 Felix-Millet rotary engined bicycle *(Courtesy of Gerard Delafond)*

FIGURE 3-48 Model of a Bently BR2 rotary engine *(Photos by the author)*

The silk scarf often seen around the necks of WWI pilots was not a fashion accessory. It was used to wipe their goggles and to wrap around their nose and mouth to reduce the amount of ingested castor oil. It is said that these pilots often brought a bottle of blackberry brandy along because drinking it tends to "bind one up." The Red Baron did not have it easy!

Figure 3-48 shows a scale, working model of a British Bentley BR2 rotary engine of WWI vintage. This model was built by master model maker Roland Gaucher of Spencer MA, and it runs. The left panel shows the engine stationary and the right panel shows it running with Mr. Gaucher at the throttle.

Radial Engines were developed in parallel with rotary engines. C. O. Manly converted one of Balzer's rotaries into a radial in 1901. Air-cooled radial engines were built, mainly for aircraft, from 1903 to the present. But, the rotary engine's popularity only lasted from 1909 to 1919, just after the end of WWI. Limitations on the power available from the rotary design and their gyroscopic problems caused them to be supplanted by radial engines. Most radials were air-cooled, a more robust design for military aircraft than a water-cooled engine, which a bullet into the cooling system could destroy by overheating in short order. But, the back rows of an air-cooled, multi-row, radial engine suffered from poor cooling due to being partially blocked from airflow by the front row of cylinders.

The radial aircraft engine became the workhorse of both civilian and military aircraft from about 1920 until the end of WWII when the much more powerful jet engine took over. The U.S. Navy decreed in 1921 that it would only buy aircraft powered by air-cooled, radial engines. Most U.S. WWII bombers and many fighters were powered by multi-row radial engines made by firms like Wright and Pratt & Whitney in the U.S., and Bristol in England, as were most commercial aircraft of the era. Lindberg's Spirit of St Louis was powered by a Wright J5, 225 HP radial engine. An example of a two-row, 18-cylinder radial engine of WWII vintage is shown in Figure 3-49 and a cutaway of a Pratt and Whitney, Wasp Major (R-4360), 4-row, 28-cylinder engine is shown in Figure 3-50. This 3500-HP engine powered the B-36 bomber (six engines), Howard Hughes' Spruce Goose (eight engines), the Douglas Globemaster, and the Boeing Stratocruiser. The cutaway engine is in the Owl's Head Transportation Museum in Owl's Head Maine.

CHAPTER 3 ENGINE CONFIGURATIONS

FIGURE 3-49   A WWII vintage, 18-cylinder, radial engine *(Photos by the author)*

During WWII, the U.S. Army asked Chrysler to build a more powerful tank engine. They took five of their standard, inline, six-cylinder flathead engines and arranged them radially around a common crankshaft at 60° intervals, to make a 1,255 cu-in, thirty-cylinder, 425-HP, radial engine for the M4A4 Sherman tank. It was completed too late to see battle in the war.

Engine Balancing

We have discussed the roles that multiple cylinders, crankshaft phase angles, and V-angles play in the balance of an engine and found that a few arrangements give essentially perfect overall balance. In other instances, additional steps can be taken to improve the overall balance condition of an engine. This section will discuss some of these techniques.

FIGURE 3-50   A WWII vintage, 28-cylinder, Wasp Major radial engine *(Photo by the author)*

Crankshaft Counterweights

Prior to 1916 no crankshafts had counterweights added to them. As described in the previous section, the 0, 180, 180, 0° crankshaft of the inline-four gives primary force and moment cancellation, and the 0, 240, 120, 120, 240, 0° crankshaft of the inline-six gives both primary and secondary force and moment cancellation. But these cancellations are strictly global, which means that these forces and moments are cancelled at the motor mounts, thus do not transmit those vibrations into the chassis. However, unbalanced forces are still present within the block and crankshaft. These are felt locally as additional loads on the crankshaft bearings. Bearing loads create friction, and friction robs power.

In 1916, Hudson introduced an improved model of its inline-six, called the "Super Six." It added counterweights to each crank throw to balance them locally. By cancelling the unbalanced forces of each piston/conrod at its source, i.e., that piston's crank throw, there were no unbalanced forces left to be cancelled globally by their neighbors. This greatly reduced bearing loads and friction.

The first automobile race quite likely was held shortly after the second automobile was sold, as humans love to compete in all things. So auto racing was common even in the early years of the car's development. Auto manufacturers viewed racing as a way to advertise the performance of their brands over the competition. The mantra became "Win on Sunday—sell on Monday." Organizations were soon formed to sanction and oversee racing events. One of these was the American Automobile Association (AAA). The AAA organized an event in November, 1915 at the Sheepshead Bay Speedway to test cars that wanted to enter their Stock Car Races to make sure that they met their rules.

Hudson entered their Super Six in this trial and cleaned everyone's clock. This engine developed 52% more power than their previous year's engine of the same displacement. This improvement was due to the exactly balanced crankshaft, which allowed them to run the engine faster with no vibration. Faster equals more power. Hudson patented the balanced crankshaft to prevent their competitors from using it and, for a number of years, Hudson dominated various racing series. They also gained a reputation in the marketplace as high-performance automobiles, a reputation they enjoyed until the 1950s when the V8 race began. By that time the company was essentially on life support and merged with Nash to form American Motors. American Motors was eventually bought by Chrysler.

The April 1917 issue of *The Automobile Engineer*, a *Journal of the Institution of Mechanical Engineers* (IMeche), Britain's equivalent of the *American Society of Mechanical Engineers* (ASME), published *A Test of the Hudson Super Six Engine*, in which they ran that engine on a dynamometer to measure its power output. They did this both with the stock engine and then a second time after they had removed the counterweights from the crankshaft. They reported that " . . . *the counterweighted engine ran smoothly with very little vibration at over 2000 rpm* (but) *the non-counterweighted motor ran very roughly above 2250 rpm . . . After it reached 3000 rpm . . . one of the center main bearings burned out.*" They measured 110 HP for the balanced engine, a very high number for the time.

Since Hudson's patent expired, all engine makers have balanced their crank throws, regardless of the number of cylinders or the inherent balance of its crankshaft. In fact, most modern engines actually slightly overbalance each crank throw, as this provides a further reduction in main-bearing forces to reduce friction and wear. Figure 3-51 shows a plot of the forces felt by the mainpin and its bearing in three modes, unbalanced, exactly balanced, and optimally overbalanced to minimize the mainpin force. The en-

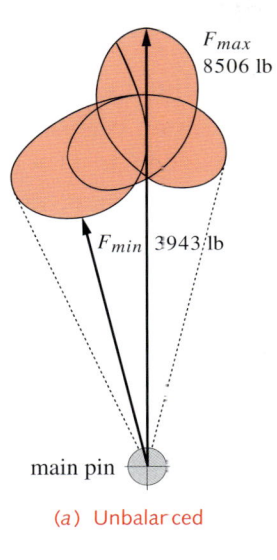

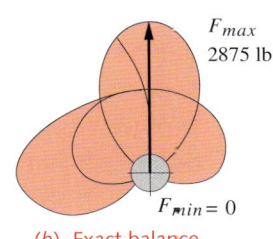

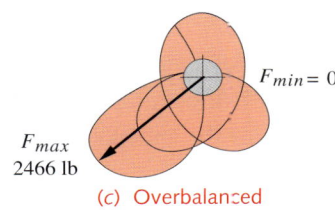

FIGURE 3-51 Force on main pin at 3400 rpm with different crank balance states. shown at same scale

gine is running at 3400 rpm and has the same size cylinders as all the example engines used in this chapter. Note the huge reduction in peak force with exact balance—from 8,506 lb to 2,875 lb. This is further reduced to 2,466 lb and made more uniform around the pin with optimum overbalancing. Note that the loops in this figure represent the tips of force vectors that run from the crank center to all points on the loops. Those vectors have been drawn in for the maximum and minimum forces in each case. Also these force vectors are turning with the shaft, meaning that

unless it is overbalanced to distribute the forces around the pin, the wear on the pin will always be on one side.

Balance Shafts

Mention was made in previous sections of this chapter of the possibility of adding balance shafts to engines to cancel forces or moments not naturally cancelled by the crankshaft or V-arrangements. This concept, called the harmonic balancer, shown in Figure 3-52a, was first proposed by the automotive genius Lanchester in 1913. Two counter-rotating shafts, driven by gears from the crankshaft, both have an unbalanced mass at some radius. The arrows shown on the masses in the figure represent the x and y components of the centrifugal forces generated by each spinning mass. Note that the vertical y components always add to, and the horizontal x components always cancel, one another. The result is a shaking force that is the sum of the two vertical y components, always acting parallel to the plane of the pistons.

If these shafts are spun at twice crankshaft speed, and arranged so their forces are the proper magnitude and always in opposite direction to that of the shaking force from the pistons, they can be made to cancel the second harmonic of shaking force. In 1976, H. Nakamura of Mitsubishi, rearranged Lanchester's balancer to put one shaft above the other on opposite sides of the crankshaft as shown in Figure 3-52b. In addition to generating the same counter-shaking force as the Lanchester device, this also creates a second harmonic of inertia torque, which cancels that imbalance as well. The Nakamura balancer was used on several Mitsubishi, four-cylinder, inline engines and was also licensed by Porsche for their 2.4 liter, 944, four-cylinder engine. Since the Mitsubishi patent ran out, many manufacturers have added this device to large four-cylinder engines in which the secondary vibrations become troublesome.

AUTOMOTIVE MILESTONES

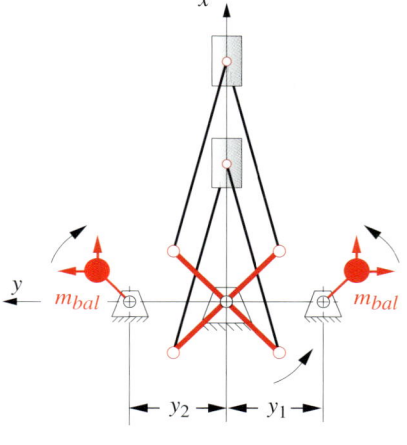

(a) Lanchester balancer

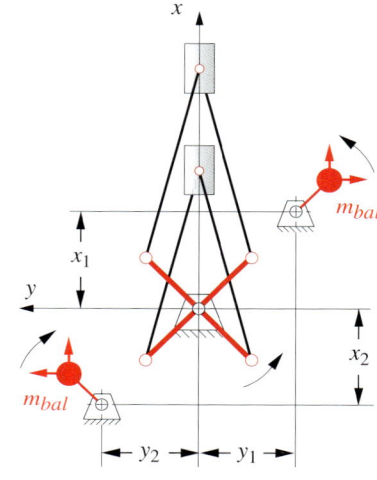

(b) Nakamura balancer

FIGURE 3-52 Two types of secondary balancer mechanisms for the four-cylinder inline engine

If an engine has a primary imbalance, such as the primary moment in a V6, manufacturers sometimes add a single balance shaft, geared to run at the same speed, but in the opposite direction, as the crankshaft. The crankshaft provides the second balance shaft. The balance shaft and the crankshaft then need two counterweights each, spaced apart along the length of the shafts to create a counter-moment at crankshaft speed that cancels the engine's primary moment imbalance.

Crankshaft Torsional Dampers

Lanchester also invented a device to reduce torsional vibration in long crankshafts. A damper is a device designed to absorb mechanical energy and convert it to heat, just as a resistor absorbs electrical energy. Shock absorbers on a car are actually dampers. Lanchester recognized the problem of torsional vibration in long crankshafts as early as 1906. His damper, designed to reduce these vibrations, is a very simple device and on modern engines is built into the crankshaft pulley that also drives a belt for the water pump. It consists of a cast iron hub, keyed to the nose of the crankshaft, and a second, cast iron ring that also forms the pulley groove. The outer ring is connected to the hub by a rubber-composite disc that has some "spring" and also provides energy absorption (damping). When the crankshaft vibrates torsionally, it sets up torsional oscillations in the damper's outer ring, which deflects the rubber disk. Deflection of the rubber causes it to heat up and this absorbs vibrational energy from the crankshaft. These devices have to be carefully engineered to match the natural frequencies of torsional vibration in the crankshaft.

Floating Power

Even when an engine is balanced to eliminate as many inertial effects as possible, there still remains an inherent torsional effect from its power pulses. The torque pulses from the engine that drive the car must be reacted against by the engine mounts; otherwise the engine would "flip over." With enough cylinders (five or more), the pulses in the gas torque function overlap to create a function that never drops below zero during the cycle. See Figures 3-16, 3-18, and 3-26 for examples. But with four or fewer cylinders, there are periods during a cycle when the gas torque drops below zero. For examples, see Figures 3-15 and 3-21 for the torque functions of four- and three-cylinder in-line engines, respectively.

If an engine is mounted rigidly to the frame of the car, as was common before the 1930s, the reaction torque pulses from a four-cylinder engine will be felt by the passengers as vibration in the chassis. Chrysler devised a clever solution to this problem for their four-cylinder Plymouth in 1931 and dubbed it "floating power." They arranged the motor mounts to support the engine at two points on an axis around which the reaction torque could rock the engine. Then they added a short, cantilever spring between a point on the side of the transmission and the frame as a torque-reaction strut that prevented any rocking of the engine about the axis. This arrangement is shown in Figure 3-53. The rear engine mount was placed under the tailshaft of the transmission for convenience. The front engine mount was then placed high on the front of the engine on a line that passed from the center of the transmission output shaft at the rear engine mount through the center of gravity of the engine-transmission combination. This made the engine into an equivalent flywheel rotating about this axis. The engine mounts contained rubber "donuts" to cushion and absorb any vibration, and the torque strut was also rubber-mounted. The car was free of discernable engine vibration and noise. Their advertising claimed it had "*The Smoothness of a Six, the Economy of a Four*." Citroën licensed this technology for its automobiles.

Summary

The saying that *they don't make them like they used to* is certainly true, but not for the reasons that this expression usually implies. The usual implication of that expression is that cars were better in the "old days." This is most certainly not true. Modern cars are much better then they ever were in the past. Technology has progressed tremendously since that time and modern automobiles and engines are head and shoulders above anything designed in the 1990s or earlier in terms of performance, economy, and reliability. Cars are now more efficient, more powerful, and more economical that they have ever been. It is now unusual not to get at least 200,000 miles of reasonably reliable service from a well-maintained engine and drive train. In the 1930s, one was lucky to go more than 20,000 miles between valve jobs and 50,000 miles between engine rebuilds with many brands. There were examples that lasted longer, but they were the exceptions. Modern cars are the best that have ever been built.

This has been a long chapter, but it has explored a complex subject in some depth. An engine is the heart of an automobile and is one of the most complex parts of the system. But, much more can be learned about engine technology than has been possible to explain here. The reader interested in finding out more about this subject is encouraged to consult the author's engineering text on the topic: *Design of Machinery; An Introduction to the Synthesis and Analysis of Mechanisms and Machines*, 5ed, McGraw-Hill New York, 2012. Chapters 12, 13, and 14 explore the mathematics and theory of the topics described in this chapter in much more depth and detail.

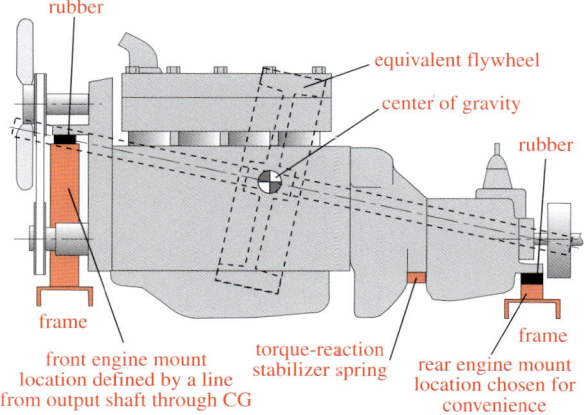

FIGURE 3-53   Plymouth's Floating Power engine mounts

Chapter 4
Valve Trains, Induction, and Supercharging

Four-stroke engines require at least two valves per cylinder, one for intake and one for exhaust. Early racing engines and many modern, production engines use more than two valves per cylinder to achieve better engine breathing. In the end of a round cylinder, four smaller valves have more total area than two larger valves and thus allow greater flow. Some engines have used three or five valves, with a larger number used for intake than exhaust. The largest pressure differential achievable across a non-supercharged intake valve is one atmosphere or about 15 lb per square inch (psi) at sea level. But an exhaust valve has the piston pushing the gas out of the cylinder and the piston can develop higher pressures. For this reason, intake valves are usually larger than exhaust valves.

Most engines use poppet valves that look something like a flat-topped mushroom with a large circular head and a long, thin stem. They are usually forced open by a cam and closed with a spring, but some engines both open and close the valves with cam action. These are called **desmodromic** (or desmo) and allow higher engine rpm than spring-closed valves. The valve springs typically limit engine speed because, if run too fast, the mass of the valve and the spring go into a condition called **resonance** and vibrate excessively. The valve does not fully close in this condition and the engine loses compression and power. This condition is called *valve float*. The "redline" on a tachometer denotes the fastest engine rpm that can be achieved without introducing valve float. There is also a danger of breaking a spring or valve in resonance. Many modern, fuel-injected engines don't show a red line on the tachometer because their engine management computer shuts off the fuel supply if the driver exceeds the redline speed. Desmodromic valves will not float as they have no spring, thus do not limit engine speed. Not all engines used poppet valves. Some used sleeve valves driven by cranks and conrods in a naturally desmodromic fashion as will be described below.

An engine's power is ultimately limited by its ability to breathe in fuel and air and to expel the products of combustion. A naturally aspirated engine is limited in this regard by the induction system—fuel supply, intake passages, and valves. Designers who want to promote engine longevity rather than engine power, will limit the size of the induction system. The original VW Beetle was

The opening photograph is of a cutaway VW, 2.0T TSI, direct-injected, turbocharged, inline, four-cylinder, OHC engine taken by the author at the New England International Auto Show in 2010.

CHAPTER 4 VALVE TRAINS, INDUCTION, AND SUPERCHARGING

fitted with a restriction in the intake manifold under the carburetor to limit the rate at which fuel/air mixture could enter the engine.[1] This controlled the maximum forces felt by the engine's moving parts and was largely responsible for that engine's famed longevity. Hot-rodders went in the other direction, substituting internally polished intake manifolds and adding more carburetors. The ultimate "hop-up" modification was to add a supercharger to force more air and fuel into the engine under pressure. This has significant effect on engine power output but can be at the expense of the health of its moving parts. Modern automobiles are moving in the direction of smaller displacement engines of fewer cylinders with turbochargers added to boost their power levels to that of a larger engine.

This chapter will explore valve trains, valve types, valve cams, and the development of valve train technology that has now reached great heights in terms of performance and complexity. The valve train is, dynamically, the most complex system in an engine. The mathematics involved in cam design is among the most complex of any subsystem in the automobile.

[1] NASCAR also requires the use of a restrictor plate to limit speeds at some of the faster race tracks.

A great deal of very sophisticated engineering has gone into bringing valve technology to its current high state of development.

Valve Actuation

Engine valves are typically opened by cams that drive a follower train that connects to the valve in some fashion. A cam is a specially shaped piece of steel on a shaft. Its shape is defined mathematically to properly control the follower's motion smoothly with minimal vibration at high speed. For part of its circumference, the radius of the cam is constant, creating a dwell in the follower motion during which the valve is closed. The camshaft containing a cam lobe for each valve is driven by gears, chain, or timing belt from the crankshaft at exactly half crank speed. The cam lobe shape is designed to open and close the intake or exhaust valve at the right point in the cycle to allow either the intake charge into the cylinder or the exhaust gases out, and then to dwell while the valve is closed.

Figure 4-1 shows cams driving followers of two types, flat and roller. Each follower (also called a tappet or lifter) is usually held against the cam with a valve spring. This spring also

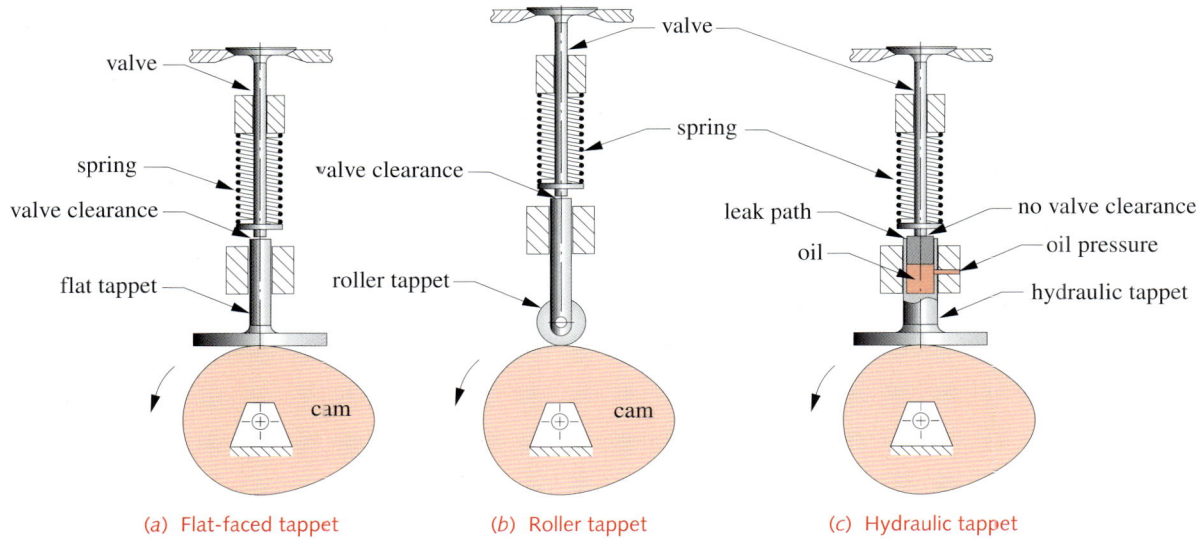

(a) Flat-faced tappet   (b) Roller tappet   (c) Hydraulic tappet

FIGURE 4-1 Cams and cam-follower trains

serves to close the valve tightly against the valve seat. Most early engines and some current ones use flat-faced followers. These are simple and inexpensive to make but their sliding friction against the cam absorbs some energy. When the energy crisis of the 1970s and 1980s made miles per gallon (mpg) an overriding factor in engine design, most manufacturers switched to roller tappets because their lower friction achieved an increase in mpg. The added cost of the roller was offset by the gains in engine efficiency. Since the advent of variable valve timing (VVT), many have reverted to flat followers.

When most materials get hot, they expand. This is true of parts in the valve train. When the valve is closed, the follower train actually must not contact the valve in order to insure that the valve closes fully on its seat to prevent leakage, which would cause loss of engine compression. There is a deliberate gap left between the tappet and valve during the dwell to allow for thermal expansion when the engine reaches operating temperature. This is referred to as **valve clearance**; it ranges, when cold, from about 0.010 in (0.5 mm) on the intake valve to 0.015 in (0.75 mm) on the exhaust valve, which gets hotter in operation. Mechanical tappets of either flat or roller type had to be adjusted periodically to maintain this clearance.

When the cam turns and takes up the valve clearance, the follower train impacts the valve, causing noticeable noise, sometimes referred to as valve clatter. (This, perhaps, explains the pseudonyms of the hosts of the popular National Public Radio *Car Talk* program—*Click and Clack, the Tappet Brothers*.) Mis-adjusted valves with too-large clearances will increase this noise to annoying levels.

Hydraulic Tappets: In 1907, Bollee (France) made an engine that had hydraulic tappets. These tappets are a close-fitting piston in cylinder that make a complete "barrel" with a passage into the side of the tappet's cylinder to allow engine oil pressure to pump it up and push the hydraulic tappet apart to exactly fill the space between cam and follower. Engine oil pressure maintains its length until thermal expansion of the cam and follower force it smaller, pushing oil out past the tappet's piston so the tappet always completely fills the gap between cam and follower. Thus, it automatically compensates for thermal expansion/contraction and eliminates the valve clearance without compromising valve sealing. The oil, being incompressible, transmits force between cam and follower.

A schematic view of an hydraulic tappet is shown in Figure 4-1c. Hydraulic tappets are more complicated than shown in Figure 4-1c and usually also contain a ball check valve. Cadillac was the first U.S. company to use hydraulic tappets (also called hydraulic lifters) in its V16 of 1930. Most present day engines use them.

Cams can be designed in what are called conjugate pairs such that one cam moves the follower in one direction and the other moves it back as shown in Figure 4-2. This eliminates the need for the valve spring and creates a so-called desmodromic cam-follower system. The lack of a valve spring eliminates the possibility of resonance in the valve train and allows much higher redline speeds. Mercedes made racing engines

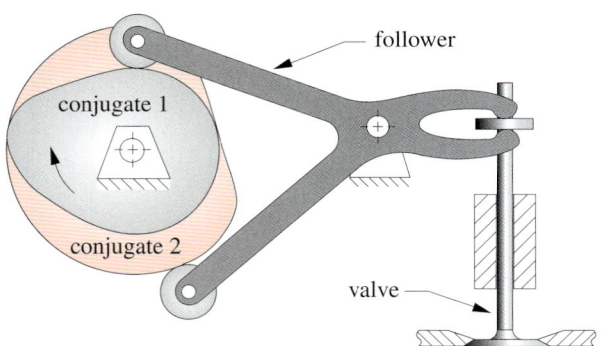

FIGURE 4-2 A conjugate cam pair for a desmodromic valve train

with desmodromic valve trains and Ducati still makes street motorcycles with them.

Valve Cam Functions

A typical set of motion functions to drive a valve is shown in Figure 4-3. Mathematically, these are called spline functions. The engine is running at 6000 rpm on the crankshaft and 3000 rpm on the camshaft. The displacement diagram shows the motion of the valve. The others show the valve's velocity and acceleration. Acceleration is the most important function as it controls dynamic force and vibration from Newton's Second Law: force equals mass times acceleration. Note that the peak acceleration on the valve is 4000 m/s^2 or about 400 G.[1] The valve-cam designer wants the acceleration pulse to be strong at the start of the rise to get the valve open as quickly as possible. The negative phase of the acceleration, which slows the valve as it approaches its maximum lift and starts its closure, is kept as small in magnitude and as flat as possible to avoid valve float. Unless it is a desmodromic system, in which the cam forces the valve shut as well as opens it, the valve spring must close the valve. Excessive negative acceleration will cause the cam to move away from the follower faster than the spring can follow and the valve will float, close late, and bounce off the seat.

Valve Arrangements

Many variations were tried in respect to valve location within the engine. Most early engines used one of several valve-in-block arrangements in which the valves were beside the piston. Others put the valves above the piston, called valve-in-head, an arrangement that is now standard. Each of these arrangements will be discussed in turn.

Valve in Block

Four arrangements were common in early engines: the T-head, L-head, F-head, and I-head or overhead valve (OHV). Despite use of the word

---

[1] A "G" is the acceleration due to earth's gravity or 9.8 m/s^2 at sea level. Four hundred G is 400 times gravity.

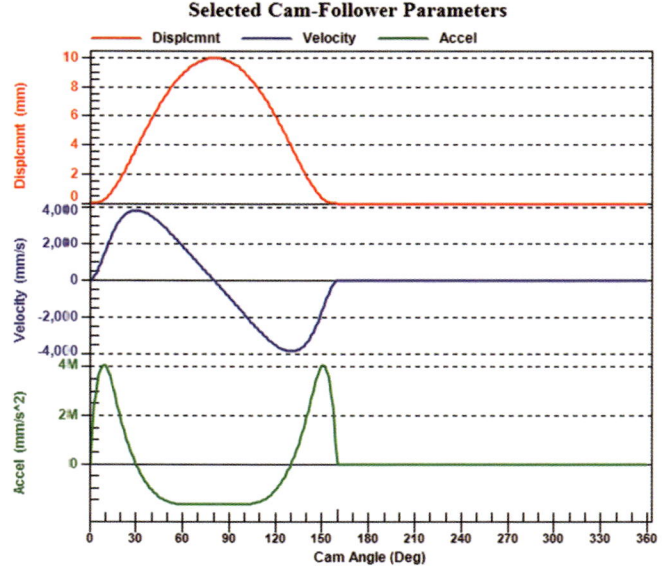

FIGURE 4-3 Cam follower mathematical spline functions

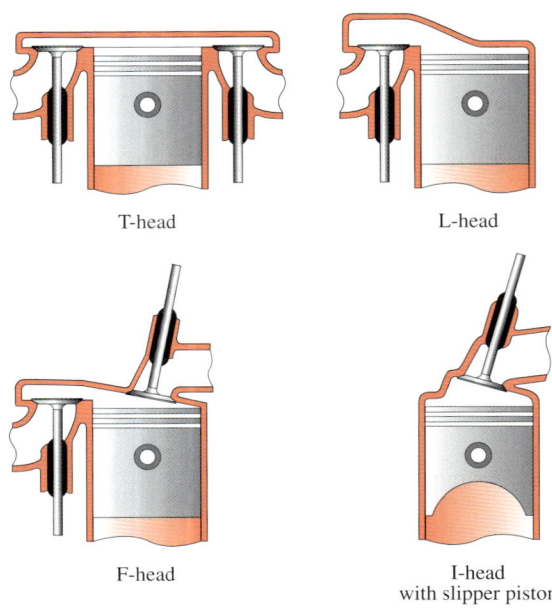

FIGURE 4-4 Various types of valve arrangements

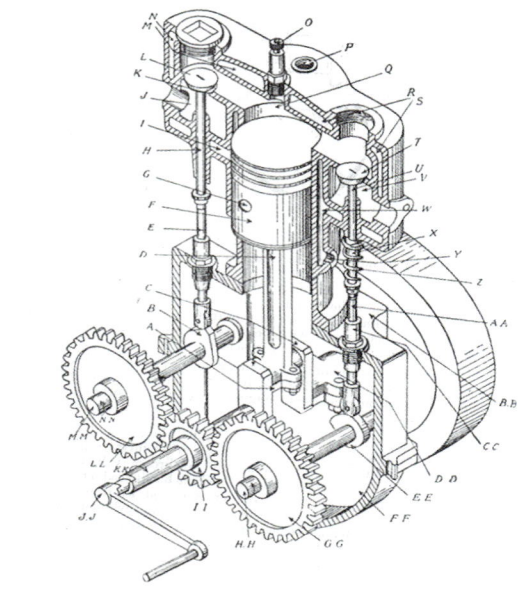

FIGURE 4-5 A T-head engine cutaway *(Public Domain top)* and a four-cyl, 417 cu in, 40-HP, T-head engine in a 1908 Locomobile *(Photo by the author bottom)*

"head" in their nomenclature, with the exception of the I-head and F-head, both intake and exhaust valves resided in the engine block. The F-head placed one valve in the block and one in the head. These arrangements are shown in Figure 4-4.

T-head

The T-head arrangement was used on the earliest engines, which used a single casting for one or two cylinders with no separate head. The intake valve was on one side of the engine and the exhaust valve on the other, making the cylinder look like the letter T in an end view. Two camshafts sat beside the crankshaft in a multi-cylinder engine, one on either side. They were driven with the standard 2:1 reduction gear from the crankshaft as shown in Figure 4-5. A camshaft must make one revolution in 720° of crankshaft rotation to achieve the four-stroke cycle. This T-head arrangement was simple to make but did not give good breathing to the engine because the gas had to do a U-turn into the cylinder. The shape of the combustion chamber was also poor for efficient combustion. This was not realized until somewhat later. The T-head engine was used only from the 1890s to WWI and then was gone.

CHAPTER 4 VALVE TRAINS, INDUCTION, AND SUPERCHARGING

FIGURE 4-6 L-head engine with a Ricardo combustion chamber *(Courtesy of Bartomiej Bulicz)*

L-head

The L-head engine (also called a side-valve or just flathead engine) puts both valves on one side of the piston and needs only one camshaft driven by the crankshaft as shown in Figure 4-6. This is a simpler arrangement than the T-head but it initially suffered from a convoluted path for the gas and a less-than-optimal combustion chamber shape. In 1919, Harry Ricardo in England discovered that turbulence within the combustion chamber increased flame speed and improved combustion. Figure 4-6 shows Ricardo's design of a combustion chamber that improved the efficiency of the L-head engine. In time, most everyone used Ricardo's combustion-chamber design and the L-head became the defacto standard for engines for many years. By 1917, L-heads, also simply called flatheads, were in about 70% of engines and comprised the majority of engine designs from then until the 1950s when they gave way to the overhead valve I-head arrangement, which was much superior in many ways. The automotive flathead engine essentially died in the U.S. in the 1950s, though small engines for lawn mowers, snow blowers, and such continued to use them.

F-head

The F-head was a combination of the L-head and overhead valve concepts. It put the exhaust valves beside the piston as in the L-head and the intake valves above the pistons in the head, operated by pushrods from the camshaft in the block. This was a hybrid of the overhead (I-head) discussed below and the L-head. It was more expensive to make and was not adopted by many manufacturers. Rolls-Royce and Rover in England used it and Hudson also used it for a short time in the late 1920s.

Valve in Head

The valve-in head or overhead valve (OHV) concept was formed very early in the development of engines. The Belgian Pipe motorcar may have had the first OHV engine in 1904. David Dunbar Buick (USA) was a pioneer in developing OHV engines from 1904 on. Every engine Buick made from 1904 to the present has been OHV. It is not clear whether Buick preceded Pipe or vice versa.

Overhead Valves (OHV)

A typical OHV arrangement (I-head) is shown in Figure 4-4 and Figure 4-7. The valves are both above the piston, opening downward and operated by a mechanism consisting of the cam-in-block driving a follower (also called a tappet or lifter) that pushes on a long pushrod, which rotates a rocker arm that opens the valve. This

AUTOMOTIVE MILESTONES

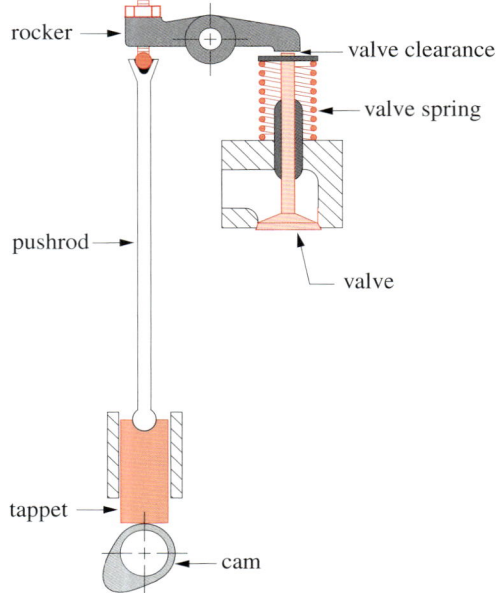

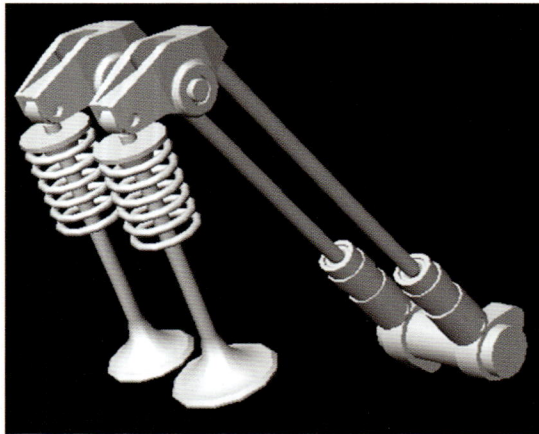

FIGURE 4-7 Overhead valve cam-follower system schematic and photo of a model
(Courtesy of Ian Brockoff)

improves efficiency. Thus OHV engines develop more power per cubic inch than flathead engines. When the U.S. horsepower race began in the 1950s, American flathead automobile engines were eventually consigned to the scrap heap in favor of OHV engines and the later overhead camshaft (OHC) engines, which are overhead valve engines with a more robust follower mechanism that moves the camshaft into the head to eliminate the long pushrod.

The design of valve cams up to WWII was rather crude and did not involve much mathematical analysis (no computers then). It was done rather more "by the seat of the pants." Engine designers got away with this because engines in the 1930s turned slowly and dynamic issues are a function of speed squared. When the 1950s horsepower race arrived and engines were expected to turn at 5000–6000 rpm, dynamic problems began to show their ugly heads. The introduction of OHV valve trains with their flexible pushrod follower mechanisms exacerbated the problem. Valve float became a big issue.

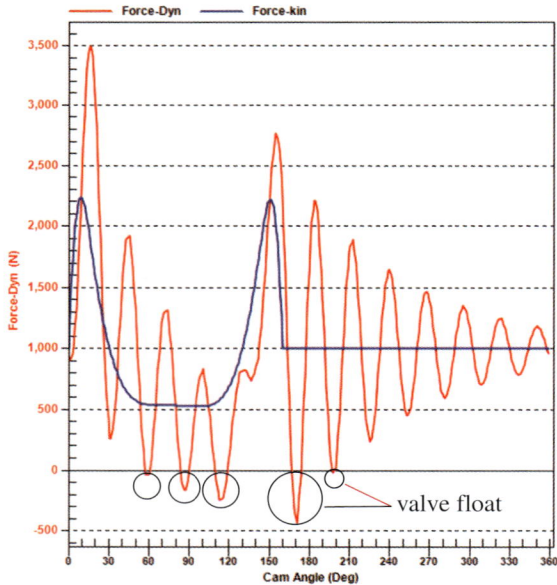

FIGURE 4-8 Kinematic and dynamic cam-follower force of an out-of-control OHV at 6000 rpm

is a rather long and somewhat flexible mechanism that can vibrate at high speeds. It limits the maximum engine speed at which valve float occurs to about 5000 rpm in most pushrod OHV engines.

Positioning the valves above the piston and having them open in line with the intake-manifold passages greatly enhances the ability of the gases to enter and leave the cylinder and this

Figure 4-8 shows a simulation of an overhead valve pushrod system running at 6000 crankshaft rpm with the follower train and valve out of control. The blue curve shows the theoretical force between cam and follower. It is always positive during motion and is a constant positive value during the dwell. The red curve shows the actual dynamic force between cam and follower including vibration due to resonance of the valve train. Note that the red curve goes negative at several points, including during the dwell when the valve should be closed. Just as you cannot push on a rope, you cannot pull on a cam joint. The two just separate if the force goes negative. This is valve float. The force continues to oscillate throughout the dwell, allowing the valve to open again when it should be closed.

In 1948, an engineer named Dudley proposed a novel solution to this dynamic problem. This solution, which is too complex to describe in detail here, was called the *polydyne cam solution*. Polydyne is a contraction of the words polynomial and dynamic. He used polynomial functions for the cam motions and combined them mathematically with the dynamics of the follower system to reduce the valve-train vibrations to zero at one engine speed. Very clever! The interested reader can find a complete explanation of this technique and its mathematics in the author's book *Cam Design and Manufacturing Handbook*, 2ed, Industrial Press, 2009, pp. 308-317.

Overhead Camshaft

An overhead camshaft (OHC) is placed in the head above the valves. This placement has significant advantages over the pushrod OHV arrangement as it eliminates the mass of the pushrods and other mechanism needed to bring the cam motion from the block to the heads. This allows the engine to run faster before it encounters the dreaded resonance problem. Either a

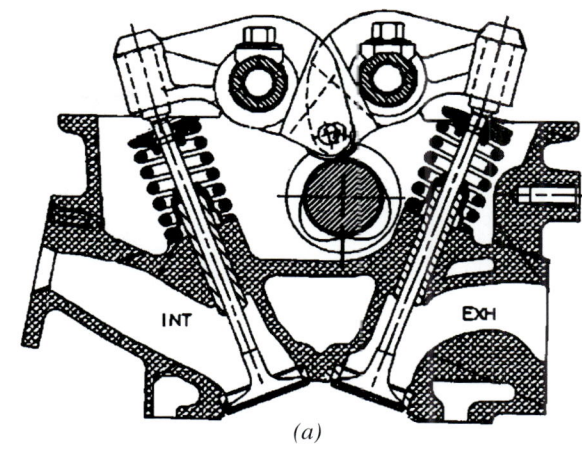

(a)

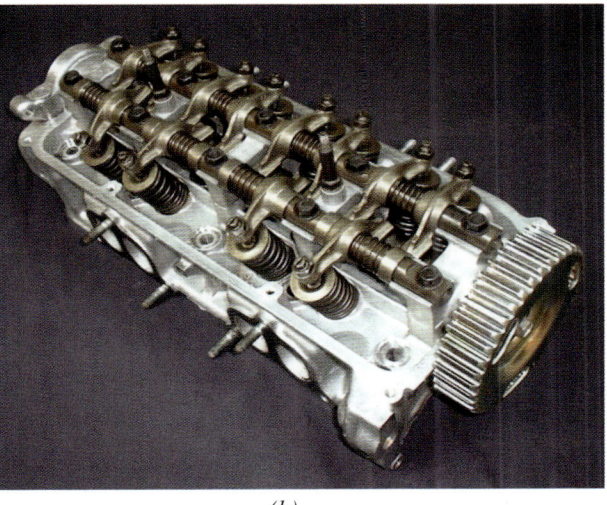

(b)

FIGURE 4-9 SOHC cam-follower system cross-section (a) and photo of a 1987 Honda CRX Si SOHC head (b) (Courtesy of Ian Brockoff,

single overhead camshaft (SOHC) can be used to operate both intake and exhaust valves, or dual overhead cams (DOHC) can be used—one camshaft operates the intake valves and the other the exhaust valves. The SOHC arrangement can have both intake and exhaust valves arranged in line under the camshaft to be operated directly, or use rocker arms to allow the valves to be angled as shown in Figure 4-9.

DOHC arrangements allow the valves to be at any convenient angles and are well-suited to a hemi-head engine. Peugeot made the first DOHC engine in 1912. The Alfa Romeo IL4s of 1928 and later were DOHC hemis as was the Jaguar IL6 of 1948–1986 used in the E-Type and XJ6 sedans. Both these engines place the valves at some angle off the vertical and put the camshafts directly above the valves driving directly on the valve stem through solid, bucket tappets. DOHC arrangements of both types are shown in Figure 4-10. This arrangement gives the lowest mass and highest stiffness follower train possible, which allows higher redline speeds. On the other hand, large mass and low stiffness in the follower train both promote the resonance problem. The OHV follower train of Figure 4-7 has larger mass and lower stiffness than those of Figure 4-9 and Figure 4-10, which makes the pushrod OHV arrangement less desirable. Figure 4-11 shows the theoretical acceleration and actual acceleration measured at the valve head in the SOHC valve train of Figure 4-9a at 7000 rpm. Dynamic force at the cam follower is essentially proportional to acceleration, and this valve was under control with no float in this test in contrast to the OHV in Figure 4-8.

As early as 1901, an engine designer, Alex Craig, apparently realized the advantages of locating the camshaft in the head to drive overhead valves directly with a minimum of mechanism. Maudslay (England) hired Craig to de-

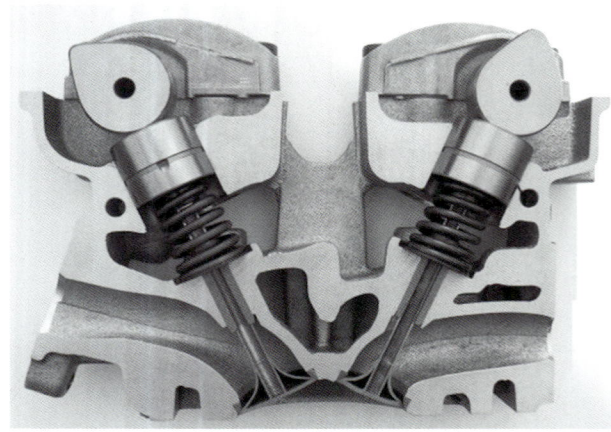

FIGURE 4-10 DOHC cam-follower system
(Courtesy of Stahlkocher)
and a Napier Lion V12 DOHC cutaway
(Courtesy of Andy Dingley)

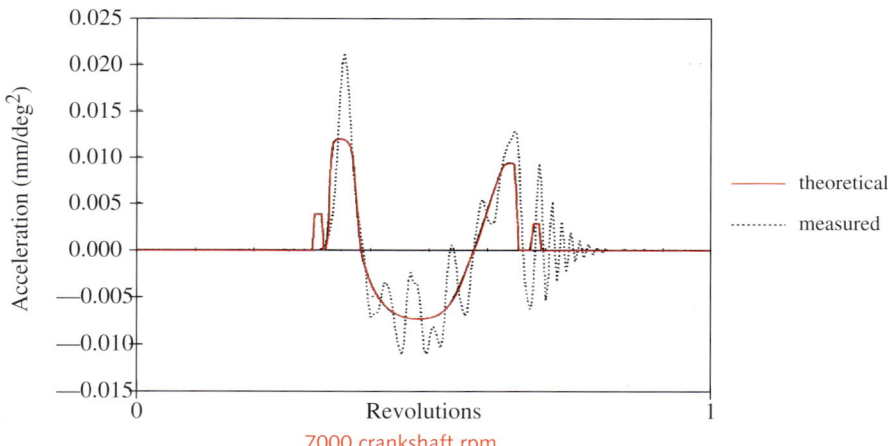

FIGURE 4-11 Theoretical and measured valve acceleration in a SOHC valve train

sign a three-cylinder, SOHC engine in 1902 and a six-cylinder SOHC in 1904. These cars were made until 1917. F.R.P. made a 454 cu in, 170-HP, four-cylinder, SOHC hemi engine from 1915-1918. In 1921, Duesenberg made a straight-8 SOHC engine and in 1928 designed a DOHC straight-8 for its SJ model that was made until they went out of business in 1937. This engine is shown in Figure 3-22. These Duesenberg engines also had four valves per cylinder—quite advanced for the time. Wills Sainte Claire (U.S.) made a SOHC V8 from 1921-1924.

Overhead camshafts were common in aircraft engines of WWI vintage and also on racing car engines from the earliest years. They were less common in production engines due to their cost and complexity. Early OHC engines used vertical shafts and bevel gearing to drive the camshaft from the crankshaft and these gear trains wore and were noisy. Chain drive soon supplanted the gear drives. Bicycle type roller chain was used on early engines but was superseded in later engines by the inverted-tooth silent chain that ran much quieter. This type of silent-chain camshaft drive is still used on OHC engines today in higher-end vehicles such as BMW and Mercedes.

The development of toothed, timing belts in 1954 offered an alternative to chains, and most OHC engines today use timing belts. Timing belts deteriorate with time and must be replaced at intervals specified by the manufacturer (often 50,000 miles) to avoid failure. Failure of a timing belt is catastrophic in some engines, known as interference engines. An interference engine is one in which the pistons and the valves, when open, occupy the same space within the cylinder at different times. When the timing belt breaks, drive and phasing between the crankshaft and cam is lost and the pistons will hit and bend the valves before the engine can be stopped, resulting in a very expensive repair as

FIGURE 4-12 Valves bent when timing belt broke at 4500 rpm *(Courtesy of Andrew Fogg)*

is implied in Figure 4-12. Non-interference engines, in which the valves clear the pistons in all positions, simply stop running when the timing belt breaks. Though timing chains do stretch, they seldom break and usually do not need to be replaced over the life of the engine.

Most modern engines are now SOHC and some are DOHC. The major exceptions are many U.S.-made V8s that still use pushrod OHV valve trains. The Cadillac Northstar DOHC V8 is an exception among U.S.-made V8s, as are the (SOHC and DOHC) Ford Modular V8 engines. BMW engines are DOHC or SOHC. Mercedes' engine line includes SOHC and DOHC models. Chrysler's IL4 and V6 engines are SOHC or DOHC, but their V8s, used primarily in trucks, are pushrod OHV.

Sleeve Valves

Dual-Sleeve Valves—The Knight Engine

In 1903, Charles Yale Knight had a better idea. After buying a Knox automobile in 1902 and finding the valve noise annoying and their unre-

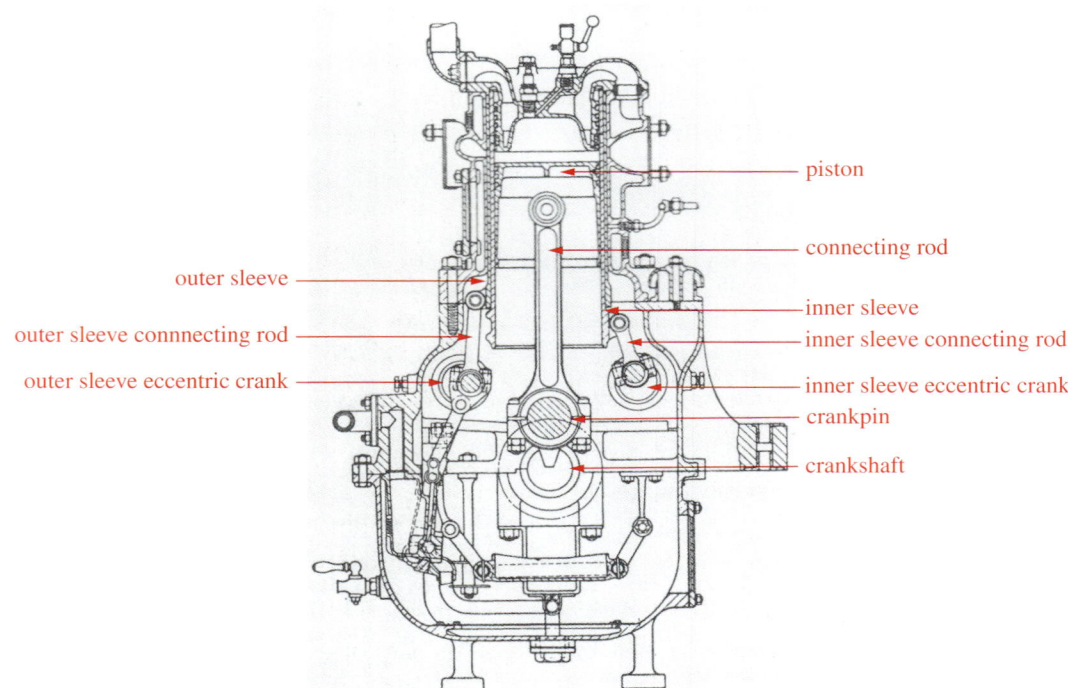

FIGURE 4-13 Daimler-Knight sleeve-valve engine *(Public Domain)*

liability frustrating, he began work on a better design for engine valves. The invention that resulted in 1905 was the "Silent Knight," dual-sleeve valve engine. Ultimately patented in eight countries, this engine, for a time, revolutionized engine valve-train technology. Instead of cam-driven poppet valves, Knight arranged two concentric sleeves within the engine's cylinder, one inside the other, with the piston fitted into the inner sleeve. The piston was driven in conventional fashion from the crankshaft and the sleeves were driven from a second, smaller crankshaft driven at half-speed from

FIGURE 4-14 1914 Stearns-Knight sleeve-valve engine cutaway in the AACA Museum, Hershey, PA
(Photo by the author)

the main crankshaft by small connecting rods. The sleeves had holes strategically placed so that, as the sleeves rose up and down, the holes aligned with one another and with the intake and exhaust ports in the block at the right times in the cycle to allow the engine to breathe in mixture and expel exhaust. Figure 4-13 shows a cross section of a Knight sleeve-valve engine and Figure 4-14 shows a cutaway Knight engine in the AACA Museum in Hershey, PA. These engines can have truly hemispherical combustion chambers uncluttered by valves.

Since there were no cams, no followers, no valve clearances, and no impacts between cam and followers, it was truly a Silent Knight. It was not a surprise that existing engine manufacturers "bad-mouthed" this engine claiming that it used too much oil (which was somewhat justified), would not last, etc. As a result, Knight could not convince any U.S. manufacturers to license and adopt his patented engine for which he demanded a $100 royalty per engine produced. Knight himself produced a car with this engine called the Silent Knight from 1905 to 1907 and exhibited it at the 1906 New York Auto show. Knight only made a few prototypes to demonstrate to prospective licensees.

Daimler (England) expressed interest and in 1908 adopted it and also added their own improvements (thinner, lighter, steel sleeves instead of cast iron) resulting in another English patent. Daimler brought out their first Daimler-Knight in 1909. Knight was as good a businessman as inventor and apparently insisted that his name be added to the model name of any car that used his engine. For reasons unknown, that was not the case for some manufacturers such as Panhard, but it was for most, which is why a perusal of any automotive encyclopedia will show a large number of makes that have the Knight name appended to them. Some authorities claim that up to 90 Knight-engined models were made.

Daimler was criticized by its competitors for using the Knight engine, so it arranged a test to be done of their Knight-engined automobile by the independent Royal Automobile Club (RAC). The club ran two Knight engines at full load on a **dynamometer** test stand for 132 hours, 58 minutes (5 1/2 days), measuring the horsepower periodically. They then put the engines in cars that were driven for 2143 miles (51 hrs) at an average speed of 41.88 mph around the Brooklands race track. The engines were then removed and run for an additional five hours on the test stand and the power again measured.

The engineers running the test were quite surprised to find that the engines actually made about 10% more horsepower after all this running than they had at the start. This was unheard of with poppet valve engines, as they typically lost up to 30% of their power over the test due to valve springs weakening and valves burning. After this test, the Knight engine was found to have higher compression than it had before the test began, which explained the higher power. While it was apparently true that Knight engines burned a bit of oil when new, it turned out that after running for some miles, the burned oil formed carbon deposits between the sleeves that actually improved their sealing and also provided lubrication (graphite, a form of carbon, is a solid lubricant). This result led to Knight's advertising slogan, *The Engine That Improves With Age*.

This result was in stark contrast to the fate of engines with conventional poppet valves that needed to be reground or replaced frequently—often every 10,000 miles or so—due to leakage. When finally disassembled after this 8-day marathon, the RAC reported "*The* (Knight) *engine was completely dismantled and no perceptible wear was noted on any of the fitted surfaces. The cylinders and pistons were found to be notably clean. ... The ports of the valves showed no burning or wear."* The RAC subsequently awarded Daimler-Knight the Dewar Trophy, Europe's most prestigious award for precision and excellence in manufacturing, in 1909. (Cadillac had received this trophy the year before for its interchangeable parts.) This demonstration of the Knight's performance by an independent testing agency quieted most of the critics.

After the Knight engine's success in England, American manufacturers began to get on board. In 1911, the Stearns-Knight appeared followed in the same year by the Stoddard-Knight, and Columbia-Knight. Between 1912 and 1927 the Edwards-Knight, Moline-Knight, Atlas-Knight, Brewster-Knight, Hadley-Knight, Sterling-Knight, Willys-Knight and Falcon-Knight appeared in the U.S. Soon, over sixty percent of New York city buses were powered by Knight engines made by the Yellow Coach Mfg. Co. They claimed that these buses went 300,000 miles without an engine overhaul. Many Knight automobile owners claimed 50,000 to 100,000 miles without an engine teardown. A poppet-valve engine was lucky to go 10,000 to 20,000 miles between overhauls at that time.

John North Willys of the Willys-Overland Motor Company encountered Charles Knight on a transatlantic voyage to England in 1914. Willys was noncommittal to Knight's sales pitch to buy a license for his company, but when Willys arrived in England, he promptly rented a Daimler-Knight with chauffeur. For fifteen days he made the chauffeur drive him over 4500 miles around England. At the end he asked the chauffeur how much work he had to do on the engine during this trip. The exhausted driver replied that he had been too busy to even lift the hood. That convinced Willys of the validity of Knight's claims.

Rather than negotiate with Knight, Willys purchased the Edwards Motor Car Co. on his return to the U.S. and renamed their Edwards-Knight the Willys-Knight. Willys later absorbed the Stearns-Knight and Falcon-Knight but kept their names. Willys and others made Knight engines in four-, six-, and eight-inline and V8 configurations. Overseas, Daimler Germany made Mercedes-Knights and a V12 Daimler-Knight. France added Peugeot, Mors, Voisin, and Panhard (who made an aero V12 among other automobile configurations). Belgium's Minerva took a license in 1908, the same year as Daimler England. They all made automobiles with Knight engines, Minerva until 1939. The last U.S. Knight-engined car was made in 1932, probably as a result of the depression. In all, over a half-million Knight engines were built between 1906 and 1939, 330,000 of these by Willys alone. Those $100 royalties per engine must have really added up for Knight.

The Knight engine did not reappear after WWII, perhaps due to the depression that caused most of its adopters to fail, but also because it could not accommodate post-war demands for increased power due to its limitations in respect to valve timing, in contrast to a cam-driven poppet valve. Cams are much more amenable to making subtle changes in the timing and shape of valve motion than is a crank-driven slider. Also, the advent of hydraulic lifters quieted poppet valve train noise, and improved materials made poppet valves last much longer.

Cam-driven poppet valves are still used in all current four-stroke engines and it is rare to need a valve job in 100,000 or even 200,000 miles with a contemporary engine. Many attempts to replace cam-activated valves with solenoid operated, piezoelectric operated, or hydraulically operated ones have been made in recent years, but so far these methods have met with only limited success. Eventually, some technology will probably emerge to replace the cam-driven engine valve, but until then, cams rule!

Single-Sleeve Valves—The Burt-McCollum Engine

In 1911, Scottish inventors Burt and McCollum patented a single-sleeve valve engine. Knight subsequently sued them for patent infringement but lost in court. In order for a single sleeve valve to accommodate both intake and exhaust ports, it had to both reciprocate up and down and also rotate to move its ports in line with the intake and exhaust openings in the cylinder. This required a more complicated mechanism than that of Knight's two crank-slider mechanisms. But engineers solved the problems and made it work.

The inventors licensed their patent exclusively to Argylls Ltd. in Scotland, and that company produced cars with this engine under their monopoly from 1911 to 1914 when they went bankrupt. The patent rights then reverted to the inventors who subsequently licensed it non exclusively to a number of other companies including, in 1926, Continental Motors in the U.S. Continental made engines (largely opposed configurations) for aircraft and automobiles. They developed a number of single-sleeve-valve engines in 6 and 8 cylinder configurations for automobiles and single-sleeve-valve, air-cooled aircraft radial engines with 6, 7, and 14 cylinders. In 1929, Peerless dropped its V8 in favor of a Continental, straight-eight, single-sleeve-valve engine. Unfortunately Peerless went out of business in 1931.

Like the Knight engine, the Burt-McCollum design was independently tested and showed similar results. After a 1000-hour test run on an engine it produced more power than at the outset of the test. Upon disassembly and inspection, no appreciable wear was found. It was as silent in operation as a Knight engine and appeared to be as long-lasting in tests. The highly respected Harry Ricardo commented in a 1926 report on the single-sleeve-valve engine

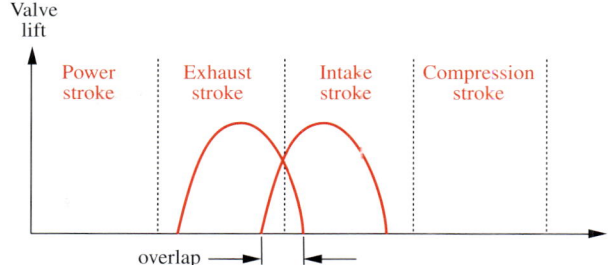

FIGURE 4-15 Schematic of increased valve overlap for high rpm

prepared by the engine sub-committee of the British Aeronautical Research Committee that "... *as the result of 9-months to 2 1/2 years' running experience on several engines, the conclusion has been reached that the single-sleeve-valve engine gives better performance than a poppet-valve engine of the same size, maintains its performance far longer, and possesses very considerable mechanical advantages.*"

The reasons that this single-sleeve-valve engine did not survive in aeronautical applications after WWII is probably due to the introduction of the turbojet engine. Its lack of impact on the automotive industry is probably due to the same factors cited for the fate of the Knight engine in previous paragraphs.

Variable Valve Timing (VVT)

As far back as the 1920s, builders of racing engines knew that more performance could be had at higher rpm by increasing the lift and duration of valve opening. Increasing duration also increased valve overlap. The intake valve would begin to open well before the exhaust valve was fully closed as shown in Figure 4-15. This allowed the engine to breathe better at higher rpm but made its performance poorer at low rpm. Hot-rodders in the 1950s also knew this fact and would replace the stock camshaft with a 3/4 or full-race cam. The 3/4 race design increased the valve duration a bit less than the full race design and so was more tractable for street use. These cams increased lift very little or not at all, as that could cause interference between valves and pistons. Both 3/4 and full-race cam modifications would result in increased performance at high rpm at the expense of a rough idle and poorer performance at low rpm. The effect of these cams on an idling engine was audible to anyone within earshot.[1]

It was obvious that a design that could change these valve parameters with rpm would be a superior solution. This approach is referred to as Variable Valve Timing or VVT. The term refers specifically to changing the time of opening/closing of the valves and does not mention changing valve lift or duration. Nevertheless the term VVT has come to include all of those changes to valve performance. The first U.S. patents for VVT were granted in the 1920s and work continued on such schemes for decades, but virtually no VVT engines reached production until the 1980s. The gas crisis of 1973 and its concomitant drive for better economy renewed interest in the topic. Between 1989 and 1990, 110 patents for VVT were issued in the U.S. alone. Fifteen basic approaches to VVT were classified by the cam researchers Dresner and Barkan in 1989. Only a few of these proposed approaches have actually been implemented in engines.

The first VVT systems changed only the valve timing by phase-shifting the camshaft as the engine reached a certain speed and had only two positions, one for low and one for high rpm. Later versions provided continuous variation of camshaft phase over a range of rpm. These systems do not change valve duration or lift. They shift only the angle at which the valves begin to open in a DOHC engine. Some are applied to both intake and exhaust camshafts and some act only on one camshaft.

<hr />

1 The author built a 3/4 race Ford flathead V8 in the 1950s and can attest to its lumpy idle that we referred to as "rump-rump."

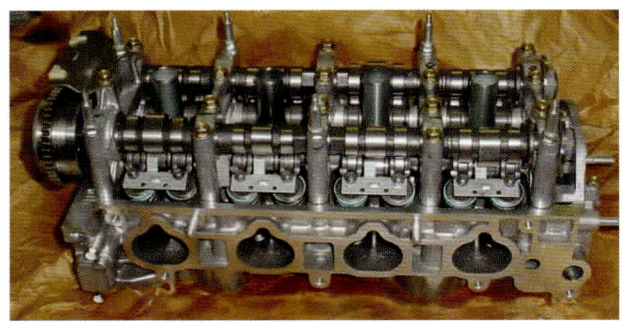

FIGURE 4-16 Honda VTEC head *(Courtesy of Joe Flores)*

ALFA ROMEO was the first to use a cam phaser (called *Variator*) in a production engine in their DOHC four in 1972. It used a helical gear that was driven axially by hydraulic pressure to rotate and advance the intake camshaft 25° with respect to the cam-drive sprocket at about 1500–2000 rpm.

NISSAN followed in 1987 with the next implementation of a cam phaser in a production engine. Its Nissan Valve Timing Control System (NVTCS) advanced the intake or exhaust camshafts or both on various models of its DOHC V6 engines. It was first used in the 300ZX (Z31) 300ZR model. It used a phaser similar to that of Alfa Romeo except that Nissan's was electronically controlled rather than mechanically as on the Alfa.

HONDA introduced its Valve Timing Electronic Control (VTEC) in 1989, which was the first system to vary valve lift and duration. This uses two cam lobes for each valve, both contacting the tappet at all times. One of the cams is free-spinning on the camshaft at low rpm and so does not drive the tappet. At low speed, the low-lift, low-duration cam lobe that is fixed to the camshaft operates the valve. At about 4500 rpm, the higher-lift, higher-duration, cam lobe is locked to the fixed lobe with a hydraulically driven pin and takes over, lifting the valve

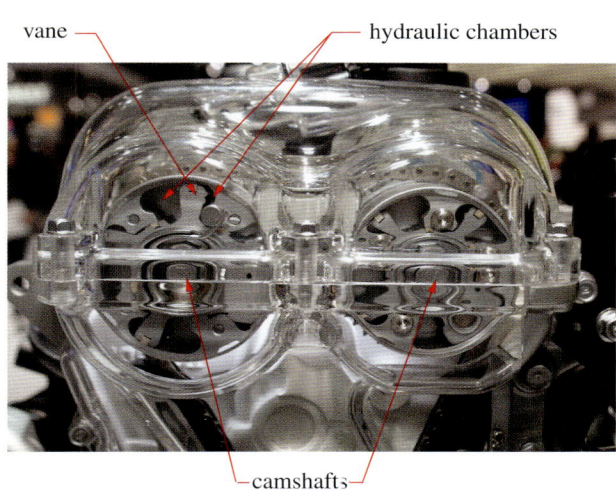

FIGURE 4-17 Hyundai vane phasor for camshaft timing variation *(Courtesy of Dmitry Ko)*

higher, opening it sooner, and holding it open longer. This system was first used on the Honda Civic, CRX, and Acura NSX, and was later extended across Honda's entire line of engines. Honda claims a 30% increase in peak power in a 1.6L engine with an increased redline of 8000 rpm (9000 rpm in their S2000 model). A Honda VTEC head is shown in Figure 4-16. Recently, a three-stage version of this system has been introduced that uses three cams, locking none, one, or the other in turn to the fixed cam.

PORSCHE introduced its first *Variocam* system in 1991, which was a cam phaser. Instead of a helical gear to rotate the camshaft, the first version used a mechanical tensioner on the timing chain to bias the camshaft sprocket versus the crankshaft. This had only two positions of phase angle. The Variocam Plus version, introduced later, used a form of vane pump on the end of the camshaft to change phase continuously over a range. Figure 4-17 shows a similar device used more recently by Hyundai. The phaser vanes rotate through an angle when hydraulic fluid is pumped from a chamber on one side of the vane to the other, causing it to rotate the camshaft versus the cam drive sprocket. This gives continuous phase adjustment over its range of motion.

Porsche *Variocam Plus* also added variable valve lift with a system similar to the Honda VTEC. It uses three cam lobes, the outer two of which are identical and have higher lift and more duration than the center cam lobe. There are two cylindrical, flat-faced tappets, one inside the other. One is a donut and the other fills the donut hole. The center, low-lift, shorter duration cam acts on the small tappet in the donut hole, driving the valve at low speeds. At this time, the outer donut tappet is being driven by the outer cams but it has no connection to the valve. At a pre-programmed speed, a valve opens to apply hydraulic pressure to a cross pin that locks the donut tappet to the inner tappet. Now the outer cams take over and drive the valve through the pin connection between the outer donut tappet and the inner tappet. It is a two-stage system like the original Honda VTEC. This system used both continuous phase adjustment and discrete, two-stage lift/duration adjustment. It is used in Porsche 911, Cayman, and Boxster models.

BMW entered the VVT business in 1992 with its VANOS cam phasing system that provided continuous phase angle variation on the intake cam over about 30° by using a hydraulically driven helical gear to rotate the camshaft versus its drive sprocket. Their Double VANOS system later added the mechanism to the exhaust camshaft as well. In 2001 they developed their *Valvetronic* system that allows continuous variation in valve lift. This uses a stepper motor that turns a separate, auxiliary camshaft through a part revolution in response to the engine controller. This auxiliary camshaft has simple eccentric cams that engage with a second rocker arm at each intake valve. Rotation of this second rocker arm changes the mechanical advantage of the main rocker arm between the cam and valve to increase or decrease valve lift. This system has continuously variable lift adjustment, unlike the Honda and other systems that only provide two or three discrete steps of lift adjustment. The main purpose of *Valvetronic* is to reduce fuel consumption rather than to increase power. BMW also claims that this VVT system eliminates the need for a throttle plate in the induction system. Instead of throttling the airflow into the engine with a throttle plate (see Figure 4-18) to control speed, the VVT system uses variable valve opening for the same purpose. Elimination of the throttle plate reduces **pumping losses** which improves engine efficiency. This is similar in concept to all diesel engines that have no throttle plate and control engine speed by varying the amount of fuel injected directly into the cylinders.

NISSAN, who had one of the earliest cam phaser systems, followed it with a continuously variable valve lift/duration VVEL system in 2007. It is similar in concept to the BMW Valvetronic but is a different mechanism. It uses the conventional camshaft to drive an intermediate mechanism that varies the lift transmitted to the valve in a continuous fashion by changing the rocker-arm ratio between cam and valve. It retains the variable phasing system and adds the ability to control lift and duration continuously.

TOYOTA introduced a cam phaser system in 1991 but in 2008 added its *Valvematic* system, which has similar function to the BMW Valvetronic and Nissan systems. Like the BMW and Nissan systems, Toyota adds mechanism between the camshaft and valve that, using links and an additional cam, can vary the amplification of the cam-lobe's lift to the valve. Toyota claims its system adds 10% power and reduces fuel consumption 5–10%.

AUDI has a two-stage, discrete, variable lift/duration system that shifts a pair of intake cams axially along the camshaft to switch from low to high lift. The axial shift motion is mechanical via barrel cams that are engaged by cylindrical followers actuated by solenoids. One barrel cam shifts the cams left and the other right. This changes the valve lift from 2 mm to 5.7 mm. It also allows the throttle butterfly to remain fully open under low load, which reduces pumping losses. Audi claims a fuel consumption savings of 7%.

FIAT introduced their **Multiair** VVT system in 2009, which is a hybrid mechanical/hydraulic system. It uses a conventional, single cam lobe for each valve, but the force path from the cam to the intake valve contains a hydraulic cylinder filled with oil. A piston acting on this oil is driven by the cam and a second piston at the other end of the chamber actuates the valve, much like a hydraulic jack or hydraulic brakes. A solenoid valve controls an opening from the hydraulic cylinder to a side chamber. The solenoid valve can block or allow hydraulic fluid to enter or leave the side chamber. When the solenoid valve is closed, there is a direct hydraulic-mechanical connection between cam and valve, and the valve responds to the full lift of the cam. When engine conditions requires less valve lift, the control system will open the solenoid valve, allowing some of the cam motion to push fluid into the side chamber, essentially stealing motion from the valve. Before the cam completes its lift, the solenoid's response is quick enough to shut off the leak into the side chamber and again provide a solid connection between cam and valve. So by manipulating the solenoid valve, the valve lift and duration can be controlled essentially continuously. Fiat claims that this system can provide 10% better fuel economy and also 10% more power.

OTHERS: Many other manufacturers besides those described here offer VVT systems of various types, It has almost become the defacto standard for engines in 2014. The improvements in engine efficiency and power output are due in large part to VVT technology, and the ultimate VVT system has probably not yet been created. Many are working to eliminate the camshaft from the system and substitute solenoid or hydraulic actuation of valves. Lotus and Eaton have been developing an electro-hydraulic system called Lotus *Active Valve Train* (AVT) since 2003 and have it working in the laboratory. It is quite complex. Double-acting (push-pull) hydraulic cylinders open and control each valve in response to servo-hydraulic valves that control the waveform of the fluid flow to the cylinders to get any valve lift or duration desired. A hydraulic pump supplies the fluid pressure and a computer controls the servo valves in real time. Engine valve motion can be changed continuously in response to engine demands. This is a very expensive system (and also heavy) and so far has not been used in a production automobile engine. MAN Diesel and Turbo does produce a large diesel engine for ship propulsion that uses an electro-hydraulic, camless valve train.

Table 4-1 lists a number of automobile makes and their current VVT technologies. In some cases, different models within a manufacturer's offerings may have different or no VVT configurations. The combination of dual overhead camshafts, VVT, electronic ignition, and electronically controlled direct fuel injection (to be discussed in a later section) have all contributed to an evolution, if not a true revolution, in the efficiency, specific power output, driveability, and reliability of engines, which have now reached a level of performance unseen before this time.

Cylinder Deactivation

It is a fact that only about 15–20 HP are needed to propel a reasonably aerodynamic automobile down a level highway with no headwind at highway speeds. The only reason that we are fond of cars with 300–400 HP is their ability to accelerate to highway speeds in only a few seconds. It is the demand for acceleration of a two-ton mass to 65+ mph that drives the horsepower craze (to which your author pleads guilty). This fact prompted some manufacturers to provide

Table 4-1 Variable Valve Timing Systems Currently Used by Various Manufacturers

Manufacturer	Cam Phasing	Variable Lift/Duration	Trade Name
Alpha Romeo	continuous I and E	no	CVVT
BMW	continuous I and E	continuous	Double VANOS/Valvetronic
Chrysler	continuous I and E	no	Dual VVT
Citroen/Peugeot	uses BMW system	uses BMW system	-
Daihatsu	continuous I and E	no	DVVT
Ferrari	uses BMW VANOS	no	-
Ford	continuous I	no	VVT
GM	continuous I and E	no	DCVCP
Honda	discrete I and E	discrete - 2 or 3 stages	VTEC
Hyundai	continuous I and E	no	CVVT
Kia	continuous I	no	CVVT
Lamborgini	uses BMW VANOS	no	-
Madza	continuous I	no	S-VT
Mitsubishi	discrete I and E	discrete - 2 stages	MIVEC
Nissan	continuous I and E	continuous	N-VCT, VVL, CVTCS
Porsche	continuous I and E	discrete - 2 stages	Variocam Plus
Toyota	continuous I and E	discrete - 2 stages or continuous	VVT (discrete), VVT-I (continuous)
Volvo	continuous I	no	CVVT
VW	discrete I and E	no	similar to original Variocam

engines that could operate on eight cylinders when rapid acceleration was demanded, but run on fewer cylinders when loafing down the highway at constant speed to save fuel. This was not a new idea. In 1905, Sturtevant offered a six-cylinder car that could also run on three cylinders. The 1917 Enger V12 also offered the ability to switch from twelve to six cylinders by moving a lever on the steering column. The first to attempt this feat again in a modern production automobile was Cadillac in 1981 when it introduced its V8-6-4 engine in most of its offerings. This engine was programmed to run on all eight cylinders when the load was high, but cut out two or four of its cylinders when less power was demanded by the driver's foot. It did this by shifting a solenoid-activated plate that changed the rocker arm pivots such that neither intake nor exhaust valves for that cylinder would open. The engine's throttle-body fuel-injection system reduced the amount of fuel injected to the amount needed by the active cylinders. Fuel injection is discussed in the next section.

With both valves in the disabled cylinders closed throughout the cycle, their pistons would simply compress the trapped air in those cylinders and the wound-up spring of that compressed air would drive the piston down on the next stroke. So the losses associated with closing off the valves in some cylinders were limited to the friction of the piston rings and the heat lost in compression of the air in the closed cylinders. Theoretically, operating the engine on fewer cylinders should give significantly better fuel economy, as it reduced both pumping losses and fuel usage.

Unfortunately, the state of the art of microprocessors to control all this technology was not up to the task in 1981. They were too slow and created hesitation when switching between cylinders. This engine suffered from many technical and reliability problems, which caused Cadillac to drop it shortly after it was introduced. Since then, computer technology has improved immensely and current manufacturers are offering V8 engines that switch seamlessly between 8- and 4-cylinder operation with no indication except a dashboard readout that announces how many cylinders are in operation and what the instantaneous mpg is. GM since 2005 achieves cylinder deactivation in its pushrod OHV V8s by collapsing the hydraulic lifters for half the cylinders. In the author's GM pickup truck with this feature, there is no indication of the switch between 8 and 4 cylinders other than the dashboard annunciator. But, unfortunately, the only time this 5500-lb vehicle switches to 4-cylinder mode is when going downhill or on the flat with a tailwind.

Chrysler introduced a similar system on its Hemi V8 in 2004. Honda added cylinder deactivation to its VTEC V6 engines in 2005 using the split rocker arms of its VTEC system. They hydraulically shift the rocker arms out of engagement with the tappets to deactivate the valves of one bank of three cylinders.

Mitsubishi was one of the pioneers in cylinder deactivation. In 1982, it introduced a system called modulated displacement (MD) on its four-cylinder engines. It was not continued for very many years due to lack of buyer interest. Perhaps performance on two cylinders was disappointing. They revived it in 1993 when they launched their VVT system, again switching from 4 to 2 cylinders. This also was dropped in 1996. Mercedes developed a cylinder deactivation system in the late 1990s for its V12 engine. This system was adapted to the Chrysler Hemi V8 in 2004 during the period when Daimler controlled Chrysler.

Engine Start-Stop

The ultimate in cylinder deactivation has to be stopping them all, and this has become popular especially in most hybrid and some non-hybrid automobiles for which fuel economy is

CHAPTER 4 VALVE TRAINS, INDUCTION, AND SUPERCHARGING

the main focus. The idea is to shut the engine off automatically after the car has been stopped for a few seconds, presumably at a traffic light or in a traffic jam to not waste fuel on an idling engine. Pressing the accelerator starts the engine again. Many manufacturers currently offer or have offered this feature. Toyota added it to its Crown model in the 1970s and Honda has had it on multiple models since 1999. Citroën offered it on some models in 2006 and BMW included it on its Mini line in 2008. Fiat first offered it on the Fiat 500 in 2008 and on the Alfa Romeo Mito in 2009 that uses the Multiair engine. The Toyota Prius and Ford's hybrid models all have it, as well as do a number of other makes. If one does mostly stop-and-go city driving, this feature can result in noticeable fuel savings.[1]

Fuel Control

One of the most difficult challenges that faced the pioneers of engine development in the 1880s and 1890s was finding a way to vaporize liquid gasoline to fuel the engine. The earliest engines used illuminating gas as fuel, which was fine in the laboratory but not very portable for an automobile engine. A large number of carburetor designs appeared in the last decades of the 19th century and into the first decade of the 20th. Eventually, the designs settled out to a few that became common until fuel injection took over in the 1980s.

Carburetors

Invention of the carburetor is often credited to the Italian Luigi De Cristoforis in 1876. Benz also developed a carburetor for his early engines and Daimler and Maybach came up with the atomizer nozzle in 1885. Some early "surface" carburetors simply passed air over the surface of liquid gasoline to pick up vapor. Lanchester experimented with wick-type carburetion in 1896. Two basic designs ultimately won out: the **fixed venturi** which came to dominate the market and the **variable venturi**, which also found many users.

A venturi is a passage, usually circular in cross section, that varies in area over its length. A venturi is typically shaped like an hourglass, larger at each end than in the middle. When air or any fluid flows through a venturi, its velocity increases as the diameter decreases and its pressure simultaneously drops. This effect is due to the principle of Bernoulli.

FIXED VENTURI CARBURETOR The venturi shape is built into the body of the carburetor as shown in Figure 4-18. Air flow through the venturi is controlled by the throttle plate which is connected to the accelerator pedal. Increased airflow increases velocity and drops the pressure in the neck of the venturi. The lower pressure sucks more fuel into the air stream through small orifices called jets placed at the smallest diameter of the venturi. There is typically an

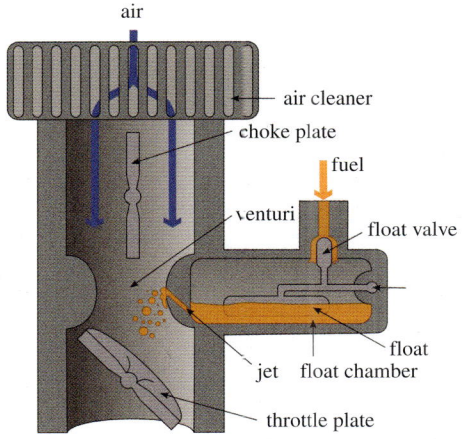

FIGURE 4-18 Cross section of a fixed venturi carburetor *(Courtesy of K. Aainsqatsi)*

---

[1] However, since most engine-bearing wear occurs during startup and stopping of the engine, perhaps this start-stop feature will have a negative effect on bearing life. When an engine is up to speed, the crankshaft journals are riding on a film of oil with no metal-to-metal contact. During the transition from stopped to idle speed (and vice versa), there is metal-to-metal contact that causes wear. (See Norton, *Machine Design*, 5ed, Prentice Hall, pp. 660-685 for details.)

In order to assure a steady supply of fuel to the jet, a float bowl is built into the carburetor body. It functions much as the toilet tank in your house. A float valve allows fuel in from the fuel pump until the level reaches a certain point at which the valve controlled by the float shuts off the supply from the fuel pump. As the float level drops, this valve opens to replenish the bowl. The jets run from the bottom of the float bowl to the neck of the venturi. Figure 4-19 shows a fixed-venturi carburetor with a transparent case on the float bowl so the float can be seen.

VARIABLE VENTURI CARBURETOR This carburetor has a tightly fitted piston within a cylinder. The piston is exposed to the vacuum generated by its venturi via a passage from the venturi to the top surface of the piston. The piston is open to atmosphere at its bottom. Attached to the piston is a tapered, conical needle in a tapered bore that meters fuel flow. When air flow through the venturi increases as a result of opening the throttle plate with the accelerator pedal, the venturi pressure drops and its increased vacuum lifts the piston, which opens the needle valve and allows more fuel flow. But the piston movement also increases the size of the venturi neck, which returns the pressure in the venturi to its nominal value. These carburetors are also referred to as a constant-depression type, where depression refers to the reduction in atmospheric pressure in the venturi.

The profile of the needle determines the rate of fuel delivery over the engine's operating range, and this gives more precise mixture control than the fixed jets of a fixed venturi carburetor. Instead of an accelerator pump, the piston's upward movement is damped by an oil-filled **dashpot** or damper. The damper initially delays piston motion. During this delay, the increased airflow temporarily pulls more fuel

FIGURE 4-19 Carburetor float bowl and float
(Courtesy of K. Aainsqatsi)

idle jet that is smaller than the high-speed jet. The greater the airflow through the venturi, the larger the pressure drop and the more fuel is sucked in. This type of carburetor is designed such that the ratio of fuel to air is fairly close to the desired stoichiometric ratio over the engine's operating range. A choke plate at the top of the venturi allows the air flow to be restricted when the engine is cold in order to achieve a richer mixture for starting.

When rapid acceleration is demanded and the throttle plate is opened suddenly, the engine will momentarily starve for fuel until flow from the jets can catch up. There is an accelerator pump built into the carburetor that also responds to the accelerator pedal. When the accelerator is opened, the pump squirts an additional amount of fuel into the throat of the carburetor to keep the engine from stumbling. This extra fuel creates a temporarily rich mixture, well above stoichiometric, and this results in unburnt hydrocarbons being emitted in the exhaust stream. This was one of the reasons that made it difficult to control emissions in a carbureted engine.

CHAPTER 4   VALVE TRAINS, INDUCTION, AND SUPERCHARGING

FIGURE 4-20   SU carburetors on a Jaguar engine
(Photo by the author)

into the engine to overcome stumble on acceleration. A variable venturi carburetor is somewhat more accurate in operation than a fixed venturi type. The variable venturi carburetor was originally developed by Skinners Union (SU) and patented in 1905 in England. Variable venturi carburetors were used extensively on English and Swedish cars until the 1980s. Hitachi in Japan also made an SU-type carburetor that was used on some Japanese cars such as the Nissan 240Z. Most other manufacturers around the world used fixed venturi carburetors. Figure 4-20 shows triple SU carburetors on the author's Jaguar E-Type engine. The three cylinders contain the pistons and their black caps allow oil to be added to the dashpots.

Increasing demands for control of exhaust emissions sealed the carburetor's doom. It could not meet the stringent emission specifications demanded by governments around the world. By the 1990s, all automotive engines had forsaken carburetors for fuel injection.

Fuel Injection

Early French engine designers experimented with fuel injection (FI) as early as 1881 (Eteve) and 1891 (Tenting) as a substitute for as yet unsuccessful attempts at carburetion. But technology of the time was not able to manufacture the precision pumps and leak-free plumbing needed to achieve fuel injection. These early attempts ultimately failed. Other attempts were made in 1898 by Deutz in Germany where Daimler and Maybach were designing engines. These attempts also failed.

The nascent aircraft industry provided an incentive to make fuel injection work, because carburetors of the time could not work in all the orientations of a maneuvering airplane. The Wright brothers actually equipped their engines (which they also designed) with a crude, evaporative form of fuel vaporization to avoid the problems with carburetors, which were not yet perfected. Their system used gravity to feed fuel into the intake manifold where engine heat evaporated it. They never flew inverted.

In France, Leon Levasasseur developed a working FI system in 1905 for his Antoinette V8 aero engine. At about the same time, Hans Grade developed an FI system for the German aircraft industry. In ten years, these two inventions would be powering airplanes trying to shoot one another down.

It wasn't until diesel engines for trucks were put into production in 1923 by Benz that the demand for high pressure fuel injection became acute. Bosch was the first to solve this problem. Diesels must use fuel injection because the fuel is not volatile enough to mix readily with air in a carburetor. Instead, diesels inject the fuel directly into the cylinder at the peak of the compression stroke. The heat of compression ignites the mixture. By 1930, Bosch had perfected the diesel FI system for mass production. Gasoline engine manufacturers showed no interest in this technology for their engines because by this time, carburetors had been developed to the point that they were doing the job satisfactorily.

Nevertheless, in the 1930s, Bosch was directed by the German government to develop fuel injection for gasoline aircraft engines. Virtually all of the Luftwaffe's fleet of aircraft were fuel injected in WWII. Experimentation showed that significant power gains could be achieved in a gasoline engine by switching from carburetors to fuel injection. Early FI systems were all mechanically controlled since electronics was yet to be invented, but electronic FI would eventually take over.

Injection Methods

Gasoline engines have used different methods of injecting fuel into the engine. The earliest systems injected it at relatively low pressure into the intake manifold at one or multiple locations. The air flow through the manifold from the throttle body through which all air enters the engine would evaporate the liquid fuel into solution with the air. The best approach is injecting directly into the cylinder.

THROTTLE BODY INJECTION The usual location for a single injection port is into the throttle body just under the throttle plate, which controls air flow in response to the accelerator pedal. This is also called a "wet manifold" system.

PORT INJECTION A better distribution of fuel can be achieved by injecting at multiple locations, typically at the manifold ports that lead to each cylinder's intake valve. The injector is often aimed at the back of the intake valve.

DIRECT INJECTION Recently, gasoline engine manufacturers have increasingly used direct injection, as in a diesel, with the fuel sprayed directly into the cylinder. WWII aircraft engines used this method as it provided the best power and efficiency. Direct injection requires higher-pressure pumps than the other two approaches which inject to a location that is actually below atmospheric pressure.

Mechanical Fuel Injection

Mechanical fuel injection (MFI) used an engine-driven fuel pump to generate pressure higher than that needed for a carburetor. In a direct-injected engine, the pressure has to force the fuel through fuel lines and into the cylinder against compression pressure. The timing of the injection to each cylinder for port or direct injection was done by cam-driven piston pumps that further increased the pressure to force fuel through the injection nozzle (injector) and into the cylinder for direct injection. The small orifice in the nozzle is shaped to make the fuel disperse in a pattern that helps atomize it.

Mechanical fuel injection was used in race cars in the late 1940s. The first use was on a port-injected Offenhauser 4-cylinder in the 1949 Indianapolis 500. This was a simple system that provided continuous flow to all cylinders. Racers don't care much about mpg. Mechanical FI began to appear in production automobiles in the early 1950s. Mercedes used port injection in its 1954 300SL gull-wing coupe. The 1957 Corvette offered a Rochester mechanical FI system of the throttle-body type. This was offered on a number of other high-performance GM models in subsequent years. Its $500 price and fears of its mechanical complexity and reliability limited its acceptance. Around 1955, port-injection became the default mode for most manufacturers. Lucas developed one such system for the Jaguar D-Type in 1956, but it never found its way into the successor E-Type, which initially used triple SU carburetors (Figure 4-20.) Mercedes also made a number of models with mechanical FI into the early 60s.

Electronic Fuel Injection

Electronic FI obviously had to wait until the electronic revolution arrived after WWII, but it has continued to improve as analog electronics evolved into digital electronics, integrated circuits, microchips, and microprocessors. Bendix, in the U.S., developed the first electronic FI system in the 1950s and presented it at the Society of Automotive Engineers (SAE) in 1957. Called the *Electrojector*, it used an electric fuel pump, electric timer, and electronic control unit (all analog). It monitored transducers measuring load, speed, coolant temperature, air temperature, and other parameters, and used these to control metering of the injected fuel with solenoid driven injector pumps instead of cams. This made it a closed-loop system. Bendix put their prototype system on a 1963 Buick V8 and showed it around to all the U.S. manufacturers. GM was not interested as they had their Rochester mechanical FI system. Ford also was not interested. AMC was the first to order it for their 1957 Rambler. Chrysler soon followed, putting it on its 300-D and other models.

By 1968, Bosch became interested in the Bendix system because VW (who used Bosch equipment) could not meet the new U.S. emission requirements with their carbureted engines. Bosch designed a system for the VW under Bendix's licensed patents. The Bosch system was ultimately adopted on many European cars. This system was designed to maximize economy rather than power.

Electronic fuel injection (EFI) has since matured to a very sophisticated state. As emission controls became ever more stringent, manufacturers improved EFI significantly. A step increase in emission restrictions occurred in 1975 forcing manufacturers to adopt the catalytic converter to meet them. This device looks like a muffler but contains a platinum-based catalyst that reacts with unburned hydrocarbons in the exhaust and converts them to water and carbon dioxide. A side effect of the catalytic converter was the need to more closely control some constituents in the exhaust stream that would poison the catalyst. One of these was tetraethyl lead, then commonly added to gasoline to increase its octane rating. Lead fouled the catalyst, resulting in the banning of leaded gasoline. The EFI designers also needed to monitor the level of oxygen in the exhaust stream ahead of the catalytic converter and use this information to control the fuel injectors to minimize other harmful constituents. All this is controlled by fast microprocessors that give essentially real-time control. The end result is an extremely sensitive and effective closed-loop control system that allows engines to be cleaner-burning than they ever have and simultaneously have high power output plus good economy.

Forced Induction

Supercharging can increase engine performance by 50–60% and is as effective as weight reduction in increasing performance. The idea of forcing more air into the intake system of an internal combustion engine was essentially born with the engine itself. Gottlieb Daimler received patents in 1885 and 1889 for four-stroke engines with devices added to pressurize the intake system. Dugald Clerk also proposed a forced induction two-stroke engine at about the same time. These early inventors may have had their hands full just trying to get their new inventions to work, so it seems that they never actually built an engine with a successful forced induction system at that time. Nevertheless, they clearly understood the advantages of forcing more air into the engine. More air means more fuel and more power.

It did not take long for inventors to implement the concept of forced induction on their engines. Around 1904, Dawson in England made a car with an unusual three-cylinder engine using "air augmentation." This was a double-piston engine with two connected pistons in each cylinder, one larger in diameter than the other. The larger pistons acted as an air compressor and the smaller ones were the engine. The compressed air was stored in a tank and could be used to start the engine and/or force more air into the engine. So it actually had six "cylinders," three for the compressor and three for the engine. A slide valve allowed the compressed air to be directed to the engine. Nothing like this was seen again after the Dawson was no more.

According to *The Automobile, A Century of Progress*, in 1907, Chadwick was making luxury cars in the U.S. and wanted to demonstrate the power of his new six by racing it. Initial speed test results were disappointing, so he added a three-stage, centrifugal supercharger running at nine times crankshaft speed via a belt. It worked. It made much more power. He beat all competitors until a mechanical failure knocked him out of the 1908 Vanderbilt Cup race. Sabotage was suspected. Because he kept his supercharger a secret, no one followed up with a similar application for some time. Chadwick went out of business after producing only 235 cars in eleven years of production.

Types of Superchargers

There are two basic types of superchargers (also called blowers) commonly used on automobiles: *positive displacement* and *positive pressure*, also called *dynamic*. Positive displacement devices deliver a controlled flow of air and the pressure that results is a function of the downstream resistance to flow. Positive pressure devices deliver pressure and the downstream flow is a function of resistance.

Positive Displacement

Common types of positive displacement superchargers include the Roots, screw-type, and vane-type. Named after its inventors, the Roots air pump was developed in 1860 for blast furnac-

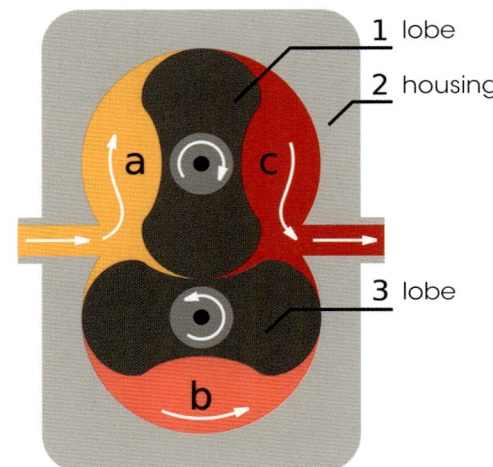

FIGURE 4-21 A two-lobe Roots blower
(Courtesy of Inductiveload at Wikimedia Commons)

es. It consists of two interleaved rotors that can have two, three, or four lobes on them. A two-lobed version is shown in Figure 4-21. Three or four lobes are more common in engine superchargers. It is only about 40–50% efficient. Screw compressors use two helical screws running together as shown in Figure 4-22. These are also called Lysholm compressors after their inventor and are more efficient than the Roots type. Vane compressors use vanes that slide radially in slots in a hub that is eccentric to the housing in which the vanes run as shown in Fig-

FIGURE 4-22 A Lysholm screw compressor
(Courtesy of Motorhead at English Wikepedia)

CHAPTER 4 VALVE TRAINS, INDUCTION, AND SUPERCHARGING

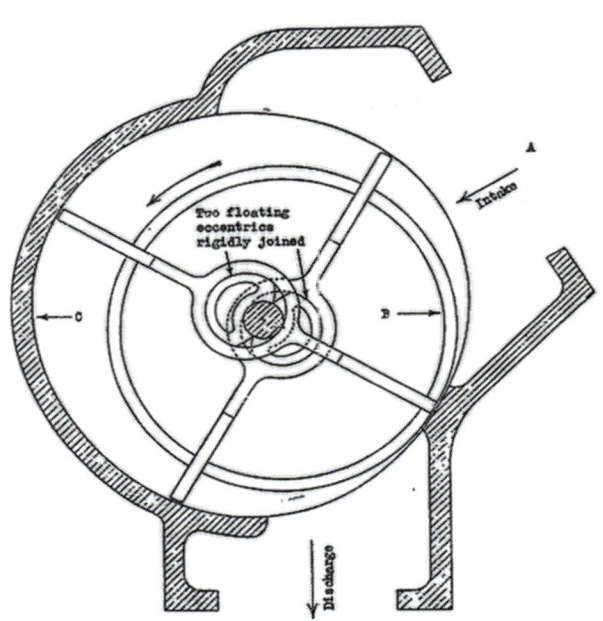

FIGURE 4-23 A vane pump *(Public Domain)*

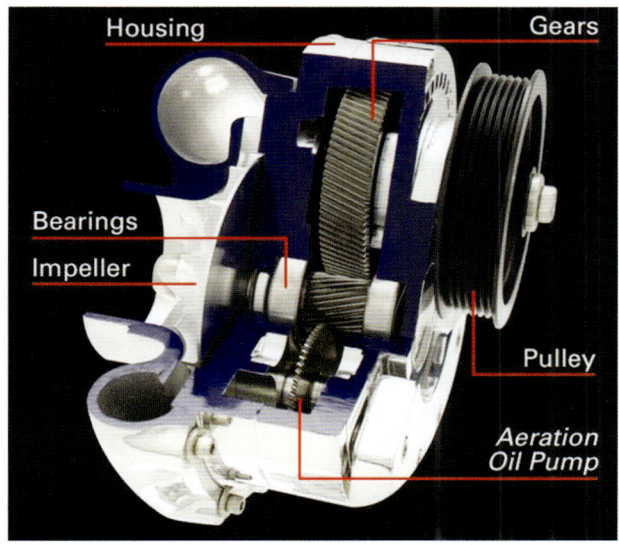

FIGURE 4-24 Cutaway of a centrifugal compressor *(Courtesy of ATI ProCharger)*

ure 4-23. As the vanes turn, they slide in and out and the volume between them contracts and expands to pump the fluid. Vane compressors are more efficient than the screw compressor.

Positive Pressure

Common types of positive pressure superchargers include centrifugal and axial flow. A centrifugal compressor has an **impeller** with vanes arranged radially like spokes, fitted in a toroidal chamber as shown in Figure 4-24. Air enters at the center and is accelerated radially by centrifugal force to exit tangentially from the housing with increased pressure. Centrifugal compressors can be 70–85% efficient, a bit better than a screw compressor.

An axial flow compressor is analogous to a window fan in that both pull air from one side of the fan and expel it axially on the other. Axial compressors have more efficiently designed blades than a window fan and run in a duct to eliminate radial losses from the blade tips. Most also have multiple blades (stages). These are more commonly used as the compressor in turbojet engines than in IC engine superchargers.

Driving the Supercharger

There are two ways to drive a supercharger: mechanically from the engine crankshaft, or by using the exhaust gas stream to drive a turbine that turns the compressor. The latter is called a *Turbo-supercharger*, or more often a *Turbocharger* or just a *Turbo*. Each of these concepts has been around as long as the other. Mechanical superchargers date from about 1904 and the turbocharger was invented by Alfred J. Buchi in Switzerland in 1905. We will discuss each in turn.

Mechanical Superchargers

Aircraft piston engines were most in need of supercharging because the pressure and density of air drops with increasing altitude. At 30,000 ft, air density is 1/3 of its density at sea level. This means that an unsupercharged engine will have only about 1/3 the power at that altitude. Some British and American WWII fighters and bombers used mechanically supercharged engines such as the Rolls-Royce Merlin, but most used turbochargers as they added less weight. German WWII aircraft relied more on mechanical superchargers, though they also had some turbocharged aero engines.

The term supercharger, used without qualifiers, usually means a mechanically driven one. Power for this type of supercharger is taken from the engine crankshaft via direct-drive, gears, chains, or belts. It typically absorbs about 20–30% of engine power to drive it. So it must add more than that to the engine for a net gain in power. The supercharged Rolls-Royce Merlin V12 engine (Figure 4-25) used in the WWII Spitfire and P51 Mustang fighter planes used 150 HP (20%) of the unsupercharged engine's 750 HP. With the supercharger, the engine produced 1000 HP. The supercharger added 400 HP and absorbed 150 for a net gain of 250 HP.

Mercedes introduced the first supercharged production automobile at the 1921 Berlin Motor Show. By the mid-twenties, Fiat, Miller (U.S.), Alfa Romeo, Sunbeam (GB), Delage (Fr), Bugatti, and Bentley were all selling them. Figure 4-26 shows a 1929 Blower Bentley. Su-

FIGURE 4-26 1929 Blower Bentley *(Courtesy of SFoskett)*

percharging of race cars began when a Duesenberg with a centrifugal supercharger won the 1924 Indianapolis 500. This motivated Harry Miller to develop the "Millercharger," which he claimed ran at 37,500 rpm and increased power from 120 HP (90 kW) to 202 HP (151 kW). Duesenberg offered the SJ supercharged model in 1929, and both Cord and Auburn also offered supercharged models.

In the 1950s, supercharging reappeared in U.S. automobiles. The 1957 Studebaker Golden Hawk added a belt-driven centrifugal supercharger to its 289 cu in V8 and claimed 275 HP. The 1957 Ford Thunderbird also offered an optional belt-driven, centrifugal supercharger, claiming 300 HP from a 312 cu in V8. Many hot-rodders of this era added superchargers to their engines for more power.[1] Figure 4-27 shows an example: a 1968 American Motors AMX set up for drag racing with a Roots-type blower. There are currently a number of makes offering mechanical superchargers. From 1998 to the present, Jaguar has offered supercharged V8s on many models. Cadillac offers a 6.2L (378 cu in), 556 HP, Roots-supercharged V8 in

FIGURE 4-25 Supercharged WWII Merlin V12 engine *(Courtesy of JAW at English Wikepedia)*

[1] In 1958, the author fitted a well-used, belt-driven, centrifugal supercharger to his 3/4-race Ford flathead engine with mixed results. It made a lot of noise but seemed to take as much power from the crankshaft as it added to the engine. It was subsequently removed.

CHAPTER 4 VALVE TRAINS, INDUCTION, AND SUPERCHARGING

FIGURE 4-27 1929 supercharged AMX dragster
(Courtesy of Christopher Ziemnowicz)

its CTS-V series. The 2015 Corvette Z06 has a 6.2L (378 cu in), direct-injected, supercharged, aluminum V8 that claims to make an SAE-certified 650 HP and 650 lb-ft of torque. That is 1.72 HP/cu in! A hot-rodder's dream!

Turbochargers

Not long after Buchi invented the turbocharger in 1905, the Frenchman Auguste Rateau began experimenting with the Buchi concept to develop a working turbocharger. He succeeded in fitting a turbocharger to a Renault aero engine in 1915 and these engines were used in WWI. In 1917, the U.S. War Department created a secret project to outfit a V12 Liberty aero engine with a turbocharger and engaged experts in gas turbine design from General Electric to work on it. By 1918 they had one running on a test stand in a wind tunnel to simulate operation at airspeed. They shipped the engine, test equipment, and dynamometer to Pikes Peak to test its performance at 14,000-ft altitude. The engine developed equivalent power at that altitude as it did at sea level without the turbo and the test was considered a success, but the war, by then, was over.

These early turbochargers suffered from their high exhaust temperatures as materials were not up to the task at that time. Heat resistant materials were not as available in the early 20th century as they were later. Modern exhaust turbines use ceramic rotor blades to resist the heat. Heat, in general, is an issue with both turbochargers and mechanical superchargers. When air is compressed, its temperature rises, and warm air is less effective than cold air for combustion in the cylinder. The solution for both types of supercharging was to add an intercooler (essentially an air-to-air radiator) between the supercharger and the intake plenum to cool the air after compression. Multi-stage compressors often have intercoolers between the stages and an aftercooler after the last stage.

Many U.S. WWII planes, such as the B17 Flying Fortress, B24 Liberator, P38 Lightning, and P47 Thunderbolt, used turbocharged engines. The large diameter of the P47 fuselage was in part to house the extensive plumbing needed to carry the air ducts from the exhaust turbine to the inter-coolers that were behind the pilot.

Not only was a turbocharged engine lighter than the same engine with a mechanical supercharger, the turbocharged engine was much more efficient. A third or more of the energy from the burned fuel in an engine escapes in the exhaust stream. A turbocharger recaptures some of that lost energy by using the exhaust to spin a turbine that drives the centrifugal compressor. So instead of stealing up to 30% energy from the crankshaft, it increases the exhaust back pressure, which increases pumping losses. This has less negative effect on engine performance than a mechanical drive from the crankshaft.

A disadvantage with turbochargers compared to mechanical superchargers is "turbo

119

lag." This refers to the fact that at low engine speeds, the exhaust turbine is spinning too slowly to provide significant boost pressure. So stepping on the accelerator at low speed results in hesitation until the turbine spins up to higher speed as the exhaust flow gradually increases. Then the power often comes on with a rush, accelerating the car rapidly at higher speeds. Modern turbo installations have reduced turbo lag significantly.

Diesel engines began using turbochargers in the 1920s. Two-stroke diesels require supercharging as they have no defined intake stroke. Four-stroke diesels also benefit from turbocharging as do four-stroke gasoline engines. A turbocharger can be used both to improve power delivery and also to improve (reduce) fuel usage for better efficiency. In recent years, increased demand for fuel efficiency by governments and buyers has driven most manufacturers in the direction of smaller-displacement engines with fewer cylinders. Turbos are often added to boost a smaller engine's power to a level similar to (and in some cases above) what the older, larger, naturally aspirated engines they replaced could do. Normally aspirated V8 engines are being replaced by turbocharged V6 engines with better performance and fuel economy. Normally aspirated V6 and inline six engines are being replaced by turbo-fours with similar effect. As mentioned earlier, the V8 engine seems to have become an endangered species except in the very high-performance segment of the market. But even those supercharged, high performance V8s are now getting better fuel economy than their predecessors.

Summary

We have mentioned this before, but the comparison is worth noting again. The saying that *they don't make them like they used to* is certainly true, but not for the reasons that this expression usually implies. Compared to this author's experience driving "high-performance" cars 50–60 years ago, and continuously since that time, the performance of the present crop of automobiles is nothing short of phenomenal. In the 1950s, the hot-rodder's holy grail was to achieve a specific power output from a hopped-up engine of 1 HP/cu in, and we never attained it.

A really fast car then could do 0–60 mph in 7–8 seconds (the 1956 Corvette claimed 7.3 sec). High-performance vehicles then were doing well to deliver 10–12 mpg in daily driving, and we did not care. After all, gasoline was only 25 cents per gallon! In 2015, the author's C6 Corvette makes 430 HP from a normally aspirated 6.2L (378 cu in), pushrod V8, which is a specific power of 1.14 HP/cu in. According to Chevrolet, it does 0–60 mph in 4.0 sec. It actually delivers 24 mpg in all-around driving and 30 mpg at a steady 80 mph on the highway (by author test). No twentieth century automobile could come close to these numbers in the same vehicle. Oh, to be 16 again in 2015!

PHOTO GALLERY

1912 Mercer Raceabout in the collection of the Revs Museum, Naples, FL.
Made in Trenton, NJ, the Raceabout was a hot-rod of its day. It set many world records and won five races.
It had a four-cylinder, 300-cubic-inch, T-head engine of 58 HP. Photos by the author.

AUTOMOTIVE MILESTONES

1913 Peugeot Type 150 Boat-tailed "skiff" in the Seal Cove, ME, Auto Museum collection. Its body is crafted from layers of mahogany. It has a 40-HP, four-cylinder engine of 7.5 liters and a four-speed gearbox. Photos by the author.

PHOTO GALLERY

The 1920s and 1930s represent one of the most significant eras for automobile body design. The 1920s were a time of great wealth, until the Great Depression ruined it, and many high-end cars were fitted with beautiful engines and custom bodies. A sampling of sixteen such automobiles follows.

1922 Rolls-Royce Silver Ghost Pall Mall Phaeton in the collection of the Heritage Museum, Sandwich, MA. Made in Springfield, MA, the Silver Ghost was left-hand drive and named for its quiet engine. Cost $10,200 for the chassis in 1922—body extra, when average annual income was $1,201. Photos by the author.

AUTOMOTIVE MILESTONES

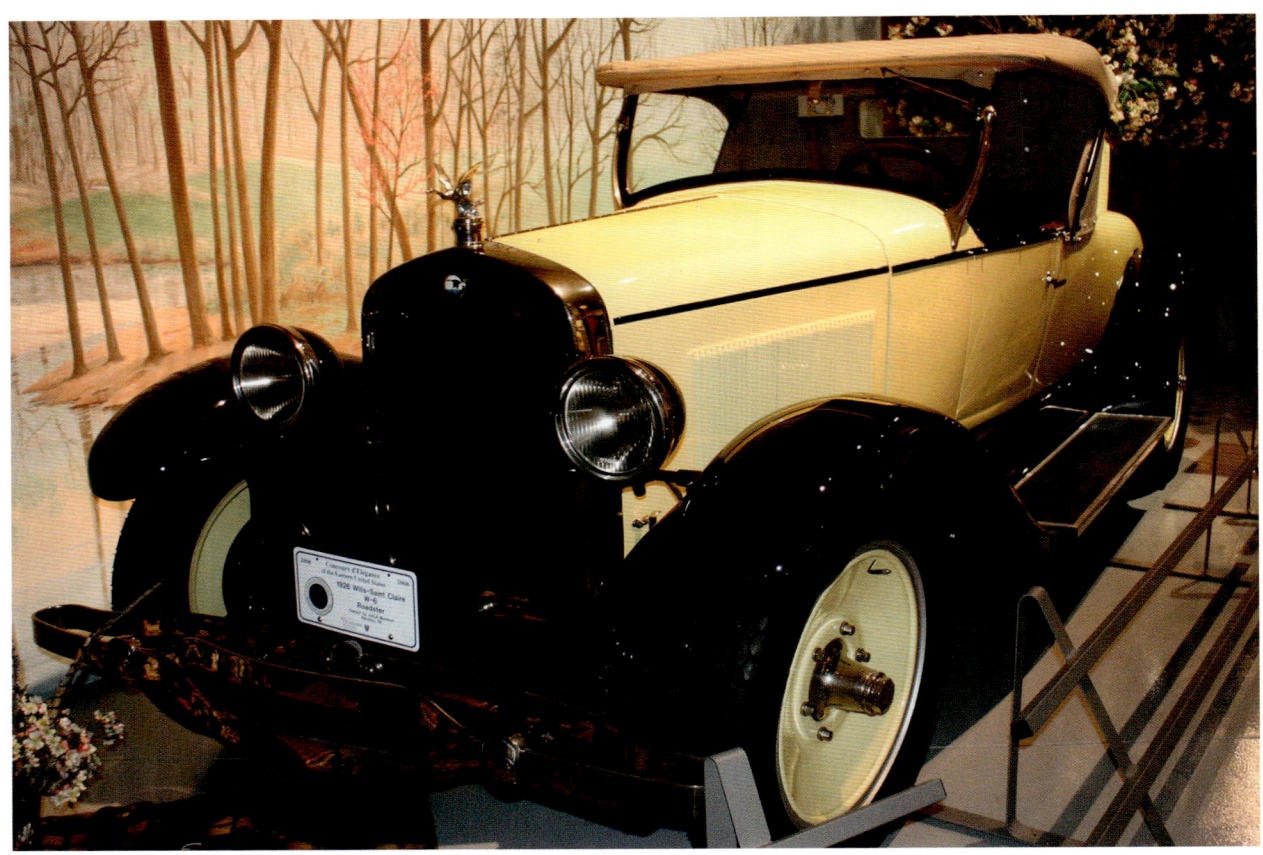

1926 Wills Sainte Claire in the AACA Museum collection. C. H. Wills had been Henry Ford's Chief Designer and left him in 1919 to make his own car. Very advanced for its time, the Wills had a SOHC V8, 66 HP engine.
Photos by the author.

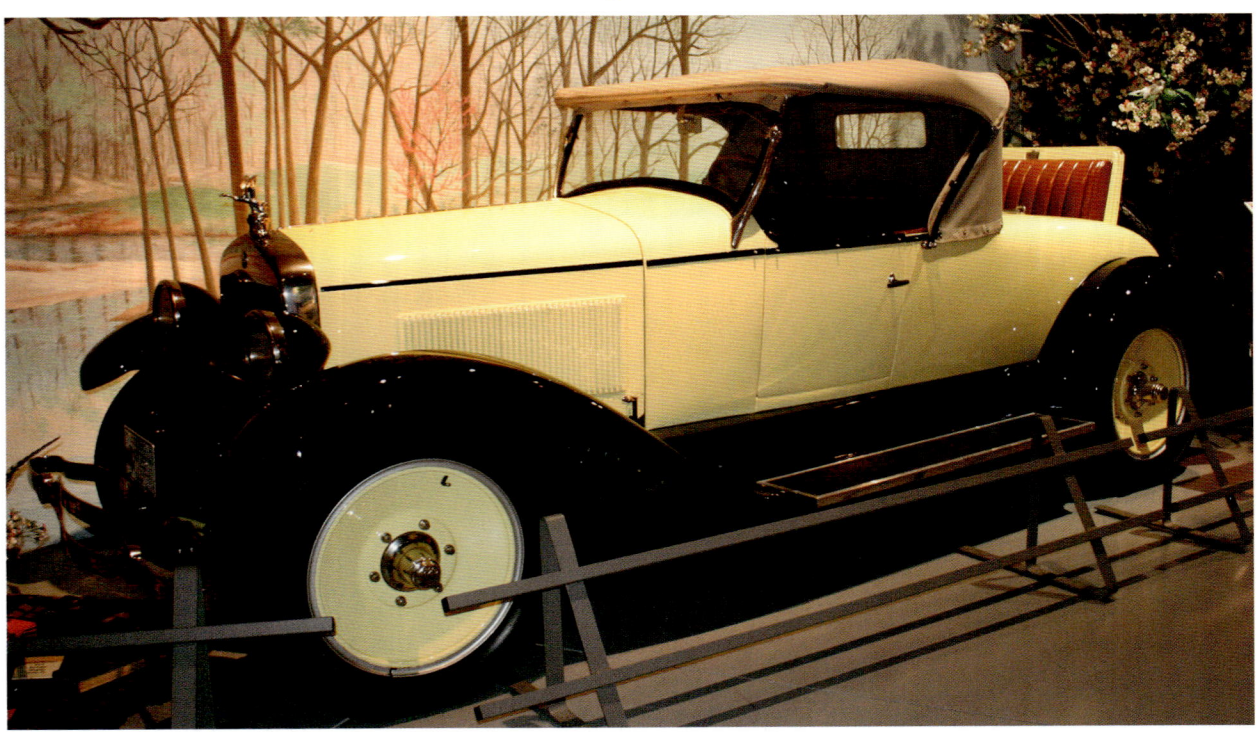

PHOTO GALLERY

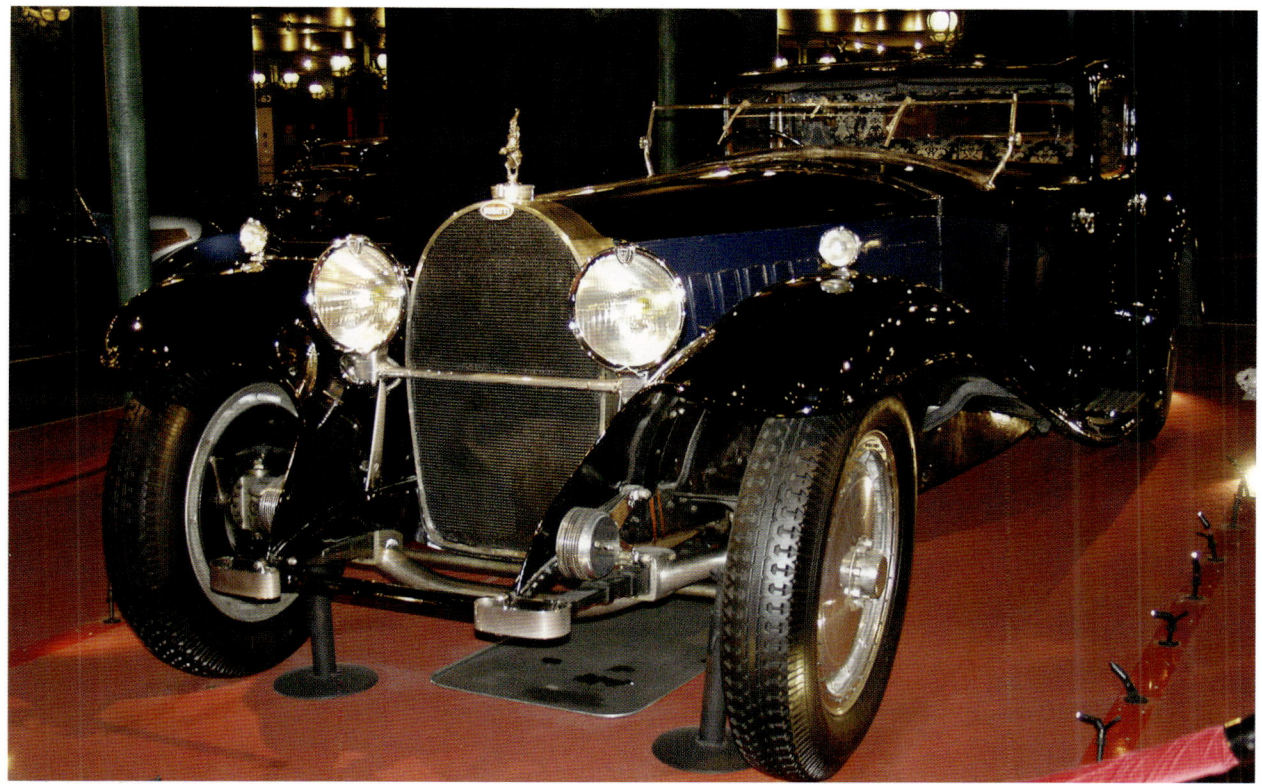

1929 Bugatti Type 41 Royale in the collection of the Musée National de l'Automobile de Mulhouse, France. Known as the Coupe Napoleon, the first of six made, remained as Ettore Bugatti's personal car. Only three of the six were sold, and all six are still in existence. Photos by the author.

AUTOMOTIVE MILESTONES

1929 Ruxton in the Tampa Bay Automobile Museum collection. The Ruxton was the first front-wheel-drive American car. It used a Continental straight-eight engine and was capable of 70 mph. Photos by the author.

PHOTO GALLERY

1929 Mercedes-Benz SSK in the collection of the Revs Institute Museum, Naples, FL. Designed by Ferdinand Porsche for both road and race track, it made 200 HP from a supercharged inline, SOHC six. Photos by the author.

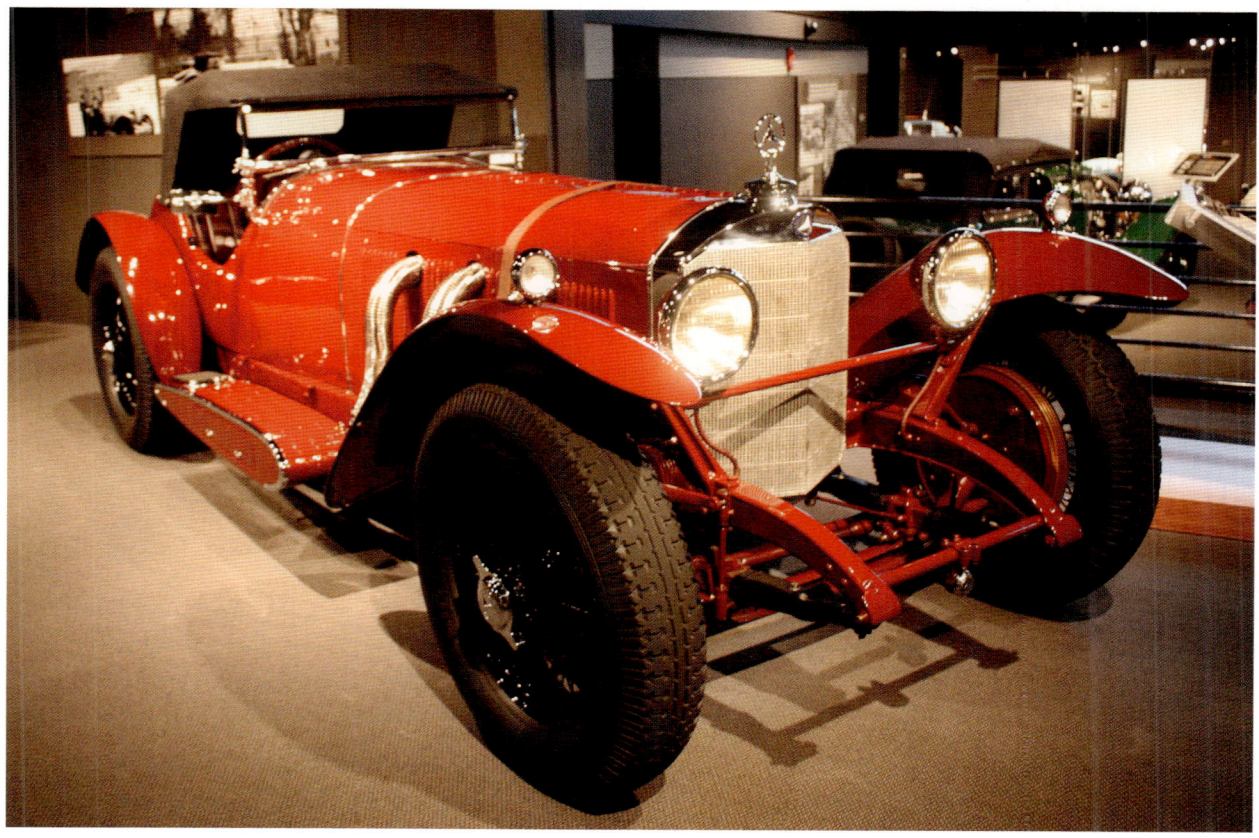

AUTOMOTIVE MILESTONES

1930 Cadillac, 165-HP, V16 Convertible Coupe in the Heritage Museum, Sandwich, MA. Collectors consider Cadillac's early V16s to be among the best ever built in America. Cost $6,900 in 1930 when average annual income was $1,388. Photos by the author.

PHOTO GALLERY

1931 Duesenberg Model J Derham Tourster in the collection of the Heritage Museum, Sandwich, MA. This car belonged to Gary Cooper. It has a straight-eight, DOHC engine with four valves per cylinder and could do 100 mph in second gear. Cost $14,000 for the chassis—body extra—in 1931, when average annual income was $1,388. Photos by the author.

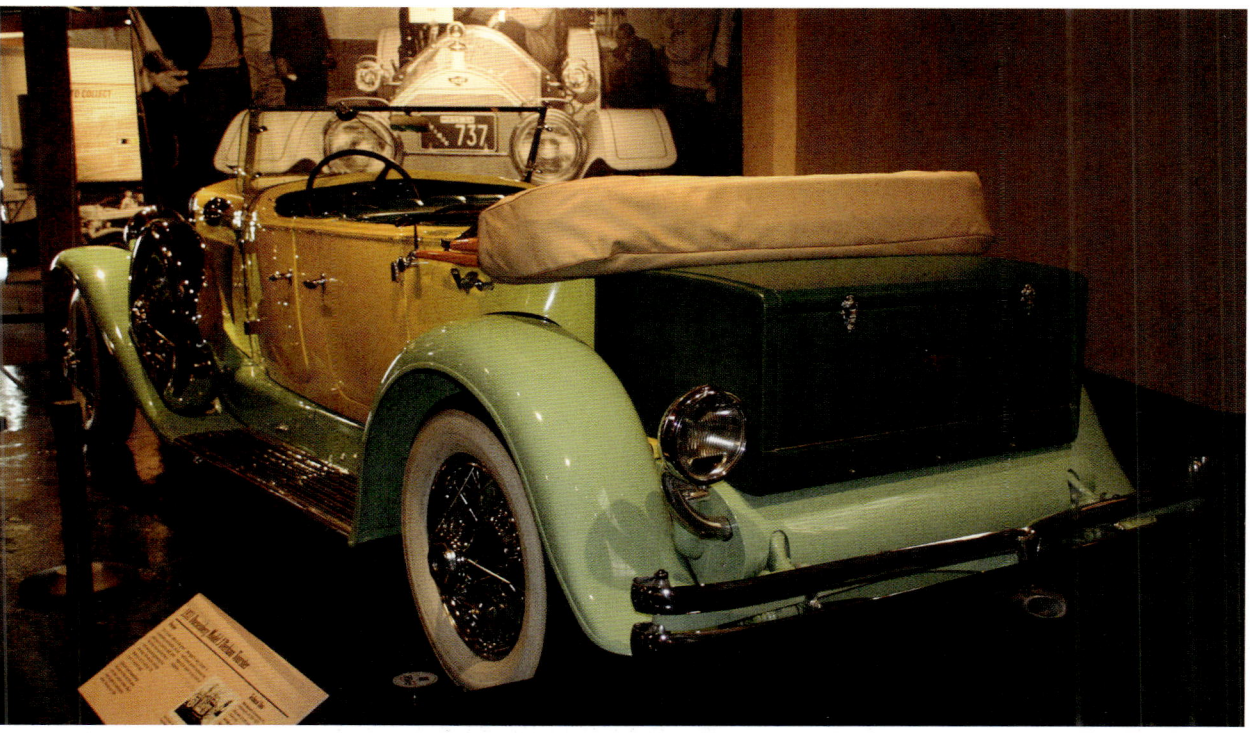

AUTOMOTIVE MILESTONES

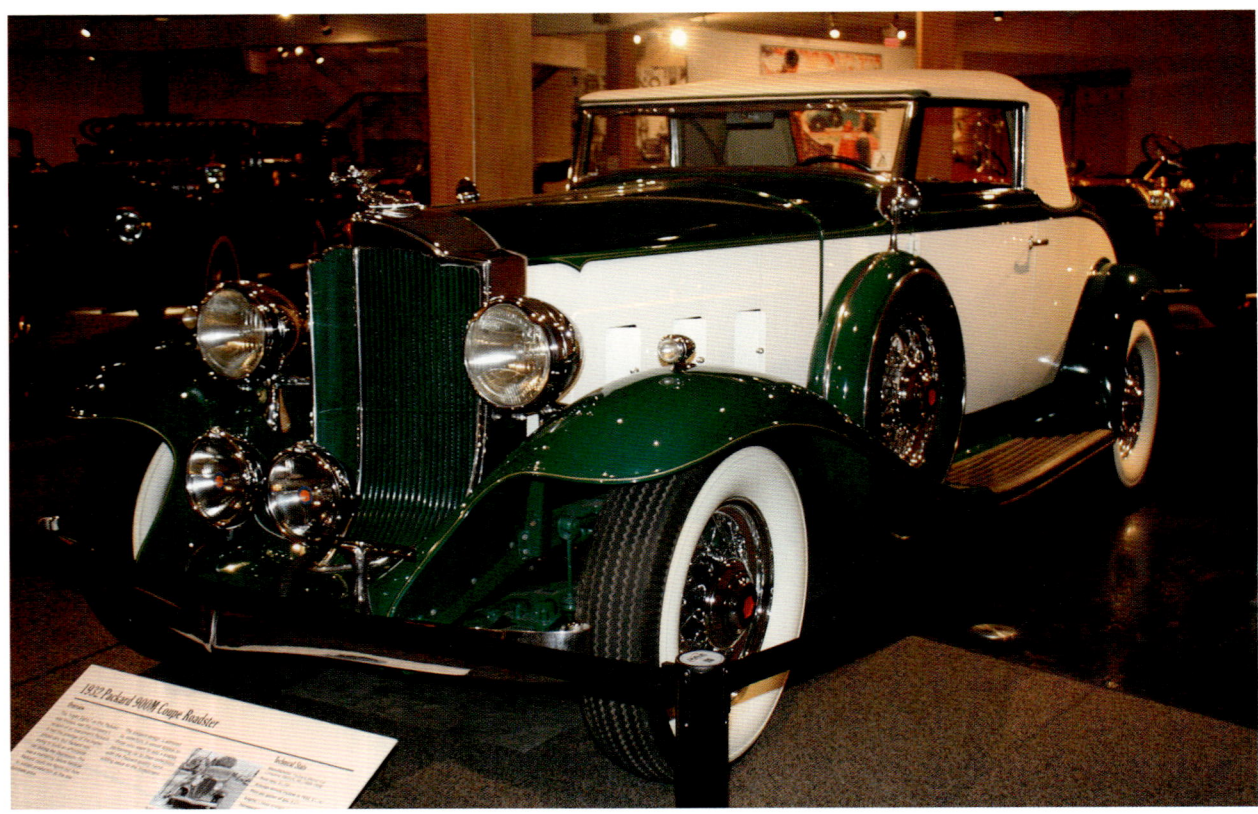

1932 Packard 900M Coupe Roadster in the collection of the Heritage Museum, Sandwich, MA. Made in Detroit, it had an L-head, straight-eight engine. It cost $1,795 in 1932, when average annual income was $1,141. Photos by the author.

PHOTO GALLERY

1932 Auburn Boattail Speedster in the Heritage Museum, Sandwich, MA.
This was the ultimate "babe magnet" driven by rich playboys in the 1930s and a favorite of Hollywood actors like Errol Flynn, who owned one. It cost $975 in 1932 when average annual income was $1,141. Photos by the author.

AUTOMOTIVE MILESTONES

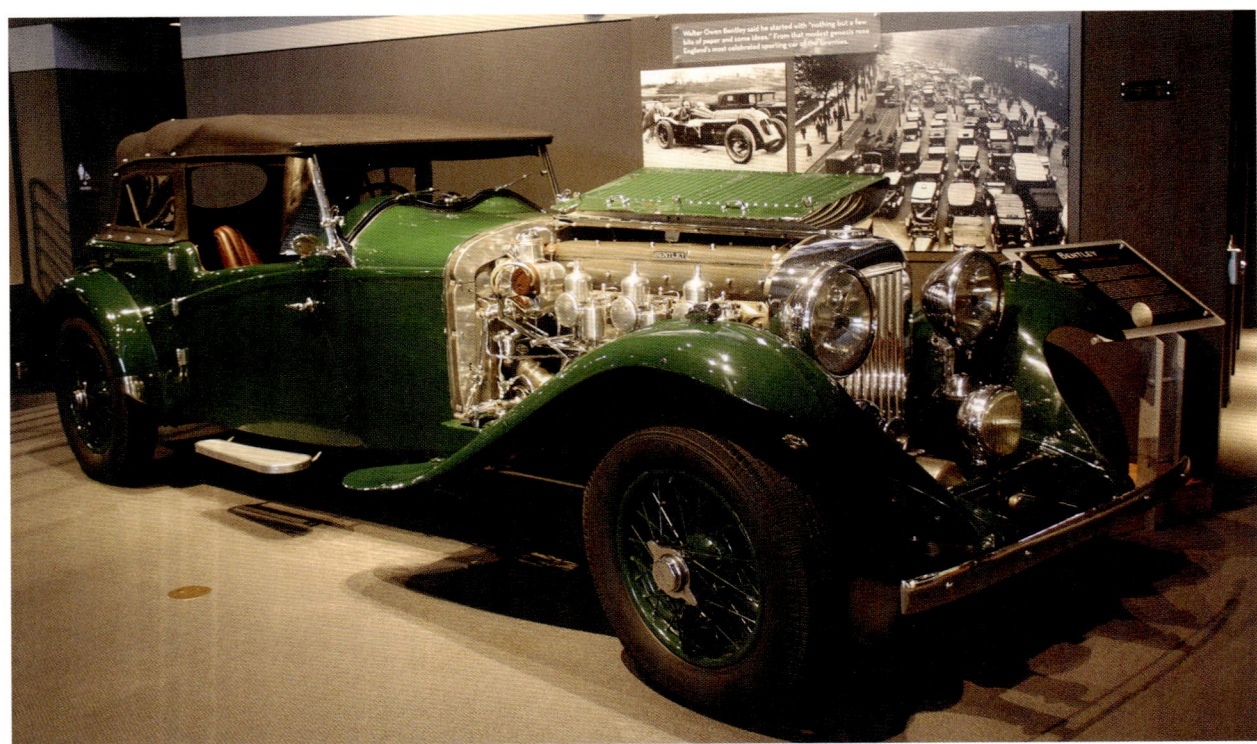

1932 Bentley Corsica in the collection of the Revs Institute Museum, Naples, FL. Designed by W. O. Bentley to compete with the Rolls-Royce Phantom II, it set a new speed record of 134.75 mph for the flying mile with its 250-HP, 8-liter, inline-six, SOHC engine. Cost $9,000 for a chassis—body extra, here by Corsica. Photos by the author.

PHOTO GALLERY

1933 Bugatti Type 55 Super Sport in the collection of the Revs Institute Museum, Naples, FL. With a 3-L, DOHC straight-eight, supercharged engine, it could do 0–60 mph in under 13 seconds and do 99 mph in the quarter mile. Photos by the author.

AUTOMOTIVE MILESTONES

1933 Packard Twelve Sport Phaeton in the collection of the Revs Institute Museum, Naples, FL. With a 67°, 446-cubic-inch V-12, L-head engine, and a wheelbase of 167 inches., it would do 100 mph in supreme comfort. Ettore Bugatti owned one. Photos by the author.

PHOTO GALLERY

1934 Alfa Romeo Tipo 8C 2300 Corto Touring in the collection of the Revs Institute Museum, Naples, FL. With a 2.3-L, DOHC straight-eight, supercharged engine, it won Le Mans four years in a row and also won the Mille Miglia. It would do 0–60 in 10 seconds. Yet it was a docile street car as well. Photos by the author.

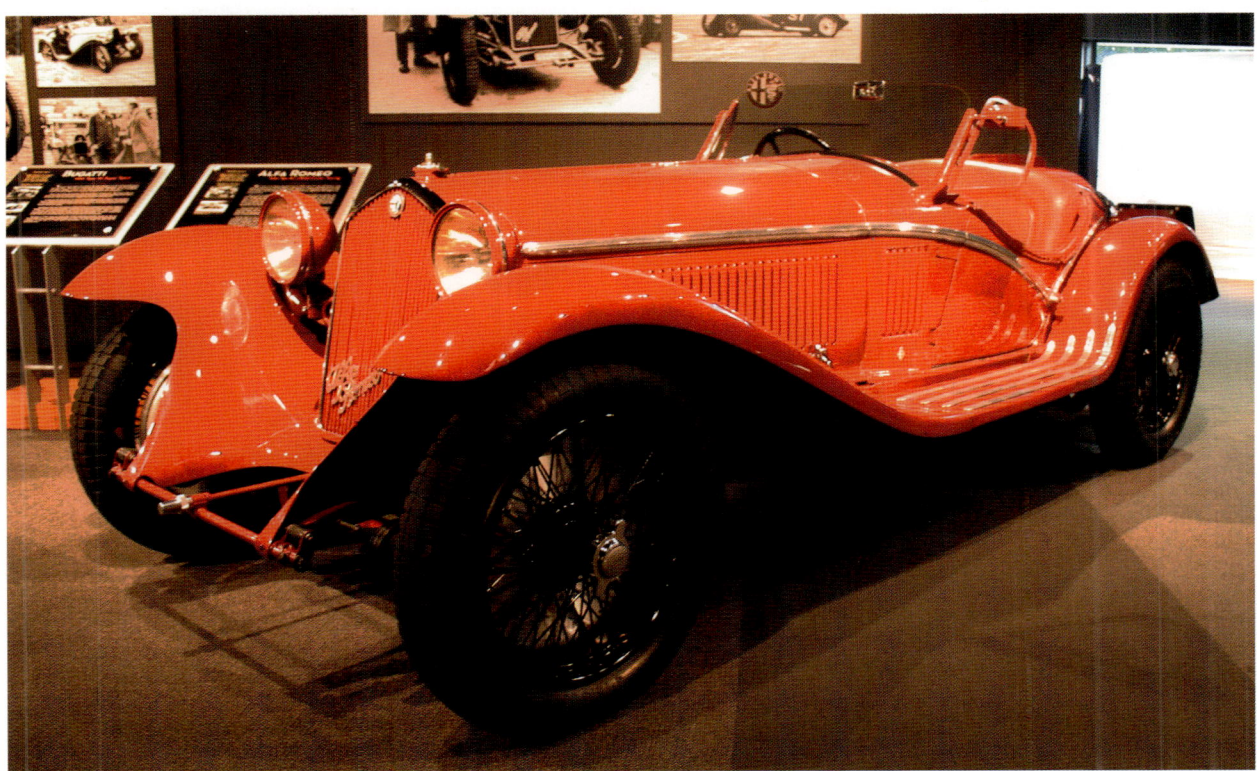

AUTOMOTIVE MILESTONES

1937 BMW 328 in the Revs Institute Museum, Naples, FL. This 3.28-L BMW was a revolution in racing sports cars with its small-displacement engine, independent front suspension, and streamlined body. Only 459 were sold to the public, but it was extensively raced and won its class at Le Mans, the Mille Miglia, and others. Photos by the author.

PHOTO GALLERY

1937 Peugeot Darl'Mart in the Tampa Bay Automobile Museum collection. It used a 2-L, four-cylinder, inline engine with a Cotal preselector transmission. It raced in Le Mans in 1937 and 1938 coming in fifth and eighth. Photos by John Perodeau.

AUTOMOTIVE MILESTONES

1937 Delahaye Type 135 Special Roadster in the collection of the Revs Institute Museum, Naples, FL. This car was the sensation of the 1937 Paris Salon. Built on Delahaye's short competition chassis, it has an aluminum body and a leather interior by Hermes. Photos by the author.

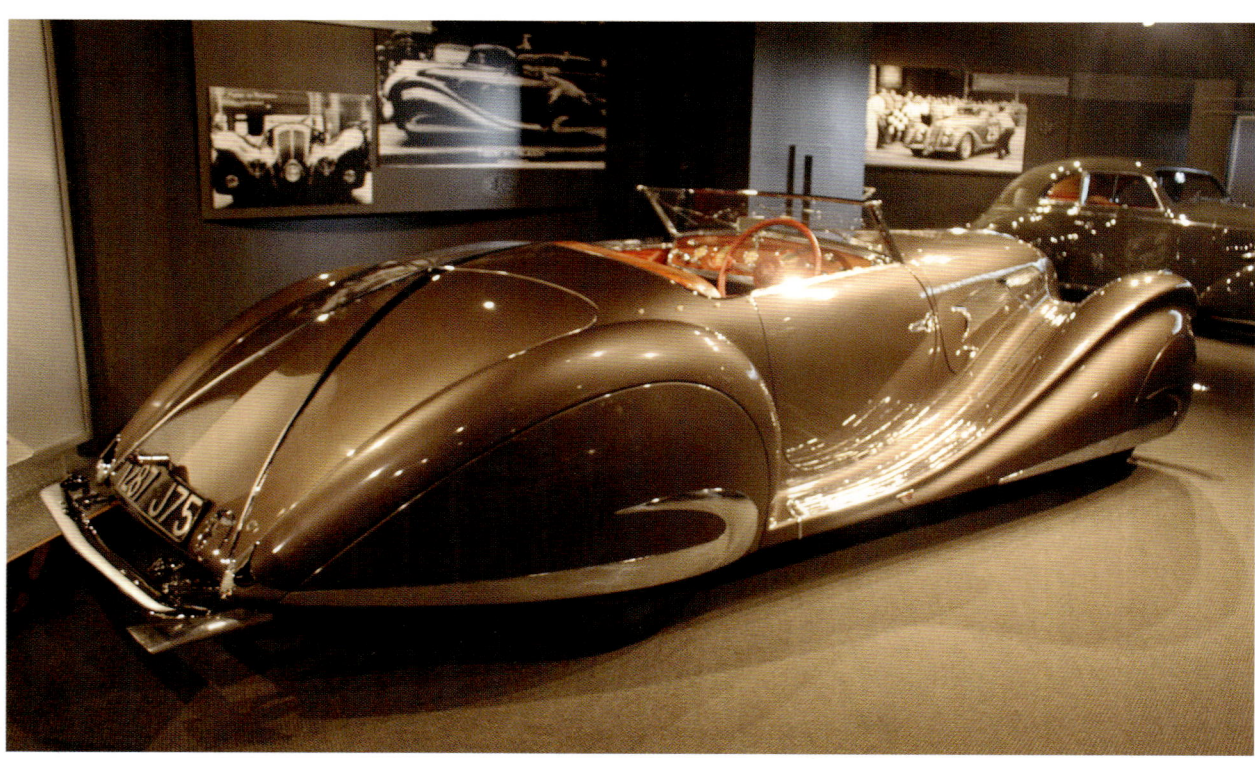

Chapter 5

Transmissions and Differentials

This chapter will describe a variety of mechanisms that have been devised to transmit an engine's power to the drive wheels. All automobiles powered by internal combustion (IC) engines need some means to match the torque and power characteristics of the engine with the load, which is the large mass of the vehicle. Many approaches have been used to transmit engine power including friction drive, belt drive, chain drive, and **gear** drive. All of these are still used in various vehicles. Motorcycles use gear, chain, or belt drive. Snow blowers use friction drive, and most automobiles and trucks use a form of gear drive.

Series hybrid vehicles use electric or hydraulic motors instead of mechanical transmissions to "couple" the engine to the drive wheels. In these cases the coupling is virtual by means of the IC engine charging the batteries or accumulators that then drive the motors at the wheels. Diesel electric locomotives are of this type as well as are some hybrid road vehicles. Electric and hydraulic motors are reversible and make maximum torque at stall (when stopped); therefore they are ideally suited to starting heavy loads and need no mechanical transmission between them and the load. Unlike electric and hydraulic motors, IC engines make zero torque when stopped and relatively low torque at idle speed and so need some means to multiply the engine torque to the level required to move the vehicle.

Power is, very simply, the product of torque and speed. Engine power varies with engine speed, from a low value at engine idle to a peak value at some rpm, and then falls off after the peak-power rpm is passed. But engine power can be thought of as constant at any given engine speed. At a given power level, you can have any combination of torque and speed between limits that you desire. The purpose of a transmission is to trade speed for torque. It does this by providing a **mechanical advantage (see Sidebar)** between engine and drive wheels. Mechanical advantage, in this context, means a torque multiplication ratio.

A transmission <u>**cannot increase power**</u> going to the wheels; in fact, some power is lost to friction in the driveline and the **power out is always less than the power in**. A transmission provides a means to select from a set of combinations of torque and speed to satisfy the immediate demands on the system. For example, to start from a dead stop, high torque at low speed is needed to accelerate the mass of the car. As the automobile's speed increases and

The opening photograph is of a cutaway automatic transmission taken by the author at the SAE International Meeting in 2004.

Mechanical Advantage

Mechanical Advantage is defined as the ratio between an output force or torque and the input force or torque required to create that output force or torque. If you can move 100 lb by applying a force of 10 lb, then you are experiencing a 10:1 mechanical advantage. Where does this advantage come from? It usually involves some type of lever ratio, as in using a crowbar to move a rock. A crowbar is a simple lever and it must sit on a pivot point, called its fulcrum. In this case the mechanical advantage is the ratio between the lever arm from pivot point to your hand and the lever arm from the pivot point to the rock. But in this example, your hand must move 10 times the distance that the rock moves. Here you are trading force for distance.

$$Force_{out} Dist_{out} = Force_{in} Dist_{in}$$

$$\frac{Force_{out}}{Force_{in}} = \frac{Dist_{in}}{Dist_{out}} = \text{Mechanical Advantage}$$

For a rotating system, mechanical advantage can be expressed as a torque ratio:

$$\text{Mechanical Advantage} = \frac{Torque_{out}}{Torque_{in}}$$

Also, the speed ratio of a gearset can be expressed as the ratio of their diameters or their numbers of teeth—and the diameter ratio works for belts or chains too:

$$\text{Speed Ratio} = \frac{Speed_{out}}{Speed_{in}} = \frac{Dia_{in}}{Dia_{out}} = \frac{N_{teeth_{in}}}{N_{teeth_{out}}}$$

If we temporarily ignore losses due to friction and other causes, then we can equate the power into the system with the power out.

$$Power_{out} = Power_{in}$$

Power is equal to torque x speed so we can substitute this in the expression above.

$$Torque_{out}\, Speed_{out} = Torque_{in}\, Speed_{in}$$

Rearrange and we get an expression relating inverse speed ratio and torque ratio:

$$\frac{Torque_{out}}{Torque_{in}} = \frac{Speed_{in}}{Speed_{out}} = \text{Mechanical Advantage}$$

The torque ratio is the inverse of the speed ratio. The speed ratio was expressed as a ratio of numbers of teeth above, and the torque ratio also can be expressed as the inverse of that tooth ratio:

$$\frac{Torque_{out}}{Torque_{in}} = \frac{N_{teeth_{out}}}{N_{teeth_{in}}} = \text{Gear Ratio}$$

The gear ratio of a gearset or transmission is the same as the torque ratio.

with it its momentum, less torque is needed to keep it moving at a higher speed. Then only the mechanical resistance (friction) within the vehicle's systems and the fluid resistance of the car moving through the air needs to be overcome. If you ask the car to accelerate to a higher speed, then you will need to increase the torque multiplication between the engine and the drive wheels by choosing a different ratio in the transmission. This may be done manually by the driver or automatically by the transmission.

Because the focus of this book is on automotive technology, we will not discuss chain and friction drives in any depth other than to briefly describe how they were used in very early automobiles and in CVT transmissions. All non-hybrid, IC-engine powered automobiles now use some form of geared transmission with the exception of the increasingly popular Continuously Variable Transmission (CVT) that typically uses some form of belt. We will focus on the many varieties of these devices and will also describe the workings of differentials that are necessary to split the power among the wheels of the car.

Gearboxes

Panhard developed the first sliding-gear transmission in 1895 and used it in its *Systeme Panhard* arrangement that became the defacto standard chassis design for many years. Their gearbox design became a standard until synchromesh was invented many years later. Synchromesh transmissions will be discussed later in this chapter. The sliding gear transmission works on the principle of compounding, which puts two **gearsets** (gear pairs) in parallel to make a **gear train** with the result that the overall gear train ratio is the product of the ratios of the two gearsets. (See the sidebar Mechanical Advantage for a definition of gear ratio.) Note that the convention is to call the smaller gear of a gearset the *pinion* and the larger the *gear*.

Figure 5-1 shows a simple, single-speed, compound gear train that demonstrates the principle. The **input shaft** has only one gear on it (gear 1) that is in mesh with a larger gear (gear 2) on the **countershaft**. The countershaft (also called a *jackshaft* or a *cluster gear*) always has more than one gear on it. The countershaft and all its gears are often made as one piece. The second gear on the countershaft (gear 3) is in mesh with

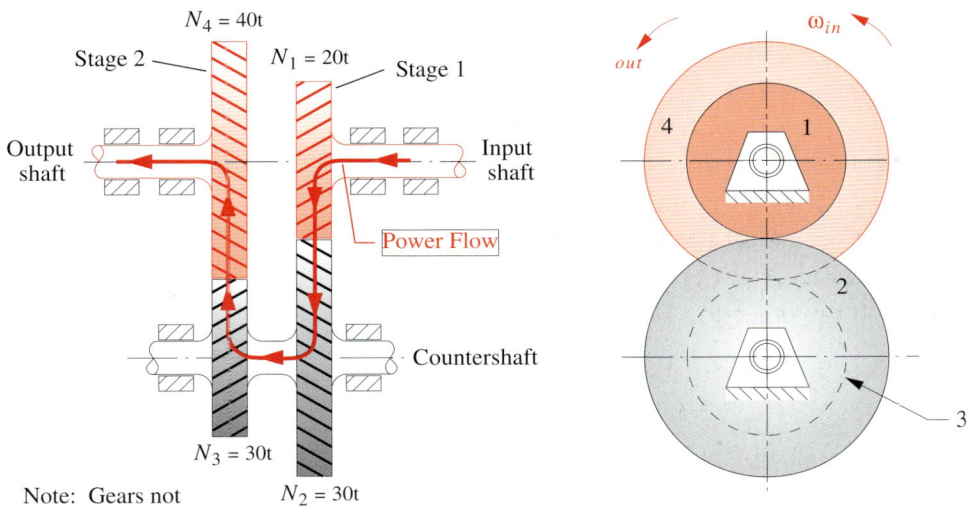

FIGURE 5-1 A two-stage, compound gear train

a single gear (gear 4) on the **output shaft**. As shown, both **gearsets** reduce speed and increase torque because the input gear is smaller than the output gear in each stage of this gearbox.

The overall transmission **torque ratio** (also called the *train ratio* or just *gear ratio*) is the product of the ratios of all gearsets through which the power flows. As an example, assume that gear 1 has 20 teeth and gear 2 has 30 teeth. The speed ratio of this first stage is then 20/30 and its torque ratio is 30/20. That is, gear 2 will turn 2/3 as fast as gear 1 and have 3/2 the torque.

Assume further that gear 3 has 30 teeth and gear 4 has 40 teeth. Gear 4 will then turn at 3/4 the speed of gear 3 and have 4/3 the torque. The overall speed ratio will then be the product of the speed ratios of its stages: 2/3 x 3/4 = 6/12 = 1/2 and the torque ratio of the train will be its inverse or 2/1. This is about the ratio of second gear in many automobile transmissions and will double the torque while reducing speed by half.

Sliding-Gear Transmissions

The gearbox in Figure 5-1 has only one fixed train ratio. A transmission needs multiple ratios and a means for the operator to shift among them. Panhard's basic design of the sliding-gear transmission put more than two gears on the countershaft and more than one on the output shaft. M. Panhard said: *It is brutal, but it works.*

The countershaft gears are typically fixed in permanent locations along the countershaft, but the gears on the output shaft are free to slide axially along the output shaft. However, these gears are locked rotationally to the output shaft with a spline. The shift collar is connected by a linkage to the shift lever, which the driver moves to engage the desired gear ratio.

A spline is a set of external teeth running along a portion of the output shaft's length. The sliding gears have a matching internal spline

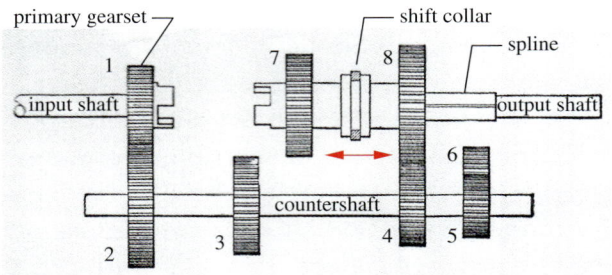

(a) First gear engaged—power path is 1-2-4-8

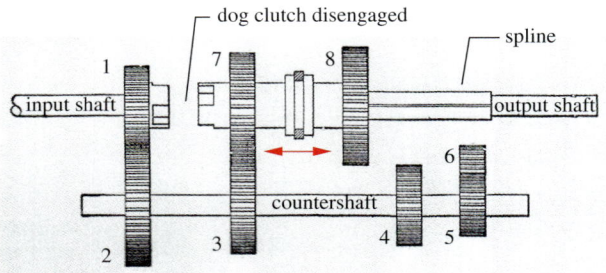

(b) Second gear engaged—power path is 1-2-3-7

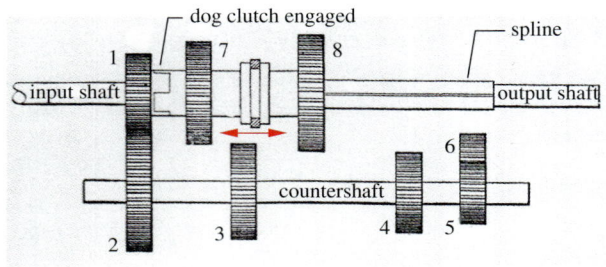

(c) Third gear (direct drive) engaged—power path is input shaft to output shaft—no gears are passing power

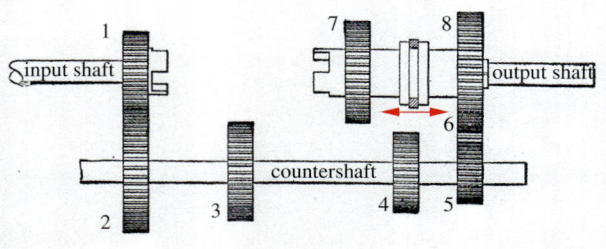

(d) Reverse gear engaged—power path is 1-2-5-6-8—gear 6 is an idler on a separate shaft to reverse direction

FIGURE 5-2 A three-speed forward, one-speed reverse sliding gear (crashbox) transmission

that transmits their torque to the output shaft at any position along the shaft. A mechanism is provided to allow the operator to slide each gear on the output shaft in and out of mesh with its mate on the countershaft to change the overall train ratio.

Figure 5-2 shows such an arrangement from an early transmission, circa 1914. Figure 5-2a shows the transmission in first gear with the shift collar slid to the right to bring gears 4 and 8 into mesh. In combination with the primary gearset 1 and 2, the overall 1st-gear-ratio typically will be between 3:1 and 4:1.

Figure 5-2b shows the transmission in second gear with the shift collar slid to the left to engage gears 3 and 7. The gear ratio between 3 and 7 is smaller than between 4 and 8 and so gives a lower overall ratio in combination with the primary gearset. Second gear is typically around 2:1 in a transmission like this.

Figure 5-2c shows the transmission in third speed or top gear. Note that, as is typical, there are no gears transmitting torque in top gear because the output shaft is now directly coupled to the input shaft by the jaws of a **dog clutch**, as shown. Since top gear is the most often used, it is desirable to have a direct mechanical connection through the transmission to avoid gear noise and wear.

When two external-tooth gears are meshed, one will turn in the direction opposite to the other. Thus the countershaft turns in the direction opposite to the input shaft. But the second set of gears reverses rotation again, making the output shaft turn in the same direction as the input shaft. To reverse the vehicle motion, an idler gear must be inserted between a countershaft gear and an output-shaft gear.

Figure 5-2d shows the transmission in reverse gear. Gear 6, which is partly hidden behind gear 5 and is on its own shaft beside the countershaft, is the reverse idler. Gears 5 and 6 are always in engagement, but only pass power when gear 8 is slid to the far right position to engage gear 6. The output shaft then turns in reverse with a ratio of about 4:1.

The transmission also has a neutral position in which no countershaft gears are engaged with any output gears. This arrangement is not shown in the figure. However, it can be understood if you mentally move the shift collar to the left in Figure 5-2a until gear 8 disengages gear 4—but not far enough to make gear 7 engage gear 3. Although the countershaft still spins, it now has no connection to the output shaft.

Figure 5-3 shows the shifting mechanism for a modern four-speed, constant-mesh transmission, schematically. Its gears do not slide into and out of engagement like those in Figure 5-2. Instead there are synchromesh clutches that lock the proper gear to the shaft for a given speed. Synchromesh mechanisms are described in the next section.

Figure 5-3a shows the transmission in neutral. The typical H-pattern of shift-lever movement can be seen, with the lever in the vertical, neutral position in Figure 5-3a. In neutral the lever can be moved left or right in the "gate" to engage one or another of the forward shift collars.

Moving the lever to the left engages the blue linkage that controls first and second gear. With the lever held left, pushing it forward slides the proper synchromesh clutch into engagement with first gear on the countershaft as shown in Figure 5-3b. Pulling the shift lever back with it still on the left side, disengages first gear, goes through neutral, and engages second gear as shown in Figure 5-3c.

To get to third gear, the lever is returned to neutral and moved through the gate to the right to engage the purple linkage that controls third and fourth gear. Forward motion of the shift le-

AUTOMOTIVE MILESTONES

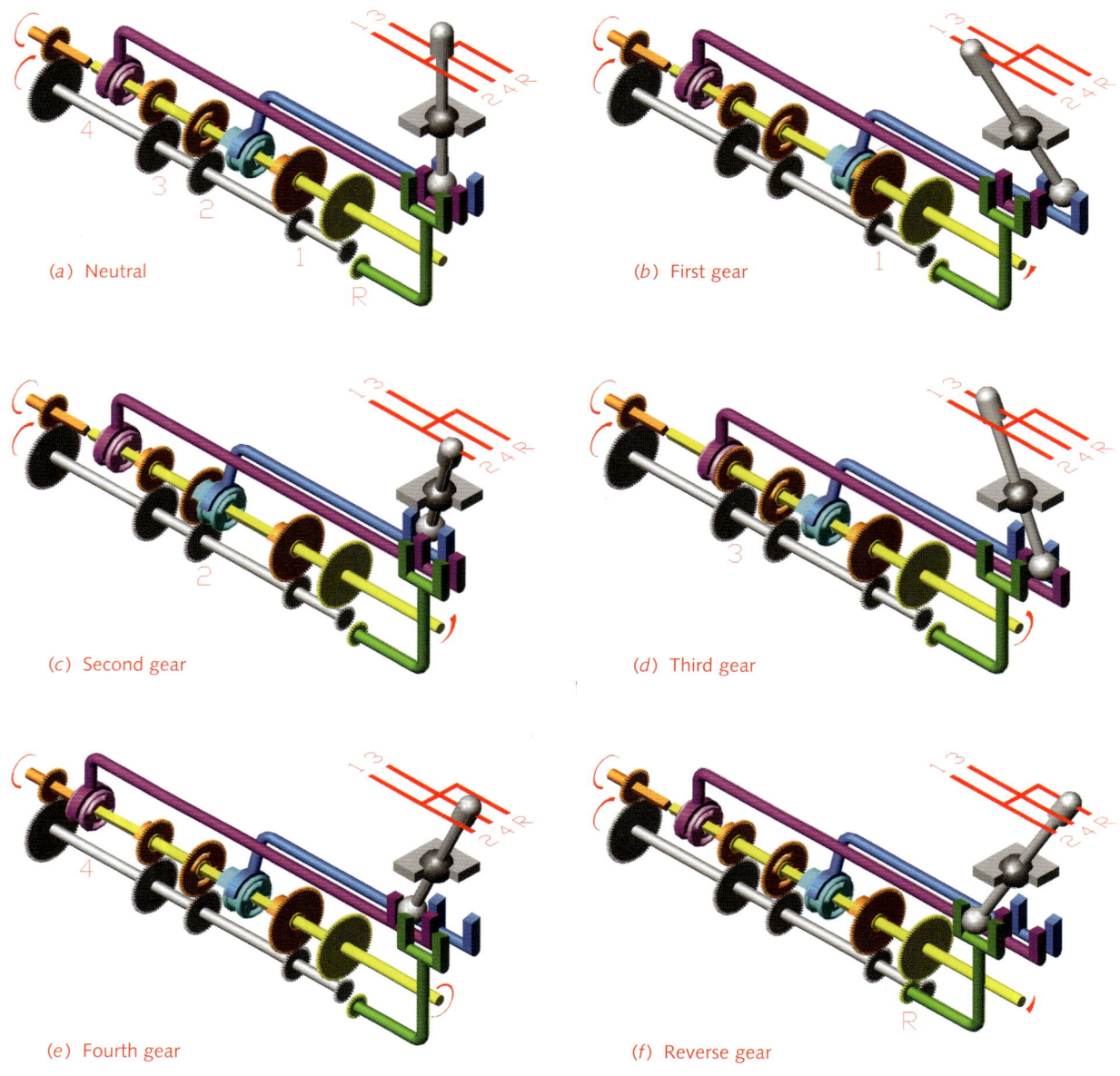

FIGURE 5-3 Shift linkage positions in a four-speed transmission with reverse *(Public Domain Courtesy of Stef Breukel)*

ver engages third gear as shown in Figure 5-3d, and backward motion disengages third, passes through neutral and engages fourth gear as shown in Figure 5-3e.

Reverse requires that the lever be pushed further to the right in neutral past the position for first gear in order to engage the green linkage that drives the proper output gear into engagement with the reverse idler gear, as shown in Figure 5-3f.

All of this gear shifting must, of course, also be accompanied by proper disengagement and re-engagement of the clutch that connects the spinning engine flywheel to the input shaft of the transmission. Power delivery must be interrupted to allow shifting gears. If they are passing torque, it will be very difficult to disengage them because of the large tooth forces.

Note that whenever the clutch is engaged, even with the transmission in neutral, the primary gearset in the transmission is spinning (see Figure 5-2), as is the entire countershaft. When shifting a conventional, sliding-gear transmission, some care must be taken to at least approximately match the tangential speed of the countershaft gear teeth to that of the output shaft gear that one is attempting to engage.

Failure to do this properly will result in some very nasty noises from the gearbox as spinning teeth run into each other. This is what gave sliding-gear transmissions their uncomplimentary appellation as "**crashboxes**." The recommended method to avoid gear clashing when shifting these transmissions is to "**double clutch**." Double-clutching involves the following steps:

1. Disengage the clutch by stepping on the clutch pedal.

2. Move the gearshift lever to neutral.

3a. If upshifting, let the engine rpm drop to match the new gearset velocity—or,

3b. If downshifting, step on the accelerator to increase the countershaft to a speed that you think (by engine sound) will properly match the expected speed of the output gear for the next ratio.

4. Disengage the clutch a second time.

5. Move the shift lever to the new gear position.

6. Hope you don't hear the sound of grinding gears.

7. Re-engage the clutch.

This technique obviously requires some practice to master. It relies on experience and some understanding of what the gears are doing in the transmission. It is not surprising that many people had difficulty achieving smooth and quiet shifts with these "crashbox" transmissions.

In fact, it is reported that Henry Ford could not master the technique of shifting a sliding-gear transmission, and so equipped his cars, through the Model T, with a planetary transmission that did not shift its gears in and out of engagement. Planetary transmissions, which are nearly as old as the sliding-gear transmission, will be discussed in a later section of this chapter.

No modern, standard-shift, passenger car uses a sliding gear transmission. They all have constant-mesh, synchromesh transmissions. However, many modern tractor trailer rigs (18-wheelers) do have multiple-speed crashbox transmissions with as many as 24 speeds. Tractor-Trailer driver training schools teach how to shift with double-clutching. A cutaway of one is shown in Figure 5-4. If you happen to be beside an 18-wheeler in traffic, listen as the driver accelerates and decelerates through the gears. You may hear him double-clutch. The telltale is to hear the engine rpm "blip" during a downshift when he or she is matching speeds in neutral during the shift.[1]

FIGURE 5-4 Transmission from an 18-wheeler truck
(Courtesy of StationNT5Bmedia)

[1] Thirty percent of new 18-wheeler trucks are now being fitted with automated manual transmissions (AMT). These transmissions are described in a later section of this chapter.

(a) Spur gears
(b) Helical gear
(c) Herringbone gear
(d) Spur gear rack
(e) Helical gear countershaft
(f) Hypoid gearset
(g) Bevel gearset
(h) Worm and wheel

FIGURE 5-5 Various types of gears *(Photos by the author)*

Another shortcoming of the sliding-gear transmission that was less serious than its crashbox nature was the requirement that the gears be of the "straight-cut," or spur variety with their teeth straight and parallel to the axis of the gear. If not, it would have been even more difficult to slide them into and out of engagement.

Spur gears are quite noisy and tend to whine when run together under load. An alternative type of gear is one with angled teeth, called a helical gear. These are stronger than a spur gear of the same size and also run much quieter. All modern transmissions, both standard shift and automatic, use helical gears mainly for this reason. Figure 5-5 shows a collection of gear types, many of which will be discussed further as this chapter unfolds.

Figure 5-5a shows two spur gears made as one piece. Note that the teeth are parallel to the gear axis. Figure 5-5b shows a helical gear whose teeth are at an angle to the gear axis and form a helix. While helical gears run quieter than spur gears, they add a component of force in the axial direction that must be countered with thrust bearings. Helical gears come with right- and left-hand helices.

Figure 5-5c shows a herringbone gear, which has both right- and left-hand helical gear teeth cut on the same part. The opposite helices counteract one another's axial forces and eliminate the need for thrust bearings. Herringbone gears are seldom used in automobiles due to their high cost. However, they are common in ship propulsion, especially in submarines, where their ability to handle high power and torque combined with very quiet running are worth whatever they cost.

Figure 5-5d shows a spur gear rack that mates with a spur-gear pinion to provide rack and pinion steering, among other uses. The pinion is on the bottom of the steering column and the rack converts the steering wheel's rotation to linear motion that moves the car's wheels left and right via linkages.

Figure 5-5e shows a countershaft for a modern, standard-shift transmission that has four helical gears cut on it. Instead of their mates being slid in and out of engagement (which would be difficult with helical gears), these gears are in constant mesh with their mating gears on the output shaft. Shifting is accomplished with synchromesh clutches, which are discussed in the next section.

Figure 5-5f shows a hypoid gearset that is commonly used in the differential/rear end of front-engined, rear-drive automobiles. These gearsets and differentials will also be discussed later in this chapter.

Figure 5-5g shows a bevel gearset that is used to take motion and torque around a corner. Early, non-chain-drive rear-drive automobiles used bevel gears to connect the driveshaft to the rear axle, but these have been supplanted by the hypoid gearset in present-day cars for reasons that will be explored later.

Figure 5-5h shows a wormset that consists of a worm (which looks like a screw) and its worm gear (or worm wheel). The worm shown has only one tooth that wraps around the worm, so very large reduction ratios are possible in a small package. With a single-tooth worm, the gear ratio is equal to the number of teeth on the worm gear. These are popular as the final drive (rear end) in heavy equipment such as large trucks. Not only do they give large reduction gear ratios, but they are very strong and can handle high loads.

Synchromesh Transmissions

Earl A. Thompson invented synchromesh and sold it to GM. Cadillac was the first to offer it in 1928. Buick followed in 1931 and made it standard in 1932. Packard also offered a synchromesh transmission in 1932 and Chrysler in 1933. Alvis (England) offered its own design in 1933. Soon all other manufacturers adopted it as it took the "crash" out of crashbox. In 1952 Porsche designed its own synchromesh using internally expanding, split-rings instead of cone clutches to match gear speeds.

Thompson's synchromesh mechanism uses a combination of a metal-to-metal cone clutch and a dog clutch at each gear on the mainshaft of a conventional sliding-gear transmission. But the gears no longer slide. They are in constant mesh with their mates on the countershaft, but are freewheeling on their shaft rather than being splined to it as before.

Figure 5-6 shows a cross-section of the 3-4 synchromesh clutch in a four-speed transmission in three modes. Movement of the shift ring to the left (not depicted) connects the output shaft directly to the input shaft for fourth gear (1:1). Moving it to the right connects a free-wheeling gear on the output shaft (shown), to its constant-mesh mate on the countershaft below (not shown) to obtain third gear. Figure 5-6a shows the synchromesh mechanism for this pair of gears in the neutral position.

The shift mechanism, instead of sliding a gear into engagement with its mate as in a crashbox transmission, now slides the shift ring that is splined to a cone clutch. In turn, the cone

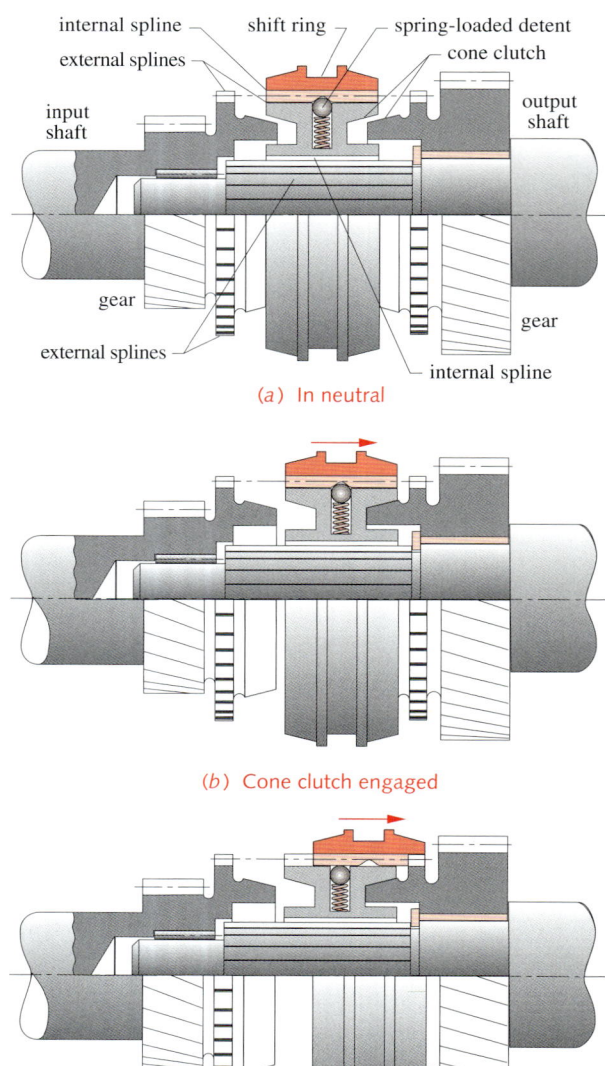

FIGURE 5-6 Synchromesh clutches

clutch." Torque then passes from the gear to the output shaft through the dog clutch, to the shift ring, to the cone clutch to the output shaft as shown in Figure 5-6c.

When shifting a synchromesh transmission, you can feel resistance as you move the lever past neutral and the cone clutch engages. This resistance decreases slightly when the cone clutch brings the two elements to the same speed. This is the signal to push a little harder on the shift lever to overcome the ball-detent and move the dog clutch into engagement for a positive connection in gear. But, you need to allow time for the speed-matching to occur.

If you rush this process, moving the shift lever too quickly, and try to engage the dog clutch before the cones have matched speed, you will hear a crunching sound. This is the sound of the edges of the dogs being worn away. The dogs are made of brass or bronze, which is much softer than the steel shifting collar. Softer metal always loses a fight with harder metal. Continued practice of this sort most likely will result in the worn dog clutches popping back out of engagement after you release the shift lever.

Shifting should be done with the clutch pedal depressed to disconnect the engine torque from the transmission. However, it is possible to "speed shift" a synchromesh transmission without depressing the clutch pedal. Drag racers sometimes did this to upshift, before automatic transmissions became the norm in that sport. Downshifting without the clutch is more difficult and dangerous to the transmission.

When the throttle is closed before shifting, the torque from the engine momentarily drops and the current gear can be disengaged by moving the shift lever quickly into neutral. Continued (careful) lever movement will engage the cone clutch and eventually the dog clutch of the synchromesh mechanism and engage the new, higher gear ratio.

clutch is splined to the output shaft. A ball detent keeps the shift ring and cone clutch together as the cone clutch engages to match speeds of the two elements as shown in Figure 5-6b.

Further motion of the shift ring overcomes the detent to allow an internal spline on the shift ring to slide over the matching external spline on the gear. The external splines on the gear are also called "dogs," and the combination of external and internal spline makes a type of "dog

However, one needs to be aware that this practice will result in shorter life for the synchronizers and can result in an expensive repair. The transmission must be removed and disassembled to replace synchronizers, which is a more expensive repair than replacing a clutch. So, speed shifting is definitely not recommended.

Figure 5-7 shows a modern 6-speed, Aisan (Japan) FWD, manual transmission cutaway with several synchronizers visible. Some countershaft gears can be seen at the lower left. Unlike a RWD transmission, in which the output shaft is on the opposite end from the input shaft, here both shafts are on the left end of the transmission.

Manual, synchromesh transmissions with more than three speeds usually build in some overdrive ratios in the higher gears. Often, 4th gear in a 4-speed and 5th gear in a 5-speed will have a ratio less than 1:1, called an **overdrive**, to reduce engine rpm at highway speeds and improve mileage. (Overdrive transmissions are discussed in a later section.) Six-speed manuals may make both 5th and 6th overdrive ratios. Modern materials and oils, along with quiet-running helical gears have made these combinations feasible. A properly maintained, unabused, modern manual transmission should give trouble-free service for 200,000 miles.

All synchromesh transmissions use helical gears on the constant-mesh gearsets to achieve their quieter running as compared to spur gears. Some manufacturers put synchromesh on reverse gear as well, but many modern, helical-gear, synchromesh transmissions use sliding spur gears in the reverse train. A few models from GM, Ford, and Chrysler, and most Lamborghinis, Hondas, and BMWs do have synchronized reverse gears.

Early synchromesh transmissions only put synchromesh on 2nd and 3rd, even when 1st and reverse had helical gears. It is not too difficult to slide spur, or even helical, gears into mesh with the transmission stopped. Chrysler used all helical gears, and the first/reverse sliding gear was on a helical spine to counteract the thrust from the helical teeth. Clever. With either a synchronized or non-synchronized reverse gear, the car should be stopped before attempting to engage reverse or significant damage could be done to the gears. If you drive a transmission with a spur-gear-reverse backwards rapidly, you will hear the spur gears whine.

Friction Drive

Friction drive was patented by Lambert in England in 1902. It is the simplest form of transmission, consisting of two disks, arranged as shown in Figure 5-8. The drive disk is attached to the engine crankshaft and the driven disk runs on a jackshaft that spans the frame rails. Chains (not shown) connect each side of the jackshaft to the rear axle outside the frame rails.

A lever, controlled by the operator, moves the driven disk left and right across the face of the driving disk to change the ratio between them. The driven disk is faced with a soft,

FIGURE 5-7   Synchronizers in a 6-speed FWD transmission
(Photo by the author)

FIGURE 5-8 Lambert friction drive *(Public Domain)*

high-friction material such as leather or cork so that the friction between the two disks transmits engine torque to the jackshaft and then to the rear wheels via the chain drives.

When the operator moves the driven disk to the center of the driving disk, the transmission is in neutral because the "dead center" of the driving disk has no velocity, even when spinning. As the operator moves the driven disk to one side of center, motion will be transmitted with a speed ratio equal to the radius of the driver at which the driven disk sits to the radius of the driven disk. If the operator moves the driven disk to the other side of center, it will drive the wheels in reverse.

So, this extremely simple mechanism is capable of an infinite number of gear ratios that vary from zero to 1:1 if both disks are the same radius. Not only that, but the same range of ratios is available in reverse! All this with no clutch, no expensive toothed gears, and only two major parts—both easy to make. A very clever design.

So why doesn't every modern automobile use this transmission? The friction material on the drive disk, of course, wears and must be replaced periodically. But modern brakes and clutches operate by friction and we have no problem replacing their friction materials periodically. But, unlike clutches and brakes, whose friction surfaces have significant area, the Lambert drive has only line contact between the disks, and this results in faster wear than experienced by modern brakes or clutches.

The main reason one does not find this transmission on modern automobiles is that the torque that can be transmitted via friction with this design is limited as compared to that which can be transmitted by pieces of strong metal (teeth) operating in an interference mode or by the CVT transmission described below.

When these Lambert transmissions were used on at least 16 makes from 1902–1917, engines then had relatively little power and speeds were low due to poor roads. As engine power and road speeds increased, friction drive of the Lambert type proved inadequate to the task. Friction drive of various designs is still used in low-power applications such as lawn tractors, snow blowers, and lawn mowers and in CVT form in automobiles.

CVTs

A different type of friction drive, invented by Daimler in 1896, has staged a comeback in modern automobiles. This is the **Continuously Variable Transmission** or **CVT** that is increasingly used in automobiles, primarily to achieve improved fuel economy.

A schematic of a sliding pulley CVT is shown in Figure 5-9. This shows a chain-driven CVT, but some use a steel or rubber belt instead of a chain. Chains and steel belts can handle

CHAPTER 5 TRANSMISSIONS AND DIFFERENTIALS

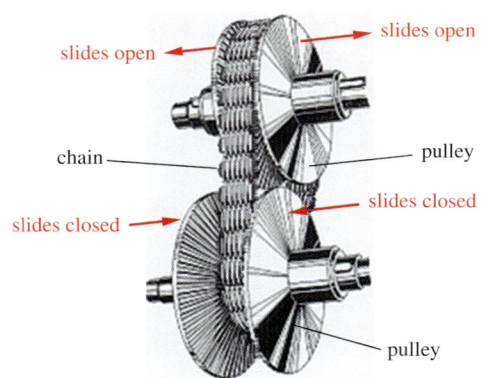

FIGURE 5-9 A chain-driven CVT transmission
(Courtesy of Büdeler Naumann)

FIGURE 5-10 Cutaway of a Toyota Super CVT-i
(Courtesy of Hatsukari715)

more torque than rubber belts. The sides of the special chain/belt are tapered to match the pulley angles.

The pulley's sides can slide open and closed to vary the radius at which the chain/belt runs. To change the ratio, one pulley is slid open at the same rate that the other is slid closed while the pulleys are turning, maintaining the chain/belt in frictional contact with both pulleys as it changes the operating radii of the chain/belt on both pulleys.

When arranged so that the drive pulley has the chain/belt at a small radius and the chain/belt on the driven pulley is at a larger radius, the transmission will increase the torque and reduce the speed of the driven pulley. Moving the pulleys in the opposite direction will result in a speed increase and torque decrease instead.

The pulley widths can be varied while the system is spinning and passing torque, making it a continuously variable transmission or CVT. It has an infinite number of gear ratios within its limits but has no dead position or reverse, as in the Lambert drive. Figure 5-10 shows a cutaway of a Toyota Super CVT-i transmission. The steel-belt assembly is at the right with the drive pulley below the driven pulley. The engine is coupled to the drive pulley through a torque converter (to be discussed in a later section) and a planetary gear train (discussed in the next section) to obtain reverse. The torque converter allows the engine to continue running when the CVT is stopped.

If the purpose of the application is to maximize fuel economy, then the CVT can be programmed to continuously vary its ratio to keep the engine running at a speed that minimizes fuel consumption.

As the load on the vehicle varies due to hills or traffic, the engine will operate at constant rpm at all times as the transmission continuously changes its ratio. As the car accelerates from a stop, the engine will rapidly speed up to its ideal rpm when the accelerator is first depressed and keep the rpm essentially constant. The CVT varies the torque ratio continuously to accelerate the vehicle.

Some people, used to the typical variation in engine sounds when driving a car with a conventional, geared transmission, find this constant droning of the engine annoying. Some also complain about the lack of distinct shift points during acceleration. Some manufacturers have responded to these customer complaints by programming their CVTs to change speed in steps rather than continuously, in order to simulate the sounds of a conventional transmission under acceleration.

CVTs are used in a variety of vehicles. Many snowmobiles and all-terrain vehicles use a rubber belt CVT. Aside from sporadic use in automobiles from the early 1900s through WWII, they first began to appear in multiple makes of automobiles in the 1980s. Many hybrid automobiles use them.

Most of these first applications were in small, economy vehicles such as the Subaru Justy, which in 1989 was the first CVT-equipped car sold in the U.S. The Ford Fiesta and Fiat Uno introduced steel-belted CVTs to Europe in 1987.

Honda put a CVT in its Civic in 1995. BMW used a belt-drive CVT in the 2001 Mini, but dropped it in favor of a conventional automatic when they offered a supercharged engine. The increased torque was too much for the CVT.

By 2006, Nissan had converted to CVTs for all its automatic transmission models of the Versa, Cube, Sentra, Altima, Murano, Rogue, and Maxima. Ford introduced a chain-driven CVT in the Freestyle, Five Hundred, and Mercury Montego in 2005. Some of these cars have fairly powerful engines, so the issues of early friction drives being unable to handle larger power levels appear to have been solved. Many other manufacturers are offering CVTs in their lineup in 2015.

Planetary Transmissions

Planetary transmissions use epicyclic gear trains. These have been known since about 1000 BC when the Chinese invented the South-Pointing Chariot. We will leave a discussion of this interesting device for the section on differentials later in this chapter. James Watt also used an epicyclic train to drive his steam engine, as shown in Figures 1-1 and 1-3, so epicyclic gear trains have been around for a long time.

Epicyclic gear trains are also called planetary gear trains (or just planetary trains). They consist of, at a minimum, one sun gear and one or more planet gears that orbit in mesh around the sun. The planets are attached to an arm, also called the carrier, that pivots on the sun axis and keeps the planets in contact with the sun as they orbit. It can make a planetary train more useful to add a ring gear, pivoted on the sun axis as shown in Figure 5-11. The ring gear provides a third element pivoted to ground to use as an input or output.

Planetary gear trains behave very differently than conventional gear trains because they require two inputs to generate a predictable output. This is referred to as having two degrees of freedom (DOF), where that term is defined as the number of inputs required to obtain a predictable output from a system. Conventional gear trains have only one degree of freedom and knowing the speed of one input gear allows quick calculation of the speed of the output gear.

Having two DOF makes it nearly impossible to intuit the behavior of a planetary train. To determine the speed and direction of the desired output requires some calculation. There is only one additional equation needed and it only involves addition and subtraction, so the calculations are not difficult. In addition, gear ratios of each gearset must be calculated. An example of those calculations is shown in the SIDEBAR: **Planetary Gear Trains** for those who are interested.

In a planetary transmission of the sort used in early automobiles, the primary input was to

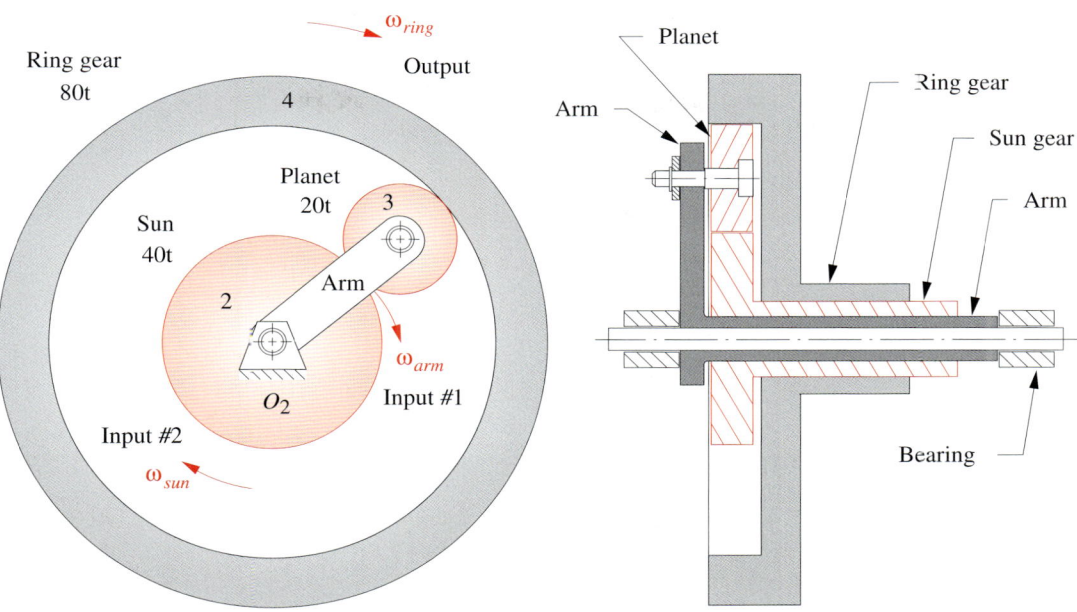

FIGURE 5-11 An epicyclic or planetary gear train

the arm, which was connected directly to the engine crankshaft. The second input was to apply a brake to a second element of the train (either a sun gear or a ring gear) to give it zero velocity, zero being a valid number. The output resulting from those two inputs was then predictable and would come from the ring gear or a different sun gear. Note in Figure 5-11 that the arm, sun and ring gear all are accessible for input or output as concentric elements on the transmission axis.

Another output, such as a different forward ratio or even reverse, was obtained by applying a brake to a different sun gear or other element to give a different ratio from the output gear. If this sounds complicated, it is. The most intriguing features of a planetary transmission are these:

1. All gears remain in constant mesh.
2. Shifting ratios only requires applying and releasing different brakes or clutches.
3. The same collection of gears is capable of delivering both forward and reverse outputs without the use of idler gears.

The manual planetary transmission was invented and patented in 1898 by Lanchester in England, three years after Panhard-Levassor invented the sliding-gear transmission. Lanchester, in typical fashion, invented a superior, three-speed plus reverse planetary transmission that he used in his 1900 automobile.

Other manufacturers settled for a two-speed plus reverse planetary. Oldsmobile was the first to put a two-speed planetary transmission in its 1901 Curved Dash Olds (Figure 2-6). Cadillac soon followed in 1903, Buick in 1904, and Willys in 1907. In 1909, Henry Ford, whom you recall reportedly could not shift a crashbox transmission, put a two-speed planetary transmission in the Model T and kept it until 1927, making over 15 million of them. Figure 5-12 shows a diagram of Ford's Model T transmission and provides an explanation of its operation.

AUTOMOTIVE MILESTONES

Planetary Gear Trains

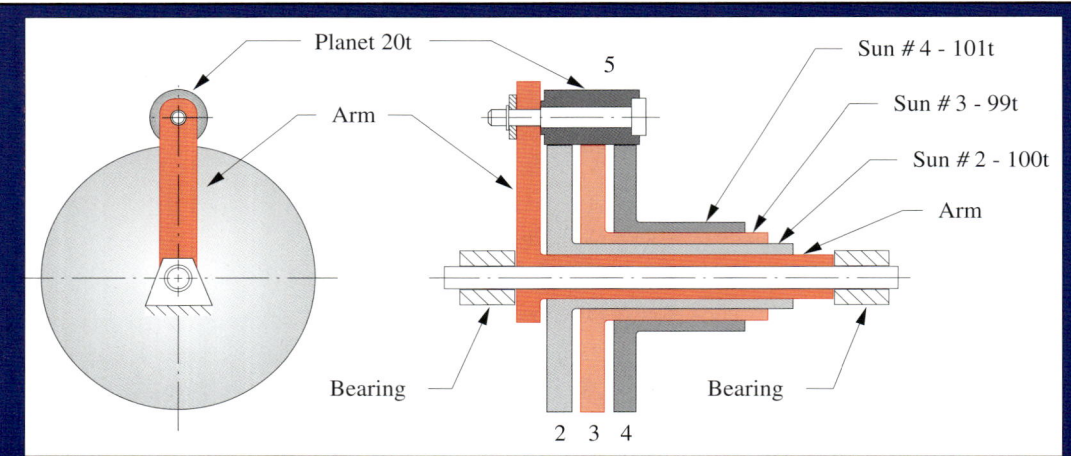

Data: Input to sun2 0 rpm N_2 = 100 teeth N_3 = 99 teeth
Input to arm 100 rpm CCW N_4 = 101 teeth N_5 = 20 teeth

Description: Sun gear 2 is fixed to the frame. The arm is driven as the second input.

Find: The angular velocities of the two outputs available, one from gear 3 and one from gear 4, both of which are free to rotate on the shaft.

Method: The equation is shown in the top row of the tables below. The left table contains only the given information. The ratios N_{in} / N_{out} are shown in the rightmost column and multiply the value in the box to the ratio's upper left to get the value for the box at the ratio's lower left. A negative ratio indicates a change in direction between the meshed gears.

The gear ratios can be applied only to the relative velocities of gear versus arm. The gear velocity is found by adding the relative velocity of gear versus arm to the velocity of the arm in each row. The answers are in the left column of the table on the right.

Discussion: Note that the two output gears 3 and 4 turn in opposite directions! Positive values denote CCW and negative values CW rotation. This planetary train is called Ferguson's Paradox, because, when it was discovered in 1858, it seemed paradoxical that the same gearbox could provide outputs in opposite directions. This feature is used in both the early manually controlled planetary transmission and in modern automatic transmissions.

Gear #	1 ω_{gear} =	2 ω_{arm} +	3 $\omega_{gear/arm}$	Gear Ratio
2	0	+100		
5		+100		$-N_2/N_5$
3		+100		$-N_5/N_3$
5		+100		
4		+100		$-N_5/N_4$

Gear #	1 ω_{gear} =	2 ω_{arm} +	3 $\omega_{gear/arm}$	Gear Ratio
2	0	+100	−100	
5	+600	+100	+500	−100/20
3	−1.01	+100	−101.01	−20/99
5	+600	+100	+500	
4	+0.99	+100	−99.01	−20/101

CHAPTER 5 TRANSMISSIONS AND DIFFERENTIALS

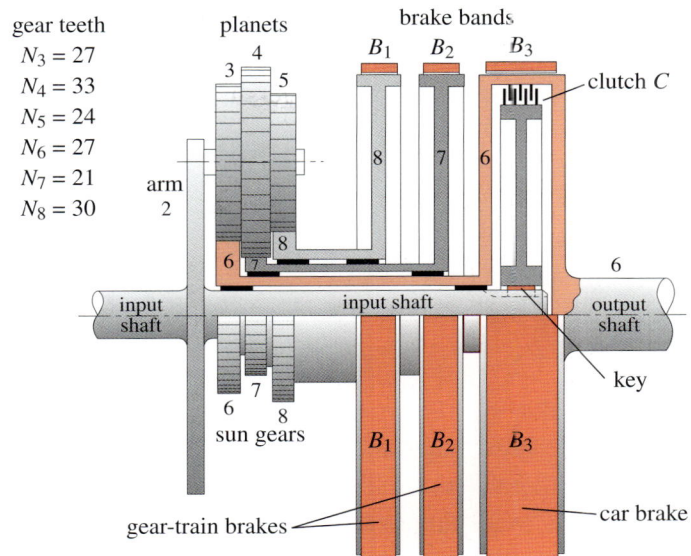

The input from the engine is to arm 2. Gear 6 is rigidly attached to the output shaft, which drives the wheels. Planet gears 3, 4, and 5 rotate at the same speed.

There are two forward speeds. Low (2.75:1) is selected by engaging band brake B_2 by depressing the "low" pedal to lock gear 7 to the frame. Clutch C is also disengaged with the clutch pedal at this time.

High (1:1) is selected by engaging clutch C by releasing all three pedals, which locks the input shaft directly to the output shaft.

Reverse (-4:1) is obtained by engaging brake band B_1 by depressing the "reverse" pedal to lock gear 8 to the frame. Clutch C is disengaged with the clutch pedal at this time.

Band B_3 is the car brake operated by a lever.

gear teeth
$N_3 = 27$
$N_4 = 33$
$N_5 = 24$
$N_6 = 27$
$N_7 = 21$
$N_8 = 30$

FIGURE 5-12 Ford's Model T planetary transmission

Except for a few hundred early examples made at the beginning of 1909, the Model T had three pedals to control motion of the car, and none of them was a gas pedal. The throttle control was a lever on the steering wheel. One pedal engaged a brake band to get low gear, a second engaged another brake band for reverse, and the third released the clutch, which, when engaged, provided a direct connection from crankshaft to driveshaft for top gear. A hand lever actuated a brake band on the output shaft to stop the car. See Figure 5-12 for a detailed explanation of their function.

Once in motion, you drove with both feet flat on the floor, controlling the speed and spark advance with hand levers on the steering wheel. The only brake was a band on the driveshaft engaged by a lever to the driver's left. These pedals and the brake lever can be seen in Figure 5-13a. Figure 5-13b shows a cutaway view of a Model T transmission showing the three friction bands, low (blue), reverse (yellow), and brake (white).

The manually operated planetary transmission became the model for semi-automatic transmissions and for the modern automatic transmission. Both use planetary gearsets and brakes/clutches in the same fashion except that the brakes and clutches are applied automatically rather than by direct input from the operator.

Overdrive Transmissions

The term overdrive refers to a gear ratio above the normal 1:1 of the typical top gear in a transmission. Overdrive originally meant a separate, optional transmission unit usually mounted behind the normal gearbox and bolted to its tail. That is the embodiment that we will discuss in this section.

Prior to the 1970s, most manual transmissions, either crashbox or synchromeshed, had only three or four speeds forward, and with few exceptions, the top gear was always 1:1. Because engines then had low power, early transmissions (1930s) also had relatively large ratios in their lower gears, typically about 4:1 in first gear.

FIGURE 5-13   Ford Model T transmission pedals (*a*) and a cutaway of the transmission (*b*)   (Photos by the author)

These cars also used a final-drive ratio in the rear end of about 4:1 to get enough torque to drive the car in top gear. Then, in first gear the overall ratio from engine to rear wheels was 16:1 (4 x 4). The engine was turning 16 times the rpm of the rear wheels in first gear.

The engines of the time were also noisy, due mainly to valve clatter. In top gear, the engine would be running four times as fast as the wheels and that resulted in a noisy ride. Noise level is a function of engine speed squared. If engine rpm could be reduced by, say, 70%, that would result in a 49% drop in noise level (0.7 x 0.7 = 0.49). This was the impetus for the invention of the planetary overdrive transmission.

As described in the earlier section on synchromesh, modern standard transmissions seldom have fewer than four or five forward speeds, many have six, and the 2015 Corvette has seven. These transmissions usually incorporate one or more ratios above 1:1 (i.e., less than 1:1) on one or more speeds in the gearbox. Some manufacturers may refer to these as overdrive ratios, but more often they make no reference to them at all. Only if you look at the spec sheet for the transmission will you know.

Planetary overdrive was invented in the U.S. by William Barnes[1] and patented in 1934. Barnes licensed it to the Borg-Warner Company and they manufactured and sold it to most U.S. auto manufacturers from 1934 to 1971. Auto manufacturers usually offered it as a low-cost option on their models. Later, around 1940, Laycock De Normanville designed a nonplanetary OD unit in England and it was sold on British cars up into the 1950s.

The Barnes/Borg-Warner overdrive unit consisted of a single-stage, planetary train with a sun gear, three planets orbiting the sun gear, and a ring gear. It looked a bit like Figure 5-11 with two additional planets as shown in Figure 5-14. Although the extra planets do not change the ratios, they balance the dynamic forces and share tooth loads.

The arm that carries the planets is connected to the manual transmission's output shaft, and this provides input to the planetary train. The ring gear is connected to the driveshaft. Be-

---

[1] Chrysler claims on its Heritage website that it invented the overdrive transmission. However, William Barnes' patent shows that he is the sole inventor, and that he assigned it to his and his wife's company, not Chrysler. Chrysler may have been the first to use it, but they did not invent it.

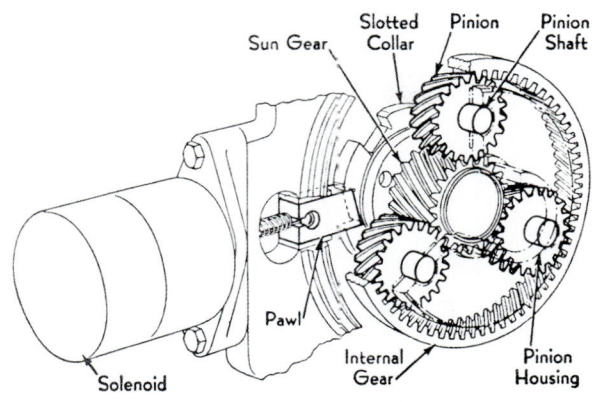

FIGURE 5-14 Borg-Warner overdrive gear train
(Courtesy of Hemmings Motor News)

tween the shafts is a one-way, or over-running clutch that provides freewheeling when the planetary train is not locked in overdrive.

Freewheeling effectively disconnects the two shafts when driving torque is not present on the input shaft, allowing the engine to drop to idle speed. The car then coasts. There is no engine braking when freewheeling.

Overdrive was unavailable at speeds lower than 25 mph. Above that speed, a switch, triggered by lifting the accelerator momentarily, engaged a solenoid on the unit that locked the sun gear to the case with a pawl, providing the necessary second input to the planetary train.

With the sun locked, the planets and ring continued to run, but at a speed 70% lower. Flooring the accelerator pedal operated a switch that took it out of overdrive for passing. Letting up on the gas pedal made it shift back into overdrive. Freewheeling was disabled when in overdrive.

Overdrive reduced noise, as described above, but also came with a bonus. Drivers discovered that their mileage increased in overdrive. Fuel usage is proportional to engine speed. As engines became quieter due to the introduction of hydraulic tappets and other improvements, improved mileage became the principal reason customers chose the optional overdrive unit. The gears above 1:1 in modern manual transmissions are there primarily to improve fuel economy, but they also reduce overall noise at highway speeds and reduce engine wear.

Semi-Automatic Gearboxes

Several designs of semi-automatic transmissions appeared in the 1930s. These were all precursors to the true automatic transmission that we know today, but were probably necessary steps to develop the technology needed for a fully automatic transmission.

Development occurred simultaneously on both sides of the Atlantic and pursued slightly different paths in each location. All would converge to a common design of automatic transmission by the 1960s. Then, in the spirit of what's old is new again, manual and friction transmissions would reappear in new, automated guises in the 21st century to compete with the automatic transmission again.

Clutches Versus Fluid Couplings

Auto companies took different approaches to the semi-automatic transmission in the 1930s. Some used a fluid coupling to connect the engine to the transmission and others used a clutch. The fluid coupling either replaced or supplemented the mechanical clutch in some early semi-automatic transmissions.

Modern, fully automatic transmissions use a descendant of the fluid coupling called the **torque converter**, which will be discussed later. Modern **Automated Manual Transmissions (AMT)**, discused in a later section, use mechanical clutches instead of fluid couplings. To understand the limitations and advantages of fluid couplings versus mechanical clutches, it will help first to explain how each of these devices works.

Mechanical Clutches

Clutches and brakes are essentially the same animal. The principal difference is that a brake has one side fixed to ground and does not move. The moving half of a brake can only move when it is not connected to the fixed half and stops when it is. Both sides of a clutch can move independently when the clutch is disconnected and will move together when connected.

Clutches come in many forms. We have already seen metal-to-metal cone clutches used in the synchromesh mechanism. Cone clutches were also the first type used to connect/disconnect engines from transmissions in early automobiles. The cone shape has some advantages as it provides a wedging action to increase the force of contact between the surfaces as compared to two disks pushed together.

Early cone clutches were faced with leather on the cone and the cup was cast iron. They tended to grab because of their wedging action and required regular ablutions of either Neatsfoot oil to make them less grabby or Fullers earth to make them grabbier. Early drivers carried vials of each potion and could often be seen under the car applying a treatment.

Disk clutches faced with asbestos soon replaced the cone clutch. Asbestos, now a known carcinogen, has since largely been replaced with less dangerous materials that have similar high-friction properties. The engine flywheel provides the driving surface of the clutch. A clutch disk splined to the transmission input shaft and faced with high-friction material similar to brake lining is the driven element.

Figure 5-15 shows a schematic of a modern automobile clutch. A steel disk called the pressure plate is spring-loaded against the outer face of the clutch disk and squeezes it against the flywheel to make the frictional

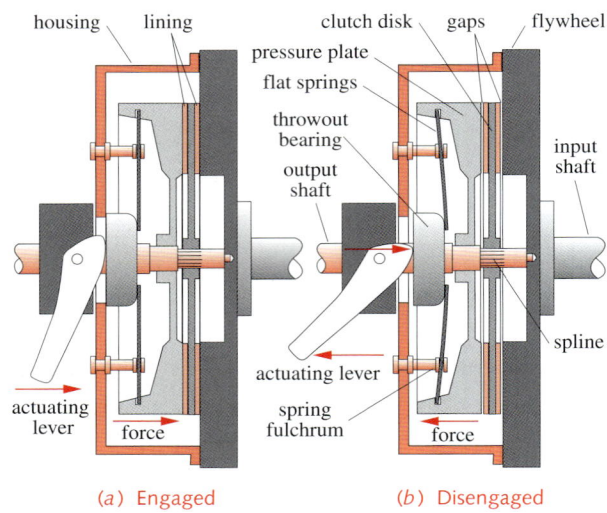

FIGURE 5-15 Automotive clutch

connection, as shown in Figure 5-15a with the clutch engaged. The high-friction material on both sides of the clutch disk is called its lining. The pressure plate and its springs are contained within a housing shaped like a shallow saucepan that is bolted to the flywheel. The flat, finger springs push between the housing and the back of the pressure plate via pins that act as fulcrums.

Figure 5-15b shows the clutch disengaged. The actuating lever that is connected by a linkage or hydraulic circuit to the clutch pedal has pushed the throwout bearing to the right, and it has bent the finger springs to pull the pressure plate away from the clutch disk to release the clutch. A small gap can be seen on each side of the clutch disk. The clutch disk is splined to the output shaft, which is also the transmission's input shaft.

The function of a mechanical clutch is essentially binary; it is either engaged or disengaged—transmitting torque or not. There is a short transient period as the clutch is engaged. During this period there is some small slippage, and its friction generates heat, which is lost energy. But this device is close to 100% efficient if properly engaged and disengaged to minimize slippage. When it is engaged, a clutch is 100% efficient and transmits torque without any losses, unlike the fluid coupling, as we shall see.

However, a clutch requires a force be applied to disengage it, and this force is typically provided by the driver's foot. The force can also be provided by other means, such as centrifugal weights, an air cylinder, hydraulic cylinder, solenoid, or magnetic field. All of these methods are commonly used in non-automotive applications of clutches, which are commercially available in all types. Many machines use clutches to interrupt their power flow.

Some current automobiles now have automated manual transmissions with mechanical clutches that use one of the methods listed above to release and engage the clutches more rapidly than a driver can with his foot. Since a synchromesh transmission is shifted internally with clutches, not only can the main clutch that connects the engine and transmission input shaft can be automated, but all the synchromesh clutches also can be.

But, this is getting a bit ahead of our narrative about semi-automatic transmissions from the 1930s and their fluid couplings. We will return to the topic of automated manual transmissions in a later section of this chapter.

Fluid Couplings

As has been pointed out many times in this narrative, most "new" ideas in automotive technology were actually invented or discovered long ago. The fluid coupling and the torque converter are no exception. Both were invented by Hermann Föttinger in Germany around 1900 and patented by him in 1905.

Some time later, Föttinger licensed the development of non-marine applications of his patent to Harold Sinclair in England. Sinclair further developed it for automotive applications and in turn licensed his technology to Daimler (England) and later to Chrysler.

A fluid coupling can replace a clutch entirely as a connection between engine and transmission. However, some manufacturers of early fluid-drive transmissions, perhaps in a "belt-and-suspenders mode," included both devices in their designs. In any event, a fluid coupling is a very simple device. As its name implies, it couples two rotating elements through a fluid, usually oil.

By definition, the term fluid includes both gases and liquids. It is easier to explain the function of a fluid coupling if we consider air as the fluid medium. Because any gas is compressible and liquids are not, a liquid will be a better choice for a fluid coupling. So please bear with us as we describe the function of a fluid coupling using air as the example fluid instead of oil.

Imagine placing two table-top fans on a surface facing one another and close together. Turn one on and leave the other unplugged. The unpowered fan spins due to the air flow across its blades from the powered fan. This is the simple principle of a fluid coupling.

A practical fluid coupling will use a liquid instead of a gas because liquids are incompressible and are much more dense than gases. More kinetic energy can be stored at any velocity in a moving liquid than in a gas because of its higher density, which means larger mass. A fluid coupling works on the principle of transferring kinetic energy.

A fluid coupling, then, consists of two shaped blades, face-to-face and running in a container full of low-viscosity oil. One, the driver, is attached to the engine's flywheel whereas the driven one is attached to the input shaft of the transmission. But, their blades look nothing like the air fans of our previous example.

You have, no doubt, stuck your arm out of a moving car at some time. If so, you know that the passing air stream exerts significant force on your open palm. But if you curl your fingers to make a cup-shape of your hand, you will feel greater force from the air.

For this reason, the blades of a fluid coupling are doughnut-shaped. The elements are called impeller and turbine, rather than "blades." The impeller is the driver attached to the flywheel which "impels," or throws, oil at the turbine attached to the transmission shaft. A drawing of a fluid coupling is shown in Figure 5-16.

The principle advantage of this device as a coupling between engine and transmission is that it allows the turbine (transmission) to be stopped while the impeller (engine) still spins. So, when the car stops at a traffic light, the engine can keep running without the driver doing anything. There is no need for a clutch or clutch pedal.

The principle disadvantage of this device is that, when the turbine is stopped and the impeller is still running, it is shearing the oil, which generates heat. Heat is lost energy. So, a fluid coupling is less than 100% efficient. It also cannot deliver all of the energy from the crankshaft to the transmission, even when spinning

When running at highway speed, the driven turbine is always turning a bit slower than the driving impeller. A fluid coupling cannot transmit torque unless there is a difference between the speeds of the impeller and turbine. The impeller must turn faster. So, there is always some slip loss in the system and this reduces efficiency. This translates to lower mileage as compared to a mechanical clutch. The difference is about 2%.

The tradeoff is convenience versus fuel economy. A fluid coupling requires less skill, involvement, and work by the driver but imposes a cost penalty. It allows the vehicle to stop with the engine running with no intervention by the driver. A mechanical clutch must be released when stopped with the engine running and the transmission in gear or it will stall the engine.

A fluid coupling also allows the driver to start from a dead stop by just stepping on the accelerator. The impeller in the coupling will gradually bring the turbine from zero speed to about 98% of its own speed, regardless of the gear that the transmission is in at the time. Power transmitted varies as the fifth power of impeller diameter, which means that small fluid couplings can handle large horsepower.

Of course, if the transmission is in a high gear, it will take a long time for the fluid coupling to accelerate the car to road speed. This is why a fluid coupling is usually used in conjunction with shiftable transmission ratios to improve a vehicle's acceleration.

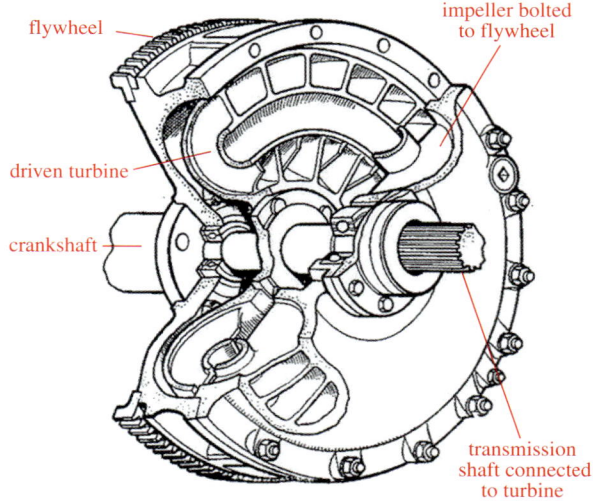

FIGURE 5-16 Fluid coupling *(Courtesy of Andy Dingley)*

Preselector Gearboxes

W. G. Wilson invented the first preselector transmission in 1928 and patented it in England. It used constant-mesh planetary trains for the first three speeds plus reverse, and fourth was direct drive. The brakes that operated on the various elements of the planetary trains to change speeds were actuated by a complicated cam-lever mechanism that was operated with a lever by the driver.

While driving in any gear, the driver would "pre-select" the gear, either higher or lower, that he wanted to engage next. Then, when ready to shift, the driver pressed and released a pedal that deactivated the current brake band and actuated the preselected brake band. The shift occurred more rapidly than in a conventional transmission and with no crashing, because of the constant mesh gears. Selecting top gear engaged a clutch that connected input to output shafts directly, like the Model T.

In a car with a conventional clutch, the gear change pedal doubled as a clutch pedal for starting off from a dead stop. The driver selected first gear, either before or after stopping, and depressed the clutch when he stopped. When ready to start, he engaged the clutch to activate first gear, and he was off.

Some cars with preselector transmissions used a centrifugal clutch between engine and transmission to eliminate the clutch pedal. A centrifugal clutch disconnects automatically below a set speed and reconnects when the input shaft speeds up. Its internal mechanism uses centrifugal force to engage friction elements that cause the output to lock to the input and spin with it above a set speed. Centrifugal clutches are used on small minibikes, gas-powered golf carts, overdrive transmissions, and on some motor scooters.

Cotal of France developed a similar system in the 1930s. Instead of the band brakes of the Wilson design, Cotal used electromagnetic clutches to engage and disengage the elements of the planetary train. These clutches were controlled by a lever in a mini H-pattern on the dashboard or steering column and directly selected the gear rather than preselected it. No pedal push was needed. It shifted when the lever was moved.

To avoid separately clutching and declutching in order to start from a stop, some cars used a centrifugal clutch or an early form of fluid coupling, which Daimler called a "fluid flywheel" (Figure 5-16). Preselector transmissions can be thought of as a "stop along the way" between the manual transmission and the true automatic transmission that was still about a decade away at this juncture.

A number of manufacturers used some type of preselector or direct-select transmission during the 1930s. Most of these companies were in Europe, but a few were in the U.S. In England, Alvis, Armstrong-Siddeley, Daimler, Invicta, Ford, Lagonda, MG, Riley, Talbot, and London's double-decker buses used the Wilson gearbox.

In France, Citroën, Delahaye, Delage, Hotchkiss, Salmson, Simca, Peugot, and Renault used Cotal gearboxes, as did Bugatti in Italy. Talbot-Lago used Wilson gearboxes. Germany used a preselector transmission in the Maybach automobile and in its WWII Tiger Tank.

In the U.S., Hudson introduced its own preselector transmission made by Bendix (who also made planetary overdrive units). An option on some models in 1935–38, Hudson called it the *Electric Hand*. The electric gear selector was mounted on the steering column in place of the shift lever. It used a conventional clutch, and like the Wilson, required the clutch be depressed to engage the preselected ratio. Unlike the Wilson, the clutches and bands within the transmission were solenoid operated. Cord, Crossley, Studebaker, and Tucker also used the Bendix unit.

Dry-Clutched, Semi-Automatics

In the mid-1930s, American companies developed semi-automatic gearboxes that did not take advantage of a fluid coupling, as did some other makes. These still required some driver involvement to release and engage the clutch and move a shift lever.

Reo's Self-Shifter

Reo was founded by Ransom E. Olds, after he was ousted from the Olds Motor Works that he had founded and was prohibited from starting another company under his own name.

Reo patented a four-speed, semi-automatic transmission in 1931. It used a conventionally shifted, sliding-gear, two-speed gearbox with a two-speed "underdrive" attached, giving four speeds. The underdrive unit was similar to an overdrive except that it had a reduction ratio of 2.07:1 instead of an overdrive's increase of around 0.7:1. Unlike the Borg-Warner OD, it was not planetary and used a centrifugal clutch to engage the underdrive ratio.

It used a conventional clutch pedal and unconventional shifter, which was a sliding lever under the dash. It functioned in a similar way to Chrysler's 1941 *Vacamatic* transmission described below, except it lacked Chrysler's fluid coupling. Instead it required the driver to de-clutch and shift between low and high range.

The underdrive in combination with the gearbox gave four speeds: 2.73:1 in Lo-Lo, 1.33:1 in Lo-Hi, 2.07:1 in Hi-Lo, and 1:1 in Hi-Hi. It was criticized for having badly spaced gear ratios. Note that Hi-Lo is lower than Lo-Hi.

With its 4.3:1 rear end ratio, Lo-Lo's overall ratio was an 11.7:1 stump-puller and not very useful. Most drivers started off in high range, which in Hi-Lo was equivalent to second gear in most cars of the time.

Reo offered it from 1933 to 1935 in their top-line Royale, but it did not sell well despite its trouble-free reputation. It was the middle of the Great Depression and the Reo was an expensive car. Despite having spent a reported six million dollars to develop this transmission, it flopped in the marketplace. Reo went into receivership in 1936.

Oldsmobile's Safety Transmission

General Motors developed the second semi-automatic transmission for the U.S. market in 1937. At that time, Oldsmobile was the most innovative division of GM. Work had been progressing since 1932 at GM's central engineering operation to develop a fully automatic transmission based on patents that GM had bought. That effort would not succeed until 1939.

When the chief engineer of Oldsmobile got wind of this transmission, he ordered it be put into production at Oldsmobile right away. But they had not yet brought a fluid coupling to a level of development that justified its production, even though Daimler in England had one since 1930. Oldsmobile decided to produce the transmission anyway using a conventional clutch and shift lever that had only three positions besides neutral—low, high, and reverse.

Their transmission design used a two-stage, four-speed planetary train as shown in Figure 5-17. They controlled the internal brake bands and clutches with hydraulic pistons. To start

CHAPTER 5 TRANSMISSIONS AND DIFFERENTIALS

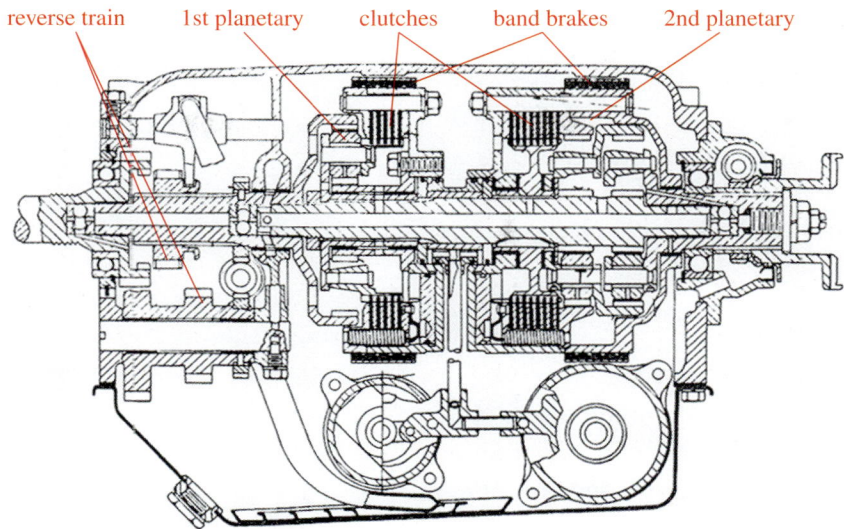

FIGURE 5-17 Oldsmobile 1937 semi-automatic transmission *(Courtesy of Hemmings Motor News)*

from a stop, they recommended that the clutch pedal be depressed and low gear selected. Once moving, the first stage of the planetary train would automatically shift to second under the control of a centrifugal governor.

The manual shift from low to high could be done without depressing the clutch or releasing the gas pedal because it only involved turning clutches and brakes on and off inside the transmission. The transmission then automatically shifted the second stage into third and later to fourth. The ratios were, from low to high, 3.16, 2.23, 1.42, and 1:1. Reverse was 3.2:1.

Drivers discovered that they could skip low range and start with the shift lever in high. The transmission then started in first and autoshifted to third and fourth without any driver involvement. Skipping second gear did not seem to slow the car's acceleration noticeably. This meant that the clutch only needed to be used when stopping, starting from a stop, or backing up. Reverse gear was a non-planetary, sliding-gear train.

This transmission was actually built in the Flint, MI, Buick plant because they had excess production capacity. But it first appeared in the Olds lineup, followed shortly by Buick. It was only offered from 1937–1939 when Oldsmobile's fully automatic transmission debuted for the 1940 model year.

Hudson's Drive-Master

Hudson had offered a vacuum-operated conventional clutch and the *Electric Hand* preselector transmission as separate options since the early-to-mid 1930s. Then in 1942, Hudson combined the vacuum-operated clutch with its *Electric Hand* to create a true semi-automatic transmission called the *Drive-Master*. Its introduction was a reaction to GM's automatic transmission of 1940 that was stealing sales from its competitors.

The *Drive-Master* offered three modes of operation at the touch of a button: *manual clutching and shifting, automatic clutching with manual shifting, and automatic clutching and shifting.* Hudson also offered the *Drive-Master* with an overdrive unit and called it the *Super-Matic.* This system was used by Hudson until 1951 when they phased it out in favor of buying GM automatic transmissions for their entire line.

Fluid-Drive, Semi-Automatics

Another technological way-station on the journey to the fully automatic transmission was *Fluid Drive*. While Europe was concentrating on the preselector transmission during the 1930s, U.S. manufacturers were taking different routes.

Daimler (England)

In 1930, Daimler applied fluid couplings first to its line of buses, and later to its top line of automobiles. They dubbed their coupling the "*Fluid Flywheel*" and it is shown in Figure 5-16 (p. 142). By 1933, all Daimler buses and automobiles were equipped with fluid flywheels and Wilson preselector gearboxes.

Chrysler's Fluid Drive

In 1939, Chrysler introduced what now appears to have been a stopgap design until they could ramp up production of the *Vacamatic*, their semi-automatic transmission (see next section) that was introduced the same year in their luxury line, the Imperial. They called this intermediate offering *Fluid Drive*. They were probably aware that GM was about to launch the first true automatic transmission in the 1940 Oldsmobile and wanted something to compete with it in the non-luxury market.

Chrysler's first *Fluid Drive* transmission simply took their existing manual, three-speed synchromesh gearbox with clutch, and added a fluid coupling in front of the clutch. If this seems redundant, it perhaps was. But the problem was that they had no transmission ready in sufficient quantity that could be auto-shifted like Oldsmobile's 1937, two-speed planetary, semi-automatic of Figure 5-17.

The only real advantage of Chrysler's *Fluid Drive* arrangement was that the operator could stop the car without depressing the clutch and the engine would not stall because of the fluid coupling. A lazy driver could then just step on the gas to resume motion and the fluid coupling would eventually accelerate the car, slowly, to a reasonable speed.

The way it was intended to be used was to depress the clutch to start in first gear and shift in normal fashion through the gears. Chrysler recommended that shifts between upper gears with the car in motion also be done by depressing the clutch pedal. But, because the fluid coupling provided a disconnect between engine and transmission, it would have been possible to shift into second or third using just the synchromesh clutches.

It was claimed that New York City taxi drivers liked this transmission because they could put it in second gear and drive around town all day without touching the clutch or gearshift lever. Since their driving was essentially stoplight to stoplight at moderate speeds, this method worked well for them. It was also very reliable, being the same clutch and gearbox that had been in use for years.

Chrysler's "Vacamatic" Transmission

By 1941, Chrysler had its semi-automatic transmission variant ready in sufficient quantity to offer it in the Chrysler and De Soto lines in addition to the Imperial. This transmission used the same fluid coupling as in their *Fluid Drive* but added a new gearbox to it. To confuse matters, they did not differentiate between these very different transmissions in their advertising and called them all "*Fluid Drive*," along with other catchy names as modifiers.

But the *Vacamatic* (so called because it used engine vacuum to shift) paired the fluid drive with a clutch (again) and added a two-speed synchromesh gearbox with a two-speed, synchromesh, "underdrive" attached, giving four speeds forward and two in reverse. The underdrive unit was similar to an overdrive unit except that it had a reduction ratio of 1.55:1 instead of the typical overdrive's increase of 0.7:1. This design was similar to the Reo *Self-Shifter* except for the added fluid coupling as shown in Figure 5-18.

The shift lever was on the steering column and had three positions besides neutral: low, high, and reverse. The manual clutch had to be depressed when shifting the gear lever. The default mode of the underdrive unit was "engaged"—the opposite of the default for an overdrive. So, manually selecting low range with the gearshift lever when stopped caused the transmission to start off in "underdrive-low" with a ratio of 3.07:1. At above 8 mph, when the driver lifted off the accelerator, the underdrive would be released, putting the combination in second gear with a ratio of 1.98:1.

Manually shifting into high re-engaged the underdrive for a ratio of 1.55:1. Lifting the accelerator pedal above 13 mph would release the underdrive unit for a ratio of 1:1. You could also start from a stop in underdrive-high (1.55:1), lift off the accelerator around 15-17 mph and get direct drive (1:1). It would automatically downshift into underdrive-high below 13 mph. These were much better-spaced ratios than in the Reo's Self-Shifter transmission.

When the underdrive was engaged in first (underdrive-low) and third (underdrive-high) the centrifugal clutch in the underdrive unit was freewheeling. The owners manual warned to not descend hills in those gears as there was no engine braking. And, when parked, those gears would not prevent the car from rolling. The parking brake always had to be set.

Most drivers started off in high range to avoid shifting and accepted the slower acceleration that resulted. In any event, these cars were not par-

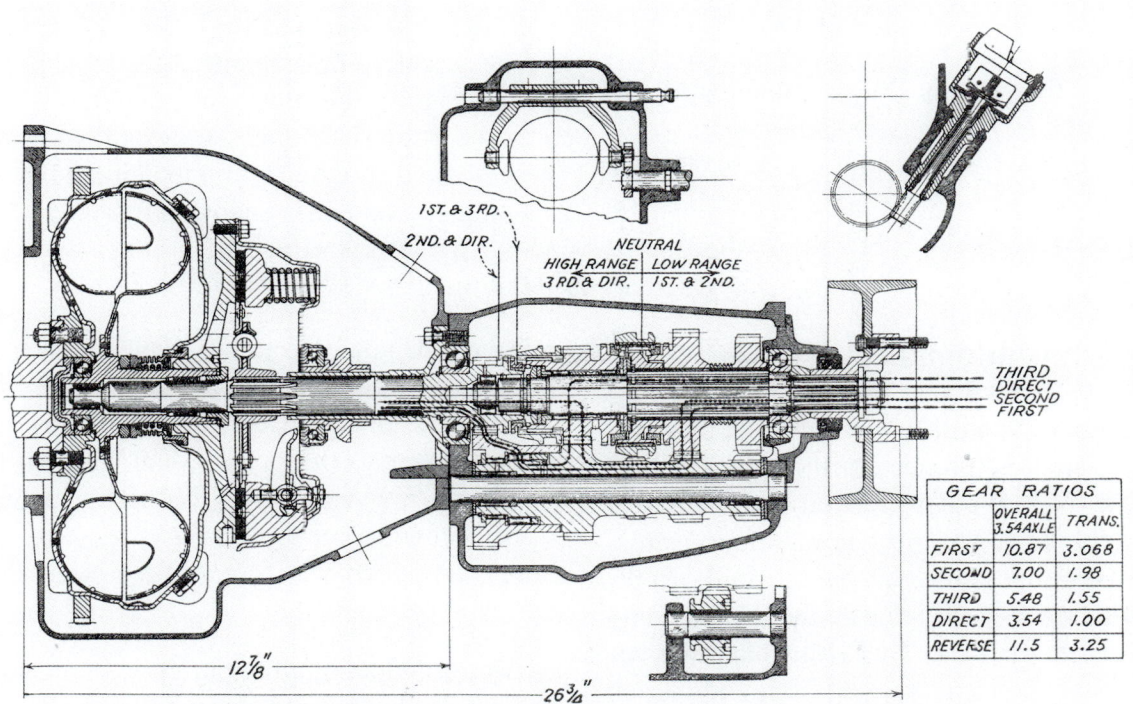

* Cross-section of Chrysler semi-automatic "underdrive" transmission, fluid drive and conventional clutch assembly, showing flow of power in four forward speeds

FIGURE 5-18 Chrysler Vacamatic Fluid Drive transmission *(Source: SAE Journal Vol. 47, No. 5, Nov, 1940, p.455)*

ticularly fast and tended to appeal to customers who valued reliability and ease of use above all. Chrysler products of this era had a deserved reputation for solid engineering and reliability. The hot-rodder set gravitated toward Fords with their V8 engines.

As is typical with all makes, Chrysler introduced its latest innovations in its higher-priced offerings. It then migrated them to the lesser marques in subsequent years; the *Vacamatic* appeared in Dodges and Plymouths shortly after its introduction in a Chrysler. They kept this transmission in production until 1953. Its name was changed for each marque for marketing purposes. But they were the same transmission.

A running change was made to the *Vacamatic* in 1946 to use hydraulics rather than vacuum to operate the underdrive train. The ratios were also changed slightly. Otherwise it was the same transmission. To improve performance, the fluid coupling was replaced in 1951 with a torque converter (see a discussion of torque converters below). They then called it *Hy-Drive*. Other names were coined for these variants such as *Prestomatic*, *Simplimatic*, *Fluidmatic*, and *TipToeShift*, among others. But the phrase *"Fluid Drive"* appeared somewhere on the bodies of them all. God bless the marketing department!

Other Semi-Automatics

Many other manufacturers offered various forms of semi-automatic transmissions before and after WWII, but we are not going to list them here. Once the fully automatic transmission was well established by the mid 1960s, semi-automatics were relegated to the economy fringe of the market. That changed when automated manuals appeared on high-end cars in recent years. We will deal with those later.

Automatic Transmissions

Again we find that seemingly modern mechanisms were actually invented long ago. This is true of the automatic transmission, which was first patented and manufactured in 1904.

The Sturtevant Automatic Transmission

The prize for the first automatic transmission goes to the Sturtevant brothers of Massachusetts, who patented their design in 1904 and manufactured it from 1905–1908 when they apparently went out of business. Their patent drawings are shown in Figure 5-19. The patent describes it as a *clutch-and-gearing mechanism ... (that) will be automatically shifted ...* .

It is quite a clever design. It does not need a conventional clutch. It is just a two-speed, plus reverse conventional, sliding-gear gearbox with automatically engaging centrifugal clutches.

Low speed uses a compound gear train. High speed is direct drive. Two centrifugal clutches control engagement of the two speeds. In Figure 5-19, the clutch housing 15 is bolted to the crankshaft and serves as a flywheel. Three pivoted weights within the clutch labeled 25 swing outward with centrifugal force to load each of the double-disk clutch plates on either side of the disk labeled 18.

When the engine is idling, both clutches are disengaged. As engine speed increases, the low-speed clutch engages due to centrifugal force, passing power to the output shaft through the four gears shown meshed. As speed increases, the second set of three weights engages the high-speed clutch that connects the input shaft directly to the output shaft.

An over-running clutch in gear 36 then disconnects the low-speed gears from the output shaft. Except for reverse, which requires that

CHAPTER 5   TRANSMISSIONS AND DIFFERENTIALS

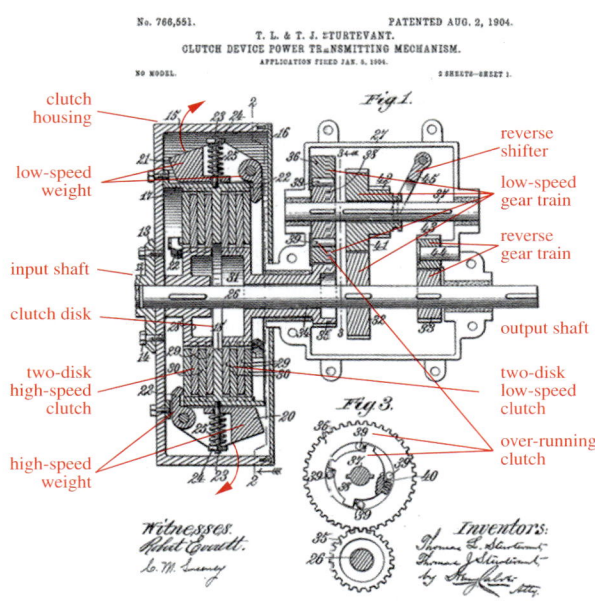

FIGURE 5-19   Sturtevant 1904 patent drawings

gear 41 be shifted manually into engagement with the reverse idler 43. the gears remain in constant mesh, and the operator need do nothing except step on the gas pedal.

One could quibble that because reverse gear required manual intervention to engage, it is not a true automatic transmission. But every modern automatic transmission requires that the driver move a shift lever from drive to reverse. Once the Sturtevant transmission was shifted into reverse, operation was automatic. Stepping on the gas caused the low-speed clutch to engage, driving the car backwards.

One interesting anomaly, not mentioned in the patent, is that if the driver could go fast enough in reverse to cause the high-speed clutch to engage, he would suddenly find himself going forward in direct drive. More likely, this would stall the engine.

It is reported that this transmission gave problems and would fail with no warning. The metallurgy of the time was, perhaps, not up to the task of withstanding the impacts from clutch engagement. Perhaps with modern materials and engineering, it could be made to perform satisfactorily now. In any event, it was a noble attempt to solve the problem of the crashbox transmission in 1904.

Other Attempts

In 1921, Alfred Munro of Canada invented an automatic transmission and patented it in Canada in 1923 and in the U.S. in 1927. This transmission used compressed air instead of oil as the working fluid. As a result, it could not handle much power and was not a commercial success. Many other interesting automatic transmission designs were invented and used on various vehicles around the world in the 1920s and 1930s. An excellent reference that describes them in detail is "Automatic Transmissions," by P. M. Heldt, *SAE Journal*, Vol. 40, No. 5, May, 1937, pp. 206–220.

The first automatic transmission to use hydraulic fluid to change gears was developed by a pair of engineers in Brazil, Jose Araripe and Fernando Lemos. They sold the prototype and plans to General Motors, and it became the basis for the first commercially successful automatic transmission.

GM's Hydramatic Transmission

As mentioned several times in the discussion of semi-automatic transmissions, GM introduced the first successful automatic transmission in the 1940 Oldsmobile in late 1939. It was called the *Hydramatic* transmission and shifted automatically through its four gears. Earl Thompson, who invented synchromesh, also had a lot to do with development of the *Hydramatic*.

GM took the four-speed, planetary transmission from the Oldsmobile semi-automatic of 1937 (Figure 5-17) and replaced its clutch with a fluid coupling. Unlike a torque converter (see next section), which would not be used until 1958, the fluid coupling did not multiply torque. It only did its basic job of allowing the engine to keep running when the transmission was stopped and in gear.

The *Hydramatic* was an instant success. There was no longer a clutch pedal or a shift lever as previously known. Now there was a quadrant and lever on the steering column with which Neutral, Drive, Low, and Reverse could be selected. Low used only the first two gears and Drive used only 1st, 3rd, and 4th. There was no Park position as on modern transmissions. With the engine off, a parking pawl was engaged in the reverse position to prevent rolling.

There were four forward speeds with ratios from low to high of 3.66, 2.53, 1.45, and 1:1. These were changed to 3.82, 2.63, 1.45, and 1:1 after the war. Both of these are similar to those of the Olds semi-automatic from which it came, but all except direct drive are a bit lower-geared than in its predecessor, probably because this car used a slightly higher (lower numerical) rear-end ratio than used in the semi-automatic to get the effect of an overdrive in fourth gear.

Reverse in the Olds semi-automatic was obtained with a manually shifted, sliding gearset as can be seen at the left side of the transmission in Figure 5-17. In the *Hydramatic*, a third planetary gearset with a 3.2:1 ratio was added to handle reverse. All clutches and brakes were hydraulically shifted as in the 1937 semi-automatic transmission. Figure 5-20 shows a detailed cross-section of the GM *Hydramatic* transmission.

In 1941, *Hydramatic* became an option on Cadillacs. It would not migrate down to Pontiac until 1948. Chevrolet and Buick developed their own automatic transmissions, which will be discussed below. During the war, the *Hydramatic* was used in Cadillac-powered U.S. tanks.

After the war, many other companies chose to buy GM's *Hydramatic* transmissions rather than develop their own. Hudson, Nash, Kaiser, Frazer, Willys, and Lincoln all used them in the 1950s. Even Rolls-Royce licensed the *Hydramatic* for manufacture in 1952. They produced it for Rolls-Royce and Bentley cars through 1967. The original *Hydramatic* continued to be used in U.S. light trucks until 1962.

The *Hydramatic* used a unique arrangement of its planetary trains and fluid coupling that made it more efficient than many later designs of automatic transmissions. Instead of connecting the fluid coupling directly to the crankshaft, in first gear it initially passed the power through the third-speed planetary (1.45:1) and then through the fluid coupling, then to the rear planetary train (2.63:1) to get their product of 3.82:1 for first gear.

This meant that the fluid coupling's impeller was turning at 70% of engine speed in first gear, reducing shearing and heating of the oil. This gave a very smooth startup because the coupling's slippage was amplified by the 1.45:1 gear ratio. Slippage diminished rapidly as the turbine came up to speed.

When it upshifted to second, the forward train locked (becoming 1:1) and the impeller ran at engine speed. This arrangement "tightened" the fluid coupling and reduced slippage, but could result in a rough 1-2 shift, especially at full throttle. The 2.63:1 ratio was now due only to the rear planetary.

Third gear locked the rear train (making it 1:1) and unlocked the forward train for an overall ratio of 1.45:1. This arrangement reduced the torque passing through the coupling from 100% to 40% of engine torque. The rest went through the forward planetary train, reducing slippage and thus increasing efficiency.

CHAPTER 5 TRANSMISSIONS AND DIFFERENTIALS

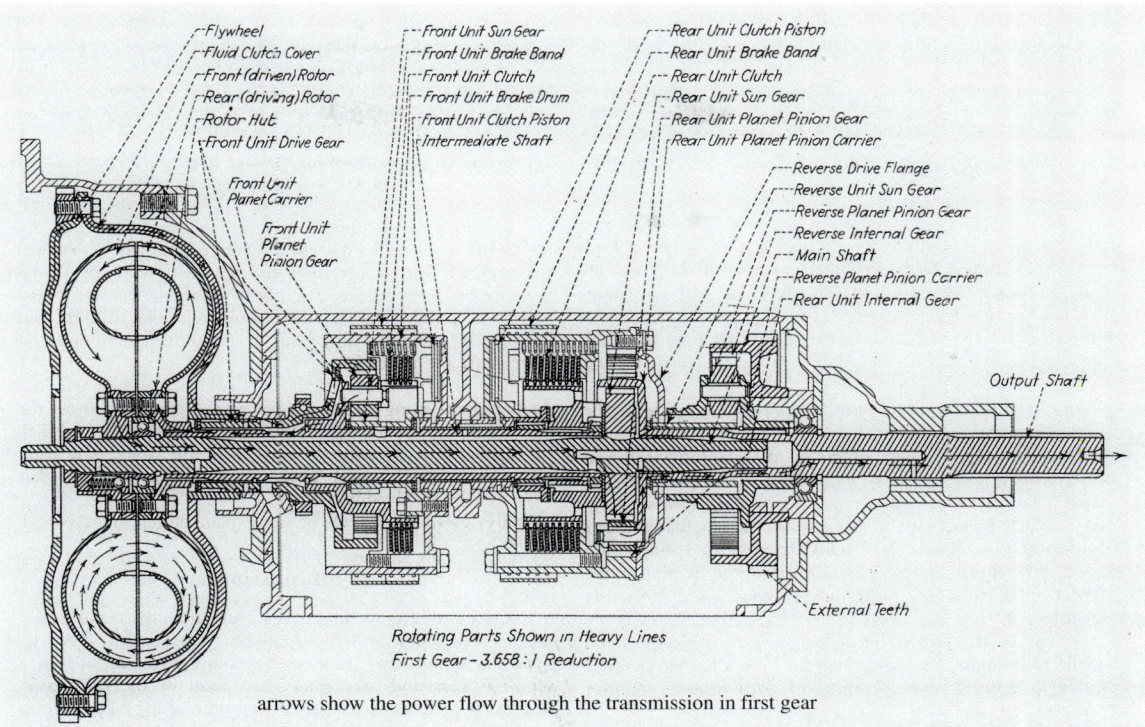

FIGURE 5-20 General Motors Hydramatic Transmission *(Source: SAE Journal, Vol. 45, No. 5, November 1939, p. 462)*

Fourth gear again locked up the forward train with the rear train still locked, giving 1:1 direct drive. Because of the torque split between the forward train and the fluid coupling, the coupling now felt only 25% of engine torque. This reduced slippage to near zero and increased efficiency further. Torque-converter automatics were not able to match this efficiency until the "lock-up" torque converter was implemented by Packard in 1949.

The *Hydramatic* transmission represents a real milestone in the development of the automobile. Not only did many manufacturers (including the exalted "best car in the world" Rolls-Royce) use it, but it was the basis for most of the geared automatic transmission designs down to the present, and it popularized automatic transmissions with the public. It was complex and expensive to build, but it was so rugged and reliable that it was used in drag racing in the 1960s. GM made millions of them before retiring it.

Nevertheless, it was not without its problems. Rough shifting was a common complaint. The 2-3 shift required the simultaneous operation of two brake bands and two clutches. Adjusting these properly was a mechanic's nightmare and some became specialists in this "art." Band brakes also wore and needed replacement fairly often. They rarely went more than 20,000 miles without needing attention.

Automatic transmissions are complicated and fairly difficult to work on. Transmission work came to be called the "brain surgery" of automotive mechanics by some. Many shops specialize in servicing them, and repairs are usually expensive.

In 1953, when most brands had automatic transmissions (many of them from GM) and

Chrysler did not, Chrysler's advertising touted the simplicity and reliability of their *Fluid Drive,* semi-automatic transmissions as compared to those troublesome automatics. Of course, by 1954, Chrysler had their own automatic transmission.

GM made the first major change to the *Hydramatic* transmission with the *Jetaway Hydramatic* in 1956. This transmission used two fluid couplings, one connected to the engine crankshaft, and another that replaced the internal clutch that controlled the forward planetary train in the original.

This change cured the rough 1-2 shift and provided smoother shifting overall. But it came at the expense of efficiency and an increase in complexity. This transmission provided a Park position on the selector. The *Hydramatic* type transmissions that used fluid couplings were phased out by the late 1960s in favor of torque-converter automatics.

Torque Converters

Recall that Hermann Föttinger patented both the fluid coupling and the torque converter in 1905. As described in a previous section, a fluid coupling has two primary elements—an impeller and a turbine. The impeller is driven by the engine and forces the turbine to spin by "throwing" oil at its curved blades.

A torque converter adds a third blade, called a stator or reactor, between the impeller and turbine. The stator is so named because, in the original implementation of torque converters, it was stationary. Modern torque converters put the stator on a one-way clutch that allows it to turn in the same direction as the impeller, but not the opposite. Figure 5-21 shows a schematic and a cutaway of a modern torque converter.

The purpose of the stator is to redirect the flow of oil returning from the turbine to the impeller and turn it in the direction of impeller motion. In a fluid coupling, the oil returning from the turbine is directed counter to impeller rotation and tends to slow it. This increases slip, reduces efficiency under slip, and generates heat. By redirecting the oil to help the impeller turn instead of opposing it, energy in the returning oil is recovered and gives a boost in torque. The one-way clutch prevents the reaction force of the oil from driving the stator backwards.

There are three stages of operation of a torque converter: **stall, acceleration,** and **coupling**. **Stall** is when the engine is turning the impeller, but the transmission is in gear and the vehicle is stopped, as is the turbine. There is then a maximum difference in velocity between impeller and turbine, and slip is 100%. In this mode, the stator is stationary and all the oil returning from turbine to impeller is trying to turn the impeller along with the engine. The result is a boost in torque of from 1.8 to 2.5 times, depending on blade design. Other transmissions, such as the *Dynaflow* and *Turboglide* described below, use multiple turbines to give a torque boost as high as 4.2:1.

This torque boost gives a distinct advantage to a torque converter over a standard transmission with clutch (or fluid coupling) in accelerating a car off the line. Assuming the gear ratios are the same in both transmissions, the vehicle with a torque converter initially has about twice the torque available for a standing start. This factor, along with quicker shifting, is why drag racers have generally abandoned standard transmissions for torque-converter automatics.

During the second stage, **acceleration**, the stator is still stationary and there is still substantial slip due to the velocity difference between impeller and turbine. There is still some torque multiplication during acceleration, but it diminishes as the turbine velocity approaches that of the impeller.

CHAPTER 5 TRANSMISSIONS AND DIFFERENTIALS

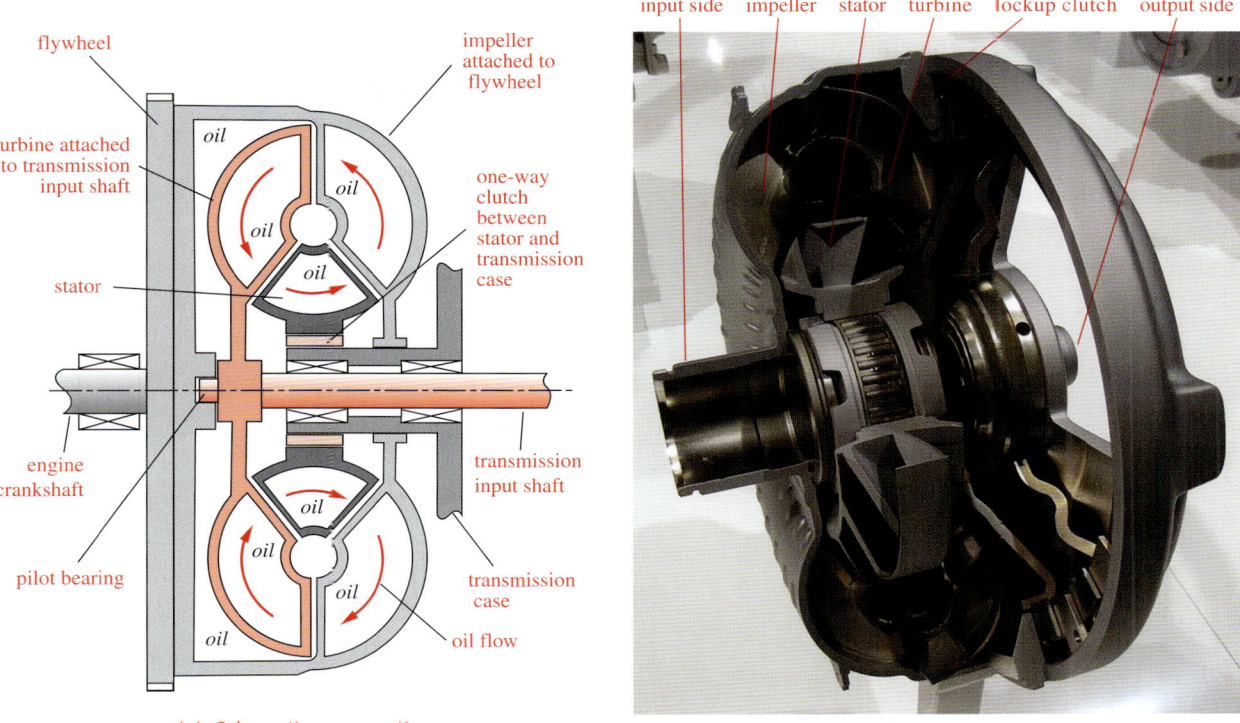

(a) Schematic cross-section (b) Modern ZF torque converter

FIGURE 5-21 Cutaways of a torque converter
((b) Courtesy of Aconcagua)

When the turbine has reached about 90% of impeller speed, the system enters the **coupling** stage. It now behaves like a fluid coupling with no torque multiplication. The oil returning from the turbine through the stator now tends to drive the stator in the same direction as the impeller and the three elements turn essentially as one.

As in a fluid coupling, the turbine can never reach the same speed as the impeller unless there is some kind of mechanical intervention. Unfortunately, in the coupling stage, the torque converter is only about 85% efficient versus the fluid coupling's 98%.

Packard was the first to add a lock-up clutch to a torque converter in its independently developed automatic transmission in 1949. This lock-up clutch engages to mechanically lock the impeller and turbine together when they reach nearly the same velocity in the coupling stage. This reduces slip to zero and the torque converter then becomes 100% efficient, as in a fully engaged mechanical clutch. When road speed drops below a set point, the lock-up clutch is unlocked to avoid stalling the engine. Since the late 1970s, most automatic transmissions' torque converters have been equipped with a lock-up clutch.

Buick's Dynaflow Transmission

In 1948, Buick not only made the first automatic transmission with a torque converter, they essentially made the transmission **be** the torque converter. Athough the transmission had a two-speed planetary train, it started from a stop in high gear. Low gear could be selected manually with the shift lever and held to about 60 mph, but it would not automatically upshift to high. You had to move the lever to high manually.

It was intended that high gear be the only one used except in some undefined special circumstances. The torque converter had five elements: the normal impeller, two turbines, and two stators. It relied on the converter's 2.1:1 torque multiplication ratio to accelerate the vehicle from stopped. It was effectively a hydraulic **constant velocity transmission**.

Stepping on the gas from a stop resulted in the engine rpm increasing to mid-range and then staying there as the car slowly, very slowly, accelerated. Buicks weighed about 4000 lb. The *Dynaflow* transmission earned the nicknames "Dynaslow" and "Dynaslush." But *Dynaflow* was very "smooth-shifting"—because it never shifted. Many customers liked it for the smoothness, which was a contrast to the *Hydramatic*, known for its rough shifts.

In addition to being slow, it was very thirsty and had poor fuel economy. This was not of great concern to most of its customers because gasoline then was cheap ($0.25 per gallon), and the typical Buick customer of the time tended to be "mature" and relatively well off. It was an expensive car, after all—one step below the Cadillac in GM's lineup.

It was an inherently inefficient design, relying on torque-converter slip to change the effective gear ratio over its entire rpm range. This was exacerbated by the dual stator arrangement, which caused added turbulence and wasted more energy than would a simpler, three-element torque converter.

It was redesigned in 1953 to eliminate one stator but kept the two turbines and the (essentially unused) planetary train. This *Twin Turbine Dynaflow* improved performance and efficiency a bit with no loss of smoothness. They added a variable-pitch stator in 1955, but these changes still left it short of its competitors in most respects.

Another change in 1958 went to three turbines and used a variable-pitch stator to increase its stall-torque multiplication ratio to 3:1. This *Triple Turbine* was canceled after 1959 due to technical problems, and they reverted to the *Twin Turbine*. *Dynaflow* was finally put to rest in 1964 in favor of more conventional two- or three-speed planetary designs that used a simple, three-element torque converter and actually shifted the gear trains, as did most of their competitors.

Chevrolet Powerglide

Powerglide was the first automatic transmission to be offered in a low-priced automobile, in 1950. This had a torque converter and a two-speed planetary train. Like the *Dynaflow*, early examples of *Powerglide* through 1952 started off in high gear and so were very sluggish. Drivers could manually shift into low to start off, but, like the *Dynaflow*, it would not automatically shift to high. When the driver moved the selector to high, the shift was rough and hard on the transmission.

One has to wonder why the Buick and Chevrolet engineers included a low gear if they did not bother to program the transmission to use it. From 1953 on they fixed this flaw in *Powerglide* (but not in *Dynaflow*), making it start in low and shift to high at a speed that depended on load and throttle setting. *Powerglide* was replaced by the non-shifting *Turboglide* transmission in 1958. It did away with low gear and, like the 1958 *Dynaflow*, used three turbines and a variable stator to change ratios over all driving speeds.

The author had the dubious pleasure of driving both *Powerglide* and *Dynaflow* equipped cars in the mid-fifties and can attest that they were terrible. They were sluggish and never seemed to be in an appropriate gear ratio no matter the circumstances. They exacted a noticeable penalty in economy over the same car with a three-speed standard transmission, and were much less fun to drive.

Both of these transmissions had a significant safety flaw in their shift quadrants. The order of selection was, from lever up to down, NDLR. There was no Park position. The flaw was the positioning of R below D and L with no interlocks between L and R to prevent R being accidentally selected while in forward motion.

The author had the unfortunate experience of hitting the gearshift lever accidentally while driving a new 1956 Chevrolet-V8 on a two-lane road at about 50 mph—knocking it from D to R. The car actually went into reverse because no gears needed to be shifted to do so in an automatic transmission. Only brakes and clutches needed to be released and applied to different parts of the train.

The rear wheels locked, accompanied by screeching tires, and the car did a "handbrake turn" through 180°, crossed the oncoming lane, stalled the engine, and planted itself in the bushes on the other side of the road. This book is only possible because there were, luckily, no other cars in sight at the time. No damage was done to the car. As for the author, if he were a cat, his count would have been down to eight. This is the first report ever made on the event. It was a well-kept secret for 58 years.

GM Turbo Hydramatic

The *Dynaflow* was replaced in 1964 and the *Powerglide* in 1969 by the much superior, three-speed, *Turbo Hydramatic* transmission, shown in Figure 5-22. This new transmission was made in several versions to match the power output of various engines across the entire GM line starting in 1964. It appeared in different makes as the *Turbo Hydramatic 350, Turbo Hydramatic 400*, and other variants.

This was a three-speed planetary with torque converter that was smoother shifting than the original *Hydramatic*. It had a Park position and adopted the now standard selector sequence of PRNDL, thus curing the above-mentioned flaw in earlier selectors. Each GM division made

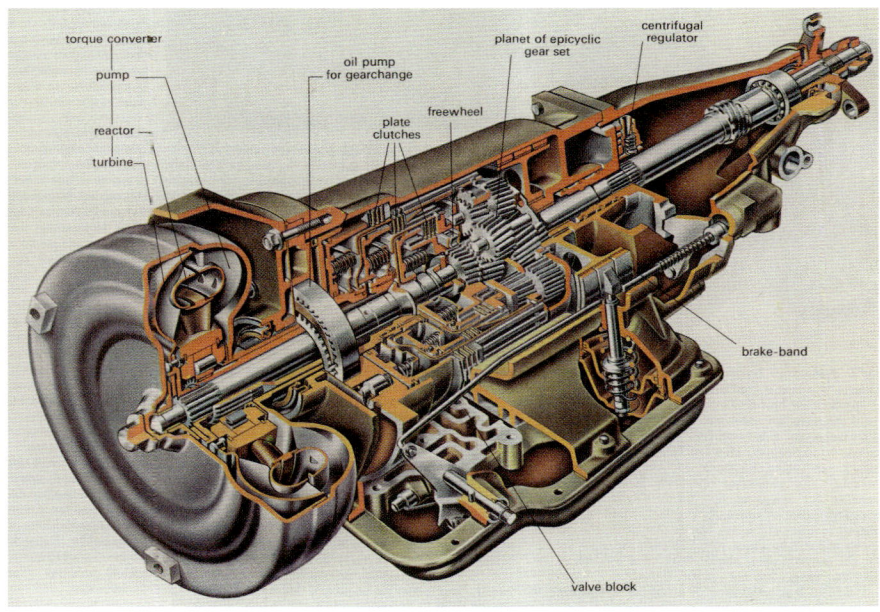

FIGURE 5-22 General Motors Turbo Hydramatic Transmission *(Source: Anatomy of a Motor Car, Ian Ward, Ed., Crescent Books, p.99.)*

changes to their versions of the *Turbo Hydra-matic* with the result that they often felt different in different makes.

By the 1970s, GM had standardized many of its offerings across brands. One could buy essentially the same car with a Chevrolet, Pontiac, Oldsmobile, or Buick label on it. They often were made on the same assembly line, but each division made changes in engine, suspension tuning, and transmission shift points, for example.

If one used rental cars frequently, this was quite noticeable. Pontiacs seemed to have crisper shifts and felt more "alive" in many respects (handling, steering) than its cousins. Oldsmobiles had good performance, but Buicks were more reluctant to change speed and felt more like an ocean liner. These are all very subjective assessments based on the author's seat-of-the-pants, rather than on any data. Each division was targeting its market niche.

Everybody Wants In On The Act!

All U.S. manufacturers scrambled to offer automatic transmissions in the late 1940s and early 1950s. Packard offered the three-speed, torque-converter *Ultramatic* in 1949 with converter lockup. *Borg-Warner* developed a series of three-speed, torque-converter transmissions in the early 1950s. They sold them to Ford, American Motors, Studebaker and others. Other independents continued to use GM transmissions.

In 1951, Ford launched its Borg-Warner *Ford-O-Matic*. Chrysler introduced a two-speed, torque-converter transmission called *PowerFlite* in the 1954 models. In 1956, they added a three-speed, torque-converter transmission called *Torqueflite*. This smooth-shifting transmission was the first to incorporate a compound planetary train and used a torque converter with two stators. By 1960, all makes offered one or more automatic transmission choices.

Modern Automatic Transmissions

In the 21st century, automatic transmissions have become much more sophisticated. Microcomputers now control their shifting, and some use "fuzzy logic" and artificial intelligence to "learn" a driver's habits and adjust their shifting algorithms accordingly. Many offer a "Sport" setting that holds upshifts to higher rpm to improve acceleration at the expense of economy.

The number of speeds in an automatic transmission has grown significantly. The German company ZF was the first to offer a six-speed automatic in the BMW 7-series. Now many marques have them. Other marques now offer seven-, eight-, and nine-speed automatic transmissions. Figure 5-23 shows a ZF, eight-speed, automatic transmission.

The quest for ever more miles per gallon has largely driven this explosion in multi-speed automatics. Having many gear ratios can make a smaller engine seem more responsive by always placing it near an optimum torque-producing rpm for all driving conditions. Taking this to its logical extreme explains the resurgence of the CVT transmission, which has an infinite number of speeds available.

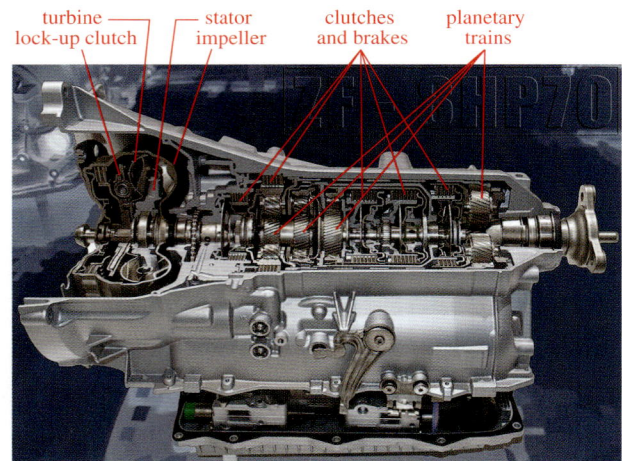

FIGURE 5-23 ZF eight-speed automatic transmission
(Photo: Stefan Krause, License: FAL)

Manumatic Transmissions

An automatic transmission just can be put into drive and it will shift perfectly well. But many automatic transmissions now offer a means to manually control their shifting. Some add paddles to the steering wheel or column that will force an upshift or downshift as long as it won't damage anything. Others allow the gearshift lever on the console to be moved to up- or downshift. These are variously called *Manumatic, Tiptronic*, and other names by various manufacturers.

Most will allow the engine to rev to the redline and will then force an upshift. Most also force a downshift when revs drop too low. So, the "electronic nannys" are continuously watching out for the "nut behind the wheel" to make sure he does not break something.

Automated Manual Transmissions (AMT)

The earliest of these was the Sturtevant, the first automatic transmission in 1904 described earlier. There also were many semi-automatic manual transmissions tried, as described in an earlier section. Some of these used a mechanical clutch to connect engine and transmission. One of them, the 1942 Hudson *Drive-Matic*, automated the clutch motion with vacuum. That transmission is a predecessor to the recent, single-clutch, automated manual transmission.

Technology has advanced a lot since 1942, and the same results now can be obtained in more sophisticated and effective ways. Modern **Automated Manual Transmissions** (also called **semi-automatic transmissions** by some) have been available in two forms: **single-clutch** and **dual clutch**. We will discuss them in that order.

Single-Clutch AMT (SC-AMT)

This transmission takes a conventional synchromesh gearbox with clutch, and automates both the engagement/disengagement of the synchromesh clutches and also that of the main clutch using hydraulic actuators. The first of these was in the Daimler/Lanchester Sprite in 1954. Now, the main clutch, normally of the dry-clutch variety in a standard transmission, is usually replaced with a hydraulically operated, multi-plate, steel-on-steel disk clutch running in oil for cooling. The transmission control computer controls all the clutch actions in response to inputs from the driver. It only allows sequential shifting like a motorcycle. No gears can be skipped. Its most common name is *Sequential Manual Gearbox* or SMG.

Most such transmissions are equipped with paddle shifters on the steering wheel or steering column, but usually they also are programmed to shift automatically. The automatic shifts can be overridden with the paddles, or it can be driven like a conventional, torque-converter automatic transmission.

The main advantage over a manual transmission is that automated shifts can be executed much faster than a human can do them. For that reason, this transmission found some of its first applications in racing. Ferrari was the first to use an SMG transmission in Formula One racing in the late 1990s. Ferrari's first SMG shifted in 60 ms.

Even before Ferrari used an SMG in racing, Izuzu introduced its NAVi5 SMG in 1984 in Japan. Initially, it had only an automatic mode, but they later added a gearshift lever in the standard H-pattern rather than paddles to manually shift it.

BMW offered a 6-speed SMG transmission in the 1992 M3 alongside choices of 5- or

6-speed manuals. Some owners who chose the SMG complained about its rough shifts. This problem plagued all SMGs, which led to their replacement by the dual-clutch AMT in later models.

Ferrari first introduced its F1 SMG transmission to a production car in its 1997, 355-F1. They subsequently stopped producing Ferraris with manual transmissions. All their models through the 599 Fiorano used SMG transmissions until being replaced with a dual-clutch AMT in their latest models. Ferrari's SMG was also used in some Alfa Romeos and in all Maserati models from 2004 to 2006. Called the Cambio Corsa, it suffered from very rough shifting in automatic mode when compared to a torque-converter transmission. Maserati replaced it in 2007 with a ZF, six-speed, torque-converter automatic and saw sales increase. The Maserati ZF transmission is a distinct improvement over the SMG, as this owner can attest.

Dual-Clutch AMT (DC-AMT)

Though the dual-clutch, automated manual transmission was reportedly invented around the same time as Hudson's *Drive-Matic*, its inventor, Adolphe Kegresse (France), never developed a working model. Harry Webster of Automotive Products (AP) in England later built successful prototypes and patented his design in 1981.

Webster's company, AP, worked in conjunction with VW/Porsche to develop what became the Porsche PDK transmission that was used in Porsche race cars starting in 1983. PDK is an abbreviation for *Porsche Doppelkupplungsgetriebe*, which is German for *dual-clutch gearbox*.

The PDK transmission was first used commercially in the 2003 VW Golf MK4 R32. It was subsequently used in many other models in the VW/Porsche/Audi lineup. A Getrag-made DC-AMT replaced BMW's SMG in 2010, and Fiat uses a Magnetti Marelli-developed version of a DC-AMT.

Figure 5-24 shows a schematic of the way a dual-clutch AMT works. It is essentially two transmissions in one case. One clutch controls the mainshaft and countershaft for the odd-numbered speeds, 1, 3, 5. The other clutch controls mainshaft and countershaft for even-numbered speeds, 2, 4, 6 plus reverse (not shown).

Having two main clutches means that torque to the output shaft need never drop to zero unless the driver selects neutral. A driver's shift-command from a paddle first preselects the requested speed. Then it simultaneously releases the clutch on the current gear's mainshaft and engages the clutch on the preselected gear's mainshaft. After the shift is complete and the previous mainshaft is released, that gear's synchromesh clutch is deselected.

The results are an essentially seamless transfer of power from one gear to the next. Shifts are as smooth as the best torque-converter automatic. The DC-AMT also has the potential for better economy than a conventional automatic because there is minimal slip in the mechanical clutches compared to a torque converter.

A DC-AMT is also capable of executing shifts faster than the best race driver. This is

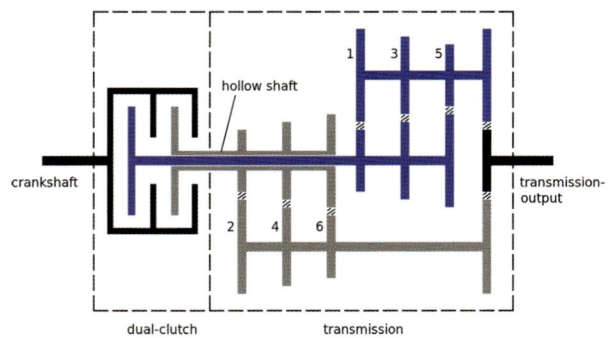

FIGURE 5-24 Schematic of a DC-AMT transmission
(Public Domain Courtesy of Xanax)

FIGURE 5-25   VW DSG, dual-clutch automated manual transmission *(Courtesy of Matti Blume)*

the reason that Formula-One race cars have converted to these transmissions. In street cars equipped with this transmission, drivers can just put the selector in drive and ignore the paddles if desired. It will behave just like a fine automatic transmission. On the other hand, if the driver wants to pretend he's Michael Schumacher, he can paddle his way to the redline at every traffic light.

Various manufacturers have given what we are calling the DC-AMT different names. The most generic of these are perhaps Getrag's DCT for *Dual Clutch Transmission* and VW's DSG for *Dual Shift Gearbox*. Other companies add a letter for their name like Porsche's PDK and Lamborgini's LDF for *Lamborgini Doppia Frizione*. BMW calls theirs M-DCT on their M-series cars. Fiat calls it ETCT for *Euro Twin Clutch Transmission*. Other companies have different names. Figure 5-25 shows a cutaway of a VW DSG transmission.

Differentials

The purpose of a differential in an automobile is to allow the outer driven wheel in a turn to rotate faster than the inner wheel, which it must do because of the difference in the lengths of the two paths. There are many possible designs of differentials, but the most common one used in an automobile is the epicyclic or planetary gear train.

Recall that our earlier discussion of the theory of planetary trains stated that they have two degrees of freedom. In a planetary transmission, this translates to applying two inputs to elements of the train to obtain a single output. In a differential, this situation is reversed. There is one input from the driveshaft and two outputs—the rotations of the two drive wheels.

But there must be two inputs for predictable motion! The second input comes from the difference in velocity of the two wheels as coupled through the pavement. When the car is traveling straight ahead, this velocity difference is zero, a valid number, and both wheels turn at the same speed. When cornering, one wheel turns faster than the other, creating a nonzero velocity difference. This causes the sun and planet gears within the differential to rotate with respect to one another to accommodate the wheels.

Invention of the differential is lost in the mists of history. The oldest reference to a planetary train being used in this (or any) manner is to the *South-Pointing Chariot* of ancient China (ca. 1000 B.C.), shown in Figure 5-26. When pulled behind a horse or camel, this device kept the little figure's arm always pointing in the same direction that it was at the start of the journey while the caravan wandered across the Gobi Desert. It served as a crude compass before the invention of the magnetic compass, also by the Chinese, in about 200 BC.

Figure 5-27 shows a bevel-gear differential as used in the final drive of an early RWD automobile. The driveshaft connects the output shaft of the transmission to the differential's input pinion. The ring gear and pinion create the rear-end reduction ratio which ranges between 2.5:1 and 4:1 depending on the vehicle. The

FIGURE 5-26 Chinese South-Pointing Chariot
(Courtesy of Andy Dingley)

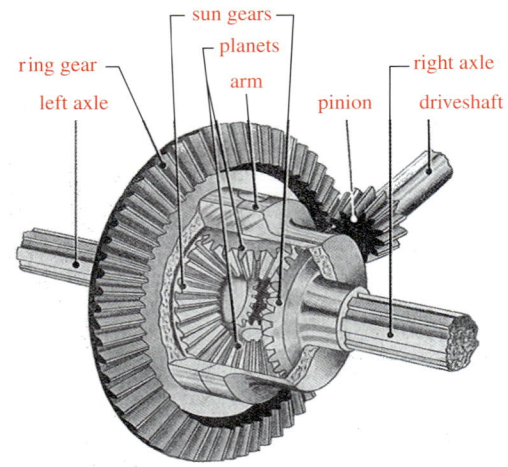

FIGURE 5-27 A bevel-gear differential
(Public Domain Courtesy of Andy Dingley)

ring gear also serves as the arm or carrier of the planetary train and provides one input to it.

The arm carries the four planets (only two are shown), which mesh with the two sun gears. One sun gear is part of the left axle and one is part of the right. The sun gears each generate one output from the planetary train. Their two outputs are coupled by the frictional contact of the tires with the ground.

An open differential, as shown in Figure 5-27, splits the torque equally between the two wheels and allows any difference in velocity to exist as dictated by vehicle motion or variation in friction between tires and road. The average speed of the two wheels always equals the ring-gear speed.

As the car rounds a curve, each wheel assumes its required speed and the speed differential causes the planets to spin about their axes to absorb that difference. When the car is moving straight ahead, the planets do not spin and serve only to key the two sun gears together,

If one wheel is on dry pavement and the other on ice, the differential will send only the amount of torque to both wheels that is needed to spin the wheel having the lower resistance. This results in the wheel on ice spinning at twice the speed of the ring gear and the wheel on solid ground remaining stopped.

To avoid the vehicle being stuck in slippery conditions, most manufacturers offer an optional **limited-slip** differential. There are many designs used to limit slip, some passive, some active. One passive type puts a clutch pack between the sun gears that adds frictional drag on them at all times. This has the disadvantage of generating heat within the differential whenever the vehicle is turning. But if stuck with one wheel on ice, the spinning wheel with no traction will transfer some torque to the wheel with traction through the clutch. Active designs use electric motors or coils to engage the clutches when slip is detected.

Viscous Couplings

A better solution than a friction-clutch pack for a limited slip differential is a viscous coupling. Developed in the 1970s, these devices use a set of interleaved, perforated plates with protruding tabs that connect two shafts within a housing. The plates do not touch one another but their tabs protrude into the space between them. The space is filled with a special, "Silly-Putty"-like, silicone-based liquid that is "shear thickening."

When the two shafts at either side of this disk-pack are turning in unison, the liquid stays cool and fluid. When there is a relative velocity between the shafts, the tabs shear the liquid, heating it and causing it to thicken in viscosity. This increases resistance to relative motion between the plates and transmits torque between the two shafts.

When this type of coupling is used to connect the sun gears in a limited slip differential, it has the advantage of low-to-zero drag when traveling straight ahead or in gentle turns, but provides torque between the sun gears when one gear is slowed or stopped and the other is spinning faster. A viscous coupling was used in the center differential of the American Motors Eagle AWD car in 1980 (see Figure 2-26).

Spiral Bevel Vs. Hypoid Differentials

Early rear ends used spiral bevel gears for the ring and pinion. A spiral-bevel gearset is a bevel gearset with helical teeth for quiet running as shown in Figure 5-28. The axes of the pinion and the ring gear of a bevel- or spiral-bevel gearset must intersect, which puts the center of the driveshaft in the same plane as the rear axle centerline. This has the side effect of raising the height of the vehicle's floor or requiring a hump in the floor.

In 1927, Walter Griswold of Packard invented the hypoid-gear rear end, and Packard was the first to use it in the same year. Instead of being based on cones as are bevel gears, a hyperboloid or **hypoid** gear is based on *hyperboloids of revolution*. The axes of rolling hy-

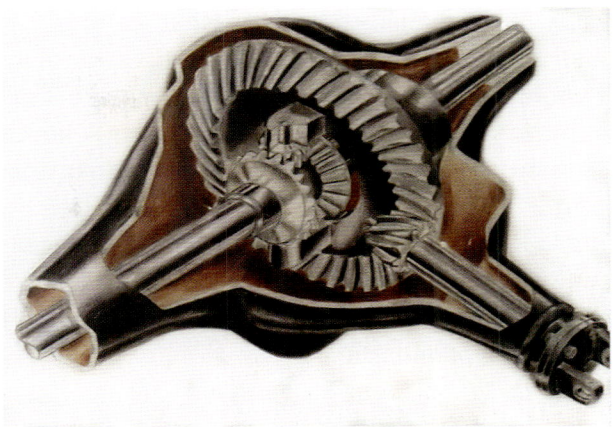

FIGURE 5-28   A spiral bevel ring and pinion
(Courtesy of DiJunge)

perboloids do not intersect, and this allows the pinion axis to be below the ring gear axis as shown in Figure 5-29.[1]

Though the hypoid gearset may have been created primarily to lower the floor of the ve-

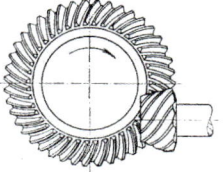

(a) A hypoid gearset

(b) A hypoid ring and pinion in a differential

FIGURE 5-29   A hypoid gearset and rear end
(Courtesy of DiJunge)

[1] For a more detailed description of hypoid gearing, see Norton, *Design of Machinery*, 5ed., McGraw-Hill, New York, 2012, pp. 500–501.

hicle and reduce the "driveshaft hump," it also had some engineering benefits. Alhough hypoid gears are more difficult to make than spiral-bevel gears, the hypoid's pinion is larger and stronger for the same gear ratio than that of the bevel gear. A hypoid pinion has 20–30% more load-carrying capacity than a spiral-gear pinion of the same size. This is why they are used in pickup-truck rear ends where driveshaft height is not an issue.

By 1936, Packard, Chrysler, Dodge, Plymouth, Studebaker, Chevrolet, LaSalle, and some Buick and Cadillac models had adopted the hypoid rear end. But wider acceptance would wait until after WWII. The major styling revolution that occurred beginning in 1946, as reported in Chapter 3, resulted in "longer, lower, wider," RWD vehicles from all makers. This required that they use the hypoid rear end to be able to lower the vehicle floor without an intrusive driveshaft hump in the back seat.

Torsen Differentials

Vernon Gleasman invented the *Torsen* (**To**rque **Sen**sing) dual-drive differential and patented it in 1958 in the U.S. This differential uses the friction generated by worm gearing to transfer torque between axles or between front and rear drives. It does not contain a planetary train. Figure 5-30 shows a Torsen center differential from an Audi Quattro.

A significant difference between *Torsen* and planetary differentials is that the slower wheel in a *Torsen* always gets more torque than the faster wheel. This makes them inherently a limited-slip differential. *Torsens* are often used as center differentials in AWD vehicles, including all Audis and many other makes. They can be designed to provide any ratio of torque-split between front and rear in a center differential.

FIGURE 5-30 Torsen differential *(Courtesy of NocturnalA6 2.7)*

Two-Speed Rear Ends

Well before William Barnes invented his planetary overdrive unit described in an earlier section, others had developed two-speed rear ends with both underdrive and overdrive ratios. Most of these were designed to be retrofitted to existing automobiles.

Walter Austin introduced a two-speed, overdrive rear axle at the 1913 Chicago Auto Show. Cadillac asked for a sample and began negotiations to license it. Before any agreement was reached, Cadillac produced one in its 1914 cars. Austin sued for patent infringement, won, and collected $10 per axle—a total of $150,000.

Later, Ruckstell made after-market, two-speed, underdrive and overdrive rear ends for the Model T Ford. The underdrives were marketed to truck owners and the overdrives to car owners. Both units were bolt-on, planetary trains. The overdrive unit gave a 0.8:1 ratio that improved mileage. A lever in the driver's compartment shifted the unit into and out of overdrive.

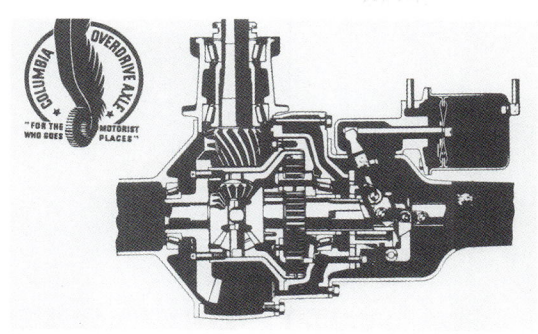

FIGURE 5-31 Columbia two-speed rear end for Ford V8s
(Courtesy of Hemmings Motor News)

In the late 1920s, Columbia built a planetary, two-speed overdrive rear end for Auburns and Cords. William Barnes, inventor of the behind-the-transmission planetary overdrive unit, was then working for Auburn—until 1931 when he was laid-off due to the Depression. By his personal account,[1] he had been impressed with the Columbia overdrive two-speed rear-end design when at Auburn, and lacking a job—or the prospects of finding one—he began to develop his overdrive transmission. This resulted in his 1934 patent, as reported earlier in this narrative.

Columbia went on to develop a two-speed overdrive rear end for Ford V8s in 1934, as shown in Figure 5-31, and this unit was eventually adopted for direct sale by Ford. They offered it as a factory option on Fords and Mercurys through 1948. It was also available on Lincoln Zephyrs until 1941.

[1] "Half-Hour History of Overdrive," *Special Interest Autos*, Mar-Apr, 1974, pp. 46-54.

Summary

Transmissions turn out to be among the most diverse and complicated subsystems of the automobile. Like engines and valve trains, they are also still evolving at a rapid rate, even well over a century after their introduction.

The latest manifestations of single-clutch and dual-clutch automated manual transmissions have only been around now for about 20–30 years, rather a small fraction of the life of the automobile. But they are based on ideas that have been around for over a century. We have also seen very old ideas (e.g., the CVT—1879) re-emerge as "new ideas" to meet current challenges. We can probably expect more "old-is-new-again" surprises in the future.

Reports in the automotive press in early 2015 detail a joint project between Ford and GM to develop a 10-speed automatic transmission. It is said to have six underdrive ratios, one direct drive ratio, three overdrive ratios, and one reverse ratio.[2] U.S. Patent 8,834,310 has already been issued to Ford for this transmission design. In September of 2014, this transmission design received a Henry Ford Technology Award. It is reputedly intended to serve all the RWD automobiles and light trucks from both automakers, perhaps beginning in 2016. A nine-speed version of the same design is reportedly slated to be used in some of their FWD models as well.[3] Hyundai has also confirmed that it is developing a 10-speed automatic, and VW has announced that they are developing a 10-speed DC-AMT.

[2] http://www.thetruthaboutcars.com/2014/12/exclusive-inside-look-fords-new-10-speed-transmission/
[3] http://www.torquenews.com/106/ford-f150-will-soon-have-10-speed-automatic-transmission

Chapter 6

Suspension and Steering

This chapter will describe the evolution of suspension systems and steering mechanisms. Early automobiles were essentially buggies to which an engine was added. A tiller mechanism was also added to steer the wheels, which only had to follow the horse before. Suspension systems can be roughly divided into two classes: *non-independent* and *independent*. Non-independent suspensions mount the wheels on both sides of the vehicle at front or rear on common axles, as did buggies. Independent suspension decouples the left wheel from the right, putting each on its own suspension system. We will see that most possible combinations of these two approaches have been tried at either end of the car at various times.

All automobiles prior to 1900 used non-independent suspensions both front and rear. One form of independent rear suspension (IRS) was patented by Rumpler in 1903 but did not come into general use until over two decades later. Christie invented one type of independent front suspension (IFS) in 1904, but this innovation did not catch on right away either. These inventors built cars with their ideas at the time, but few followed their examples until much later. Ultimately, several designs of both IFS and IRS would be invented and eventually they came to dominate automotive design. Today, virtually all passenger vehicles have IFS, and with the major exception of light trucks and their truck-based derivative SUVs, most passenger automobiles also have IRS. IFS and IRS have become the standard for good handling and riding quality. Large trucks still use non-independent suspensions both front and rear because of their higher load-carrying capacity.

All suspension systems need some type of elastic members to support the car body and isolate its passengers from bumps in the road. These usually take the form of springs. Buggies also had springs and the earliest automobiles simply carried over the buggy spring for this purpose. This approach was soon realized to be less than optimal for an automobile, and many other spring systems were introduced. All will be discussed in the course of this chapter. Tires are another consideration in a suspension system. The earliest vehicles had no tires as we now know them. They used wagon wheels made of wood and rimmed with iron for wear. The pneumatic tire was an early invention (1896) and partially solved the problem of the rough ride from solid wheels but introduced other problems that will be discussed below. Modern tires play a large role in the ride and handling equation for automobiles.

The opening photograph is of a c.1957 Talbot-Lago chassis with an early form of independent front suspension taken by the author at the Tampa Bay Automobile Museum in 2014.

CHAPTER 6 SUSPENSION AND STEERING

Springs

From an engineering standpoint, every physical part is a spring because no material is truly rigid. So when force is applied to any part made of any material, the part will deflect. Its **spring rate** or springiness is defined as the force applied divided by the deflection that results. The units are lb/in or N/m.

A well-designed automobile will have high spring rates in all of its parts except for those designated as springs. In other words, we want the body and frame to be as stiff as possible so the physical springs that support the body do essentially all the deflecting. Physical springs come in many types and all are designed to have a desired spring rate.

Leaf Springs

As the name suggests, leaf springs are made of a number of leaves, most often of hardened steel. Each leaf is made of a flat, thin piece of metal, bent to form an elliptical shape. Leaves of varying length are stacked and clamped together to form the complete spring. Two of these curved stacks can be joined at their ends to form a full-elliptic leaf spring as shown in Figure 6-1. Figure 6-2 shows a 1904 Stanley Steamer suspended on four full-elliptic springs, just as horse-drawn buggies are. The center of the top half is clamped to the frame, and the center of the bottom half is clamped to the axle.

The earliest automobiles used full-elliptic springs because that's what buggies and wagons had used. Most likely in the interest of cost reduction, they soon switched to half-elliptic springs as shown in Figure 6-3. Others used only quarter-elliptic springs as shown in Figure 6-4. Some automobiles mounted the axles on a single, transverse leaf spring instead of a longitudinal spring at each side. Fords from the 1909 Model T down to the 1948 Ford used this arrangement at both front and rear. A Model T spring arrangement is shown in Figure 6-5.

Leaf springs are relatively inexpensive compared to some other types and have an advantage in respect to doing double duty as locating links for the axles as well as springs. Their springiness is primarily in the transverse direction and they are quite stiff axially. In the ar-

FIGURE 6-2 1904 Stanley with full-elliptic leaf springs
(Photo by author at Boothbay RR Village Museum)

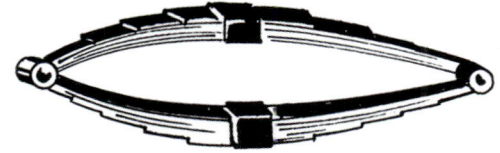

FIGURE 6-1 A full-elliptic leaf spring
(Public Domain Courtesy of Scott Foresman)

FIGURE 6-3 Semi-elliptic rear spring *(Public Domain)*

165

FIGURE 6-4 Quarter-elliptic rear spring *(Photo taken by the Author at the Tampa Bay Automobile Museum)*

rangements shown in Figures 6-2 and 6-3, they serve to locate the axles fore-and-aft and side-to-side as well as to provide compliance for vertical movement. However, their main disadvantage is friction between the leaves as they slide on one another. This can degrade ride quality when compared to other spring types.

Torsion Bars

A torsion bar is a very simple spring. It is just a round, straight, solid bar of steel that is clamped at one end and loaded in torsion (twist) at the other. Figure 6-6a depicts a torsion bar schematically in both undeflected and deflected mode. The open end always has a link fastened to it to apply the twist and this link is typical-

FIGURE 6-5 Model T transverse leaf springs *(Photo taken by the Author at the Owls Head Transportation Museum)*

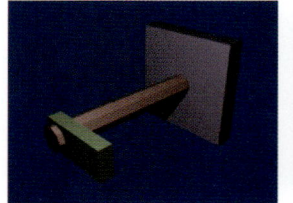 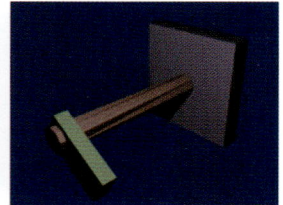

undeflected   deflected
(a) Schematic

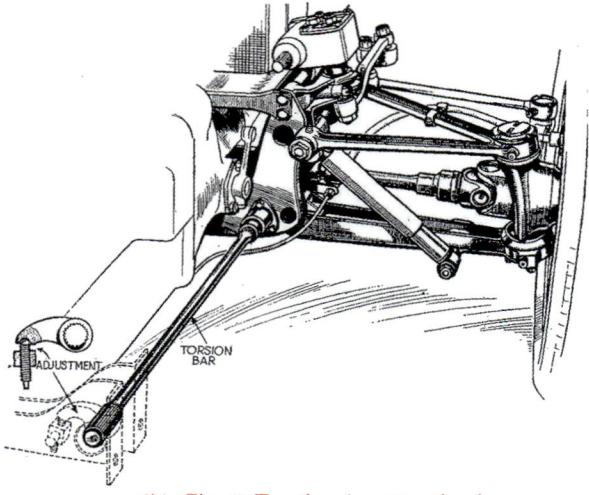

(b) Citroën Traction Avant torsion bar

FIGURE 6-6 Torsion bars *(Public Domain)*

ly part of the car suspension. But Figure 6-6a is missing a crucial bit of detail. To keep the torsion bar from also bending in the direction of the force applied to its end link, a grounded bearing needs to be placed around the torsion bar adjacent to the twist link to counter the link force. This results in a pure torque being applied to the bar. There is little friction on the torsion bar, unlike in a leaf spring.

One of the earliest cars to use a torsion bar suspension was the 1934 Citroën Traction Avant as shown in Figure 6-6b. Torsion bars were used in the 1930s Czech Tractas on which the VW Beetle was based. They were also used on all 21 million original VW Beetles. Hudson used them on its 1934–1936 models. Packard used them in the 1950s and Jaguar on its E-Type in 1961 through 1972. Packard had a unique design in 1955 called "Torsion-Level suspension" that included electric motors to automatically adjust the car's ride height and attitude both front to rear and side to side by rotating individual torsion bars at their mountings. Chrysler used torsion bars in its TorsionAire suspension from 1957 through 1989. The Oldsmobile Toronado and Cadillac Eldorado used torsion bars in the late 1960s. Many other manufacturers have used them with good success.

Torsion bars have many advantages over other types of springs for auto suspensions. They are inexpensive to make, package very compactly, and take up little space. They are also lighter than other springs. Vehicle ride height can be easily adjusted at the fixed end of the torsion bar by rotating its clamp, a distinct advantage in production as it allows adjustment of ride height for different engines and optional equipment.

Coil Springs

A coil spring is made by twisting a round bar of metal, usually steel, into a helical shape as shown in Figure 6-7. Forces are applied at either end of the spring, compressing the coils to closer spacing. Its force increases in proportion to its deflection. Despite the fact that this is called a compression spring, its coils are actually loaded in torsion as shown by the torques labeled T in Figure 6-7. The helical coil shape converts the compressive forces at either end to torsion or twist in the coils. Thus, a coil spring is actually a torsion bar wound into a helical shape.

Coil springs have an advantage over leaf springs in that they have no friction between the coils. Their disadvantage versus leaf springs is that they cannot stand much lateral force. They must be used in combination with lateral links that control wheel position in all directions. However, coil springs typically give a softer and more comfortable ride than leaf springs. They were first used in 1905 in the earliest independent front suspension (IFS). With few exceptions, modern cars use coil springs in their front suspension, and most use them in the rear also.

While some early independent front suspensions used leaf springs, virtually all converted to coil springs or torsion bars. There is still at least one modern independent suspension that uses transverse leaf springs front and rear—the Chevrolet Corvette, though its springs are made of a single composite-plastic leaf, not of steel. Others use coil springs or torsion bars.

All passenger cars made in the U.S. and most of the world abandoned leaf-spring, non-independent rear suspensions some years ago, but many trucks did not. Some non-independent, truck rear suspensions, such as those in GM and Ford light trucks and their SUV derivatives, still use leaf springs at the rear, but the Dodge Ram converted to rear coil springs in 2009. Heavy duty trucks continue to use leaf springs and axles at both ends.

Air Springs

William Humphreys patented an air spring system in 1901 but it was unlike any modern system. It used a long, sausage-like air bladder that ran along the length of each frame rail. The sau-

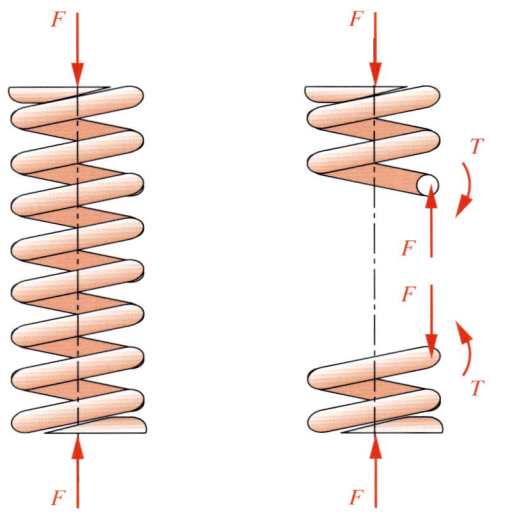

FIGURE 6-7 Coil spring forces and torques

sage was trapped in a metal trough that ran the length of the frame rails. This trough supported the ends of the transverse leaf springs of the car. No record seems to exist of any cars being built with this system. Around 1920 George Messier, in France, invented a system that used gas-filled bladders at the four wheels. Most modern systems use variations of this approach.

During WWII, the U.S. developed air suspension systems for aircraft to save weight. After the war various auto companies experimented with pneumatic and hydro-pneumatic systems. Air bladders of various types, when filled with an amount of gas, provide a spring with a nonlinear spring rate. As the bladder is compressed, the volume of the gas decreases and its pressure increases. Each additional increment of reduced volume requires a larger incremental force to further compress the gas, making it a nonlinear spring. In contrast, a metal spring has a linear spring rate, giving equal force for equal increments of deflection over its range. Plots of the force-deflection behaviors of these two springs are shown in Figure 6-8. Air springs tend to give a better ride than mechanical springs, and their spring rate and height can also be adjusted, even while in motion.

Cadillac introduced optional air suspension on its *Eldorado* in 1957 using compressed air from an engine-driven compressor. It also provided a self-leveling function. Buick followed Cadillac's design in 1958. The 1958 AMC *Rambler* station wagon offered an optional *Air-Coil Ride* with air bladders inside its coil springs.

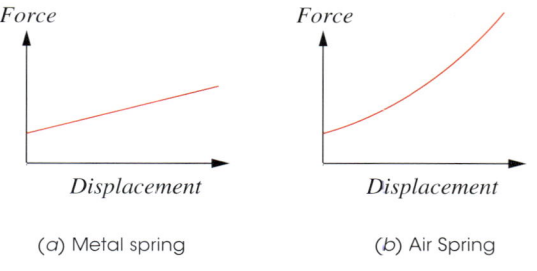

FIGURE 6-8 Rates of linear and nonlinear springs

Lincoln was the first U.S. company to standardize air suspension on the *Continental Mark VII* in 1984. Current automobiles that offer air suspension include Audi, Cadillac, Ford, Jeep, Lexus, Lincoln, Maybach, Mercedes-Benz, Porsche, Ram, Rolls-Royce, Rover, Subaru, Tesla, and Volkswagen.

In addition to the truck's conventional springs, air springs are used as after-market helper springs on trucks to increase load carrying capacity. These look like several wheelbarrow tires and wheels connected side by side. Kits are also available to convert conventional truck suspensions to air suspension. Some 18-wheeler trucks with sleeper cabs mount their cabs on air springs to isolate the driver from road shock. If you are driving beside one on the highway, you can see the air springs under the cab clearly and also see the cab bouncing softly with respect to the truck's chassis.

Hydro-Pneumatic Suspension

Citroën was the first company to develop a hydro-pneumatic system. It was based in part on an air suspension system developed by George Messier in 1922. They used it on their *Traction Avant 15H* in 1954 and later on the *DS* and many other models down to the present. It combines a gas (nitrogen) with hydraulic fluid to provide air suspension along with self-leveling. The gas is stored under pressure in accumulators (spherical metal containers) and the hydraulic fluid (oil) is pumped into and out of the accumulators and cylinders that support the suspension.

The accumulators each have two chambers separated by a rubber membrane with nitrogen on one side and oil on the other. The gas provides the spring in the system. The oil is used for damping by being forced through small orifices between the accumulators and the suspension. The oil serves also to raise and lower the car body versus the wheels by adding or removing it from the cylinders.

CHAPTER 6   SUSPENSION AND STEERING

This system allows the ride height of the vehicle to be kept constant regardless of load. As more luggage is added or passengers pile into the car, sensors detect the suspension's deflection and pump oil into the cylinders to restore the proper ride height. The driver can also control ride height manually either to increase road clearance to avoid obstacles, or to lower road clearance on smooth roads to reduce wind resistance. An engine-driven pump provides pressure to move the oil around.

Figure 6-9 shows the system under-hood of the Citroën-Maserati SM made from 1969–75. The green spheres are the accumulators, one per wheel plus a master. The Maserati V6 engine is set well back in the chassis and drives the front wheels. The shaft coming out of the engine drives the hydraulic pumps for the suspension. When parked, the hydraulic system "bleeds down" resulting in the body nearly touching the ground. When the engine is started, the car pumps itself up to ride height. This system can also jack the car up to change a flat tire.

A hydro-pneumatic suspension system is quite complex and can be troublesome. Nevertheless, other manufacturers licensed it from Citroën or developed their own variants. Rolls-Royce licensed the Citroën system for the Silver Shadow in 1965. Borgward developed their own system in 1959. Mercedes first used a variant of the Borgward system in 1962 but in 1964 developed their own system designed to avoid the Citroën patents and used it on the 600 and on later models. Citroën still uses their own design, updated with modern electronic controls and other improvements.

Tires

The tire's job is primarily to provide support and traction between car and road. But it is also a spring and so contributes to the overall dynamic behavior of the vehicle. Its carcass is made of a composite of rubber and various cords of polymer and/or steel and so is flexible. It is filled with pressurized air and is an air spring. Its spring characteristics combine with those of the actual springs to determine the effective spring characteristics of the vehicle. The tire is an important factor in ride and handling.

It is interesting to consider the evolution of the tire from its earliest examples to its present design. Early automobiles used large diameter wheels, anywhere from 24- to 36-inches at the rim. The wheels were also narrow width. Tires were correspondingly small in cross section compared to a modern tire. As cars evolved into the 1930s and suspension systems became more sophisticated, wheels were incrementally reduced in diameter and increased in width to improve the ride. A smaller diameter, wider tire was found to give better cushioning over bumps.

A 1930s era car had wheel diameters between 17 and 20 inches with widths of about 3 to 5 inches. The trend to smaller and wider rims and tires continued through the 1940s and 1950s, with 13-inch diameter by 5- or 6-inch wheels becoming common. Then the trend reversed with the result that present automobiles seldom offer rims smaller than 16-inch and many have 17-, 18-, 19-, and 20-inch rims with very wide, low-profile tires. So we seem to have come full circle with respect to tire diameters, if not width. These large, low-profile tires tend to

FIGURE 6-9 Citroen SM engine compartment *(Photo taken by the author at the Tampa Bay Automobile Museum)*

169

improve cornering and handling but can do so at the expense of ride quality in many cases, because the low-profile tire tends to have a higher spring rate than a taller one.

In any event the tire, as an air spring, has a nonlinear spring rate as shown in Figure 6-8 and its characteristics have to be taken into account when designing the suspension system. It is a spring in series with whatever type of physical springs are fitted to the car. Automobile manufacturers work closely with tire manufacturers during development to define the right tire for the suspension system. Often, the tire manufacturer designs a new tire specifically for the vehicle. This is particularly true of tires for high-performance sports and luxury cars. It is not unusual to have only one choice of tire make and size available as a replacement on a particular vehicle.

Sprung Versus Unsprung Weight

Sprung weight is made up of all parts of the vehicle that have the springs between them and the road. Unsprung weight is all the other parts that do not have a physical spring between them and the road. The springiness of the tires does not count in this context. So the tires, wheels, brakes, suspension parts, and steering gear all contribute to the unsprung weight. It is desirable to minimize the amount of unsprung weight to improve ride and handling.

When a wheel hits a bump or a pothole, it results in an impact force that can be quite large. Without going into the methods for calculating impact forces, which are quite complicated, we can say that in a crude sense there are three factors that affect the magnitude of an impact force. One is the speed at impact, another is the spring rate of what is impacted, and the third is the mass of the impacting object. In this case, the spring rate is defined by the tire. The velocity is whatever it is when the bump is struck, and the unsprung mass is what impacts the pothole. The sprung mass is isolated from the impact by the car springs. So for a given tire and car speed, the larger the unsprung mass, the worse the impact force. This is one reason why the unsprung mass should be minimized.

As will be discussed below in more detail, a non-independent suspension system tends to have larger unsprung mass than an independent suspension. This is because the heavy axle of a non-independent suspension is unsprung, but an independent suspension has no axle. With an independent suspension, heavy elements such as the differential and brakes can optionally be placed on the frame, which is sprung, rather than be part of the unsprung weight. The best-handling automobiles use all these tricks to keep the sprung/unsprung weight ratio as high as possible.

Non-Independent Suspension

All the earliest suspension systems were non-independent. Despite some pioneering attempts at independent suspensions as early as 1905, they did not become common in the U.S. until the late 1930s. They began to be adopted in Europe in the 1920s. During the intervening years, cars had both front and rear axles. While these were satisfactory when engine power and road speeds were low, they soon proved troublesome.

Non-Independent Front Suspension

A typical non-independent front suspension is shown in Figure 6-10 on a 1910 Otto. Two longitudinal, semi-elliptic leaf springs, whose ends support the frame, are clamped to a beam axle. Each end of the axle provides pivots for a kingpin about which the steering wheels

FIGURE 6-10   Typical non-independent front suspension
(Photo by author at AACA Museum, Hershey, PA)

pivot. Cars of this era did not have front brakes. This is an early example of the genre in that the kingpin axes appear to be nearly vertical. If you project those axes down to the ground (dotted lines), you will see that they are a significant distance inside the tire's contact patch. Since the wheel assembly pivots about the kingpin axis, the tire's contact patch must scrub through an arc when turned with the car stationary. Obstacles, when hit while moving, will steer the wheel violently outward with that large an offset. This is one form of bump steer.

The horizontal distance from the point where the kingpin axis (also called the steering axis) intersects the ground to the center of the tire's contact patch is called the **pivot radius**. Later designs angled the steering axis so that its projection to the ground landed closer to the tire's contact patch to minimize the pivot radius. The pivot radius is not made zero because that would eliminate "road feel" at the steering wheel. Instead, it is kept about 1 to 1.5 inches inside the center of the contact patch. This radius is referred to as a positive pivot radius.

If the pivot radius is positive (contact patch center outside the steering axis), then when the wheel hits a bump, it will steer the wheel to the outside. If the pivot radius is negative (contact patch center inside the steering axis) then a bump will steer the wheel inward. When the pivot axis is zero, it eliminates this bump-steer but also eliminates all steering feel. But, if the pivot radius is too large in either direction, the resulting bump-steer can cause loss of steering control.

Caster, Camber, and Toe

There are three angles that affect the dynamic behavior of a wheel: **caster**, **camber**, and **toe**. Collectively, they define the vehicle's wheel alignment. These are illustrated in Figure 6-11. Caster defines the longitudinal inclination of the steering axis and is positive when the steering axis meets the ground ahead of the center of the contact patch. A typical value for a passenger car is zero to 5°. Positive caster makes the wheel want to steer out of the turn, helps to center the steering when the wheel is released, and keeps the vehicle in a straight line in the absence of steering inputs. Interestingly, the effect and value of caster was recognized as early as 1896 when Arthur Krebs patented the concept in England.

Camber defines the lateral inclination of the wheel with respect to vertical. If the top of the wheel tips outward, it has positive camber; if inward, negative camber. Camber is typically set to zero or slightly positive when the vehicle is fully laden. Positive camber makes the wheel steer away from the center of the car. With both wheels equally positively cambered, each cancels the other's attempt to steer while traveling in a straight line. With independent suspension, the camber changes as the wheels move through their suspension travel. With a beam axle, camber angle is fixed by the kingpins. When a car turns, body roll puts more weight on the outside wheel. If that wheel maintains positive camber in the turn, it will try to straighten the car out and generate understeer and fight the turn. If

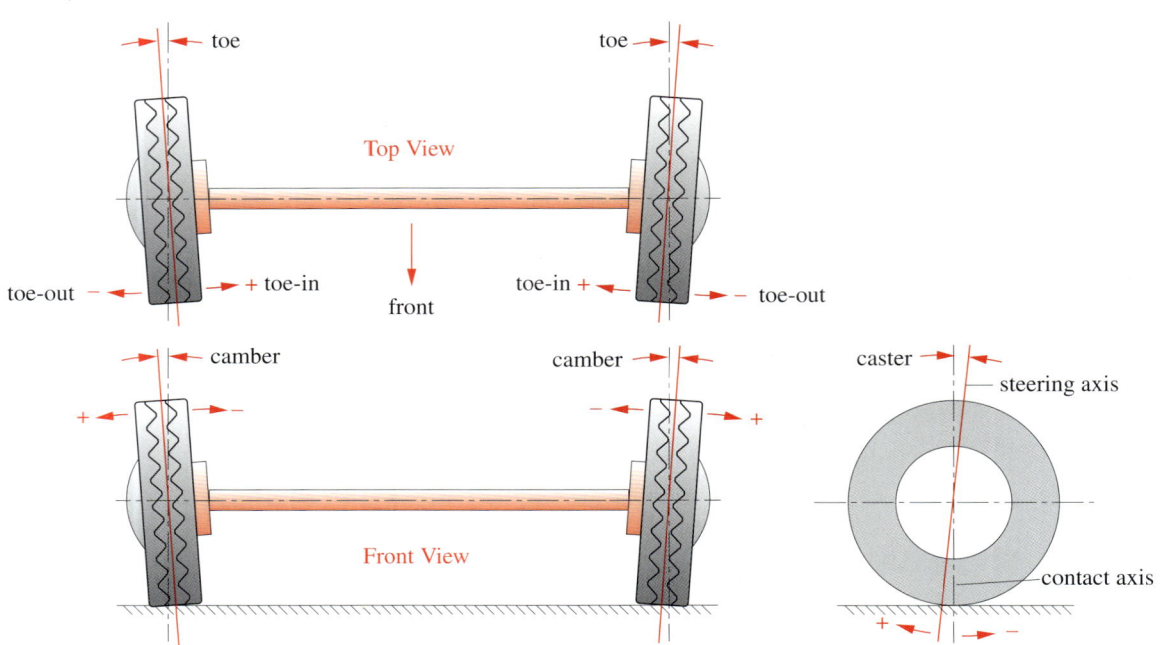

FIGURE 6-11   Caster, Camber, and Toe-in

the outside wheel has negative camber, it will promote oversteer and want to turn sharper. To minimize understeer, good independent suspension geometry will induce negative camber in the outside wheel as the car rolls in a turn and the suspension is compressed on that side. Suspension designers favor a setup that creates a bit of understeer rather than oversteer. Oversteer is considered potentially dangerous with an unskilled driver, as it can promote skidding.

Toe is usually set slightly positive, called toe-in. This points the fronts of the wheels toward one another in a "pigeon-toed" attitude. Along with caster, this helps prevent wandering when traveling in a straight line as each wheel is trying to steer the car toward the center. Toe-in also has the effect of countering the opposite turning effect of positive camber and is sometimes used for this purpose. The toe angle is only at the correct setting when the car is traveling in a straight line because the steering geometry is designed to turn the inside wheel through a greater angle than the outside wheel due to the different radii of their paths. This is discussed further in this chapter in the section on steering.

Some manufacturers of modern cars no longer provide adjustments for caster and camber, though most older cars did. So a wheel alignment now sometimes involves only adjusting the toe. All of these angles have the potential to affect tire wear as they determine the degree of scrubbing that the tires experience. Thus out-of-alignment wheels can result in rapid tire wear as well as poor handling behavior.

Shimmy, Wobble, and Bounce

When road speeds were low in the early years, the beam axle seemed to work well. After all, it had done so on horse-drawn conveyances for centuries. But as speeds increased, this suspension system began to show problems. A great deal of engineering effort was expended on these problems in the 1920s and 1930s. Dozens of research papers were published in engineering journals in the U.S., England, and France on the subject, and it generated some controversy.

CHAPTER 6 SUSPENSION AND STEERING

There was much discussion even over the terminology used to define the phenomena. The terms **shimmy**, **wobble**, and **bounce** were used quite generally, but there was some professional disagreement as to their exact definitions. It will not be of interest for us to engage in, or even describe, that controversy in any detail, but it is interesting to note that it existed. Various solutions were proposed to fix these problems and some helped, but it was only fully solved by eventually abandoning non-independent front suspension entirely in favor of independently suspended front wheels.

Taken together, the three terms, *shimmy*, *wobble*, and *bounce* describe a set of problems that plagued all non-independently suspended cars of the time. When driving on even a fairly smooth road above about 15–20 mph, if one front wheel hit a bump, it could start the front wheels oscillating left and right and make the steering wheel hard to hold onto. A rough road made it worse. In its most severe state, drivers variously reported the front end of the car violently twisting left and right, the wheels on either side alternately leaving the pavement for as much as a 9-foot horizontal distance, and the car threatening to swerve into the ditch. It did not help that car frames of the time were quite flexible.

F. W. Lanchester did extensive experimental and mathematical studies of the problem and published several papers on it. Others also published their analyses. These experts agreed that the basic culprit that started these oscillations was the gyroscopic effect. This is a physical phenomenon well known for centuries before. An explanation can be found in the Sidebar: **Gyroscopic Forces**. A simple description of its role in shimmy will be attempted here.

When one wheel at the end of an axle (say the right), hits a bump, this causes it to lift up some distance. Assuming the opposite (left) wheel remains on level ground, this upward motion tilts the axle counterclockwise from the driver's perspective. This counterclockwise rotation of the spinning wheel's axle generates a **gyroscopic couple** about the wheel's vertical axis that turns the wheel left. Because both wheels are on the same axle and their steering arms are coupled with the tie rod, this rotation of the right wheel to the left is transferred to the left wheel and the car lurches to the left.

In some circumstances, which were never fully defined to the collective satisfaction of the various experts of the time, these gyroscopic disturbances at certain speeds could set up a resonance between the tires acting as springs and the actual springs. This resonance then launched the car into the types of gyrations described above. The flexible chassis probably also joined the party as its flexibility made it an effective spring. Some witnesses reported seeing inch-wide gaps open between the hood and cowl when the car was shaking, indicating that the frame was far from stiff.

Most experts seemed to agree that these phenomena could be reduced by careful attention to proper adjustment of the steering and suspension links, eliminating any slop in pivots, and balancing the wheels. Some suggested adding damping to the kingpins. Lanchester actually designed a damping system that he thought was the cure. Others disagreed.[1] A few preached that this beam-axle system was doomed, and the only cure was to scrap it in favor of independent suspension. It turned out that they were correct. The ultimate cure for these problems was to break the connection between the wheels at either side so that a disturbance to one wheel would not transmit directly to the other. By the mid-1930s, most manufacturers had switched to independent front suspension. By 1941, the only one still using a beam axle in the front of a U.S. passenger car was the perennial holdout against new ideas, Ford. They did not switch

---

[1] Reading the technical papers of this era is always delightful and sometimes amusing, especially the long discussions among the conference attendees that follow each paper. Some of these express very polite but strong disagreement with the speaker's points.

Gyroscopic Forces

When something, such as a wheel, is spinning on an axis, any attempt to rotate the axis of spin about a transverse axis will generate a pair of forces (called a couple) that will rotate the axis in a different direction than it was moved initially.

You can do a simple experiment to experience this phenomenon. Remove the front wheel from a bicycle and hold the ends of its axle in your two hands. Extend your arms in front of you with the wheel between your arms.

As a first step, tip the axle and wheel by moving your right hand up and left hand down simultaneously. You will feel very little resistance to this motion.

As a second step, holding the axle horizontal, have an assistant spin the wheel as fast as possible. Now repeat the hand motion—left hand down and right hand up to tilt the wheel. Your right hand will be pulled away from your body and the left pushed back toward you. You are feeling the gyroscopic couple that was generated by your disturbance of the axle's equilibrium. Rotating the axis about an axis perpendicular to the spin axis causes a gyro to rotate about the axis that is 90° to the one you moved. It is as if you had steered the wheel to the left. This is called gyroscopic precession.

If you ride a bicycle or motorcycle, you have experienced the gyroscopic effect. It is more noticable on a motorcycle because of its higher speed. On either cycle, when moving very slowly, you steer by turning the handlebars in the direction you want to go. But when moving at speed, you steer differently. To go left, you either lean left or push down gently on the left handlebar and vice versa.

If you try to turn the handlebar left at speed, the bike will try to tip to the right and turn right, because the wheels are now spinning fast enough to generate a gyroscopic couple. Pushing down on the left handlebar or lifting up on the right generates a gyroscopic couple about the vertical axis that turns the front wheel to the left. All you need to do is to lean left to rotate the spinning wheel axis counterclockwise and the wheel turns left about its vertical axis, just like magic. "Look, Ma—no hands!"

until 1949. It would take longer for independent rear suspensions to take over in the 1980s. But these tales will be told later in this chapter.

Non-Independent Rear Suspension

Dead Axles

Like non-independent front suspension, non-independent rear suspension also has an axle. Some very early automobiles had a so-called "**dead axle**" at the rear. That term means the axle only supports the vehicle and does not contain any shafts. A front beam axle is a dead axle. Cars with dead rear axles typically used chain drives on both sides that turned sprockets attached to each wheel. The wheels turned on stub shafts (also called axle shafts) in bearings at the ends of the dead axle. Such an arrangement is shown in Figure 6-12 on a 1903 Mercedes Simplex.

Live Axles

With a conventional rear-drive arrangement using a driveshaft and differential in the rear axle, the axle becomes a "**live axle.**" It still has a nonrotating case (the axle) that serves to

CHAPTER 6 SUSPENSION AND STEERING

FIGURE 6-12 Chain drive on a dead axle 1903 Simplex
(Photo by author at Owls Head Tranportation Museum)

mount the springs and wheel bearings, but there are also rotating shafts within the axle case that carry torque from the differential to the wheels. In proper engineering terminology, **axle** refers to something that supports a rotating wheel but does not itself rotate, as in the case of a front axle. If something rotates and also transmits torque, it is properly called a **shaft**. A shaft in a machine may also do double duty by both transmitting torque and supporting side loads, in which case it lives a tougher life and is more highly stressed than if it had only to transmit torque.

Full-Floating Versus Semi-Floating Axles

The shafts in an axle (often called axle shafts in automotive parlance) can be arranged to be either semi-floating or full-floating. These terms refer to the way the shaft is supported in the axle

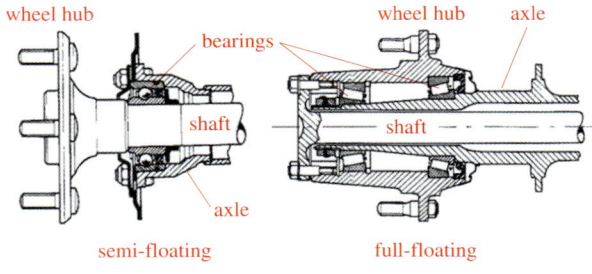

FIGURE 6-13 Semi-floating and full-floating axles

as shown in Figure 6-13. Note that there are two axle shafts, one right and one left. Only one is shown.

SEMI-FLOATING A semi-floating shaft has a single bearing at each end, one in the differential case and one at the axle end. The shaft extends through the outer bearing and the wheel is attached to it. In this case the shaft supports the weight of the vehicle at that corner in addition to transmitting torque to the wheel. This combined loading creates larger stress in the shaft than would the torque alone. This arrangement is used on passenger cars.

FULL-FLOATING A full-floating shaft has a single bearing at the differential, but at the axle ends it is splined or otherwise rotationally attached to a rotating wheel hub. The wheel hub sits on two bearings and rotates on the axle housing. The wheel mounts on the wheel hub rather than on the shaft directly, so the vehicle weight is supported by the axle housing instead of the shaft. The shaft now only has to transmit torque. This arrangement is used on heavy trucks to avoid overloading the shaft.

De Dion Axles

De Dion and Bouton patented their axle design in 1896. They recognized then that unsprung weight was bad. To reduce it, their design mounted the differential on the frame and used axle shafts with two U-joints each to drive the rear wheels. The wheels were supported and located by a tubular, dead axle that is bent in the top view to get across the car behind the differential. The body and frame were suspended on leaf springs on each side. Their 1896 patent drawing is reproduced in Figure 6-14.

De Dion axles have been used in a number of cars including several models of Alfa Romeo, Mazda Cosmo, Lancia Aurelia and Flaminia, Volvo 300-series, Rover P6, Dodge Car-

AUTOMOTIVE MILESTONES

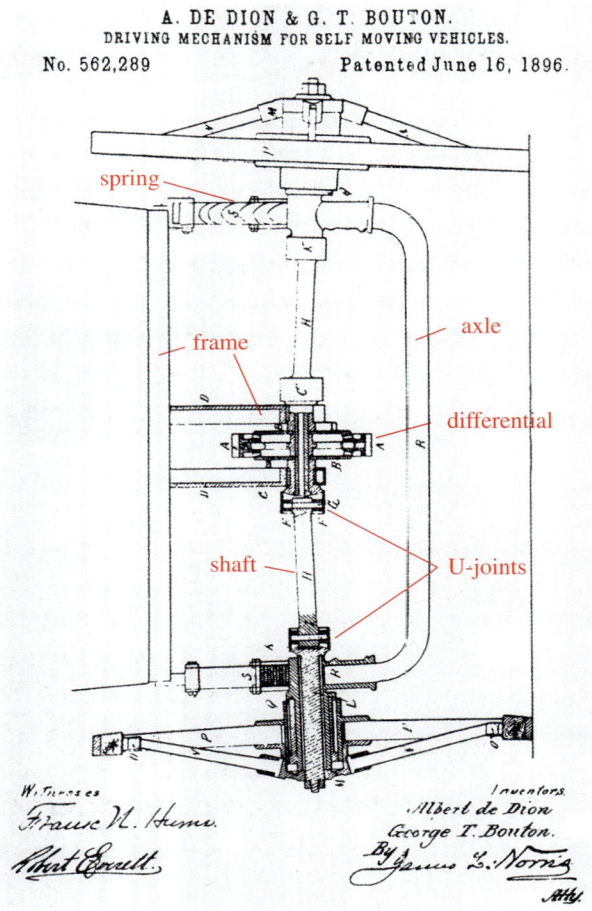

FIGURE 6-14 De Dion 1896 axle-patent drawing

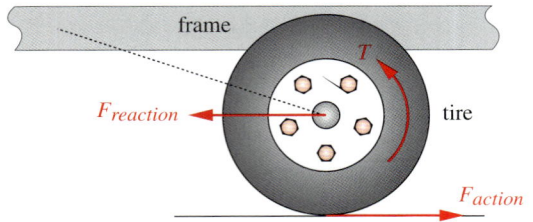

FIGURE 6-15 Forces on the wheel under acceleration

avan AWD models, Aston Martins from 1967 to 1989, and some models from Ferrari, Maserati, and Mercedes. Current models using it include the Smart ForTwo, Mitsubishi *i kei* car, Caterham 7, and the AWD version of the Honda Fit. Many of these use coil springs rather than the leaf springs of the original design.

Axle Control

Three designs are used with non-independent, live-axle rear suspensions: **Hotchkiss, torque-tube**, and **multi-link**. Each of these serves to control axle location and movement in different ways. Hotchkiss is used with leaf springs and the other two with coil springs, though some torque-tube suspensions have also used leaf springs. There are several requirements for axle control. It needs to keep the axle in the proper relationship to the frame in the horizontal plane while allowing the necessary suspension movement in the vertical plane. The suspension system also has to transmit wheel loads to the frame. Cornering applies side loads to the wheels. Acceleration and braking apply longitudinal loads.

Figure 6-15 shows a simple diagram of the acceleration forces from the drive torque. When drive torque is applied to the axle shaft to turn a wheel, it creates a pair of forces, called a couple, that are equal in magnitude, opposite in direction, and separated by some distance. One force can be thought of as the action and the other as the reaction of the wheel pushing against the ground. The drive torque is shown as T and the couple of that torque is the pair of forces F_{action} and $F_{reaction}$, one pushing on the ground and the other pushing on the axle. During deceleration, the torque and forces are reversed. Something is needed to take the force on the axle $F_{reaction}$ back to the frame in some fashion. This could be a connection along the dotted line, for example.

Leaf springs are capable of resisting loads along their length and are used for that purpose in Hotchkiss suspension systems. Coil springs are not capable of handling transverse loads, so other means must be provided if they are used. In some cases, coil-spring, non-independent

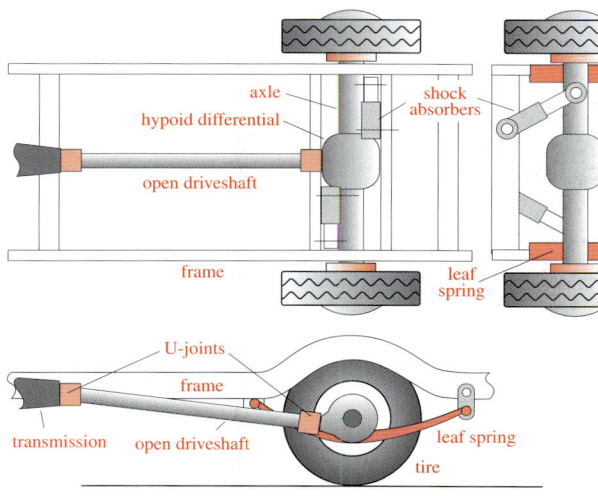

FIGURE 6-16 Hotchkiss driveline

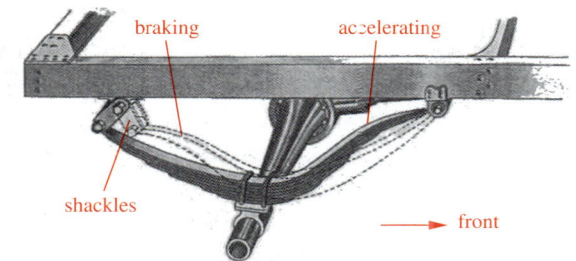

FIGURE 6-17 Leaf-spring windup in accelerating and braking
(Public Domain Courtesy of Andy Dingley)

rear suspensions use a torque tube and others use an arrangement of multi-links to control axle motion and transmit forces to the frame.

Hotchkiss Suspension

Figure 6-16 shows a Hotchkiss rear suspension. This suspension uses longitudinal leaf springs to locate and support the rear axle. They pivot on the frame at the front and attach through shackles at the rear to allow spring-length change when it deflects. The leaf springs also carry the torque reaction force from axle to frame. The shock absorbers are angled inward to provide some lateral force resistance and placed one in front and one in back of the axle to reduce axle tramp. The driveshaft is open and has Cardan (Hooke) universal joints at each end to accommodate suspension travel. Two are needed to cancel the velocity error in Cardan joints discussed in Chapter 2. The driveshaft is splined to the transmission shaft to allow its length to change as the springs deflect. Hotchkiss was the most commonly used suspension on RWD U.S. cars with leaf springs from the 1920s through the 1970s when coil springs and independent rear suspensions began to take over. Hotchkiss suspension is still used on light trucks, and their SUV derivatives that use leaf springs.

TRACTION BARS: Leaf springs are quite stiff in a longitudinal direction and will pass the torque reaction forces to the frame. But, under hard acceleration or severe braking, they can "wind up" and form an S-shape along their length as shown in Figure 6-17. This can cause "axle tramp" in which the wheels vibrate vertically under hard acceleration and cause the tires to lose traction. Severe braking can wind the springs up in the other direction and cause "wheel hop" that reduces braking effectiveness. Drag racers often add "**traction bars**," which are links pivoted to the frame in front of the axle and to the axle near the spring attachment. They look a bit like the lower links of the multi-link suspension in Figure 6-20 and serve as a stiff force-transmission "helper" for the front half of the spring without compromising suspension travel. The leaf springs are left in place.

Torque-Tube Suspension

The term torque-tube as it is being used in this context is actually something of a misnomer. We earlier defined torque correctly as a twisting force about the long axis of a shaft, like a crankshaft or a driveshaft. In the torque-tube context, torque is used to refer to the force needed to constrain the center of the wheel from moving longitudinally with respect to the chassis when the actual torque is applied to the wheel to either accelerate or decelerate the car, shown as $F_{reaction}$ in Figure 6-15.

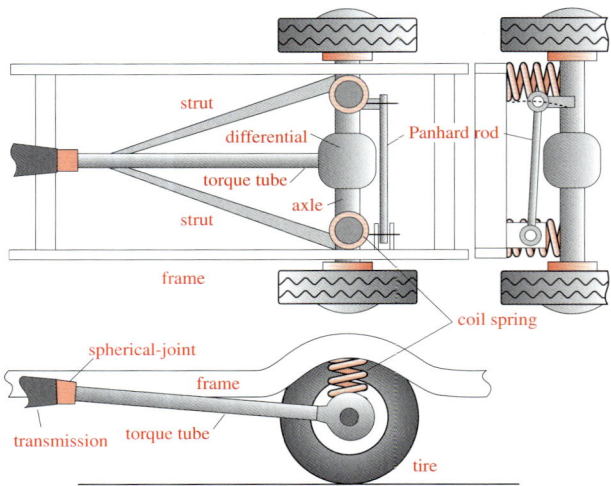

FIGURE 6-18 Torque-tube driveline with coil springs

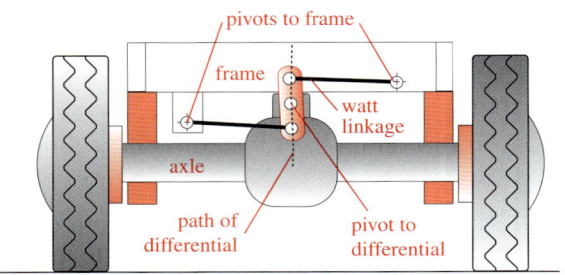

FIGURE 6-19 Watt Linkage

Since we already have a driveshaft connecting the differential on the axle to the transmission on the frame, why not use this route to take the force from axle to frame? That is what is done for a torque-tube drive line. The name comes from the fact that the so-called torque tube takes the reaction force of the torque's couple back to the frame, not because the torque tube is transmitting torque as its name implies. It is not. It simply transfers a force.

Figure 6-18 shows a sketch of a torque-tube drive line. The torque tube is rigidly connected to the differential housing and surrounds the driveshaft. At its front end, it connects to the transmission tailshaft housing with a spherical joint that allows the axle and torque tube to follow the spring deflections. Two struts rigidly connect outer points on the axle housing to the torque tube to prevent the axle from rotating about the differential in the top view. The driveshaft lives inside the torque tube and connects to the transmission shaft with a U-joint, preferably of the constant velocity type. Its other end connects to the pinion in the differential. All the reaction force from acceleration and deceleration is transmitted to the chassis through the torque tube and its struts and thence to the chassis through the motor mounts.

Coil springs do not provide resistance to side forces in cornering. So, a Panhard rod (first used by Panhard-Levassor) can be added to control lateral axle motion. This is a link, pivoted to the frame at one end and to the axle at the other as shown in Figure 6-18. The axle end of the Panhard rod travels in an arc over its vertical motion and that introduces a small lateral motion to the axle. A superior arrangement is to use the Watt straight line linkage from Figure 1-2 to force the axle to move up and down in a straight line and also provide lateral resistance to cornering forces. This arrangement is shown in Figure 6-19.

Some torque-tube suspensions in the 1920s and 1930s used torque tubes with longitudinal leaf springs such as in the Hotchkiss. These did not need a Panhard rod or Watt linkage as the leaf springs were stiff enough laterally, though a few also added a Panhard rod. Ford used a torque tube on all its cars from 1909 through 1948. But Ford used transverse leaf springs at front and rear. These gave reasonable lateral stiffness between body and frame, and so did not need a Panhard rod.

Multi-Link Suspension

Many live axles that use coil springs use the torque tube of Figure 6-18 with a Panhard rod or Watt linkage arrangement as shown in Figure 6-19. Buick has had rear coil springs with that arrangement since the 1930s. However, some

CHAPTER 6   SUSPENSION AND STEERING

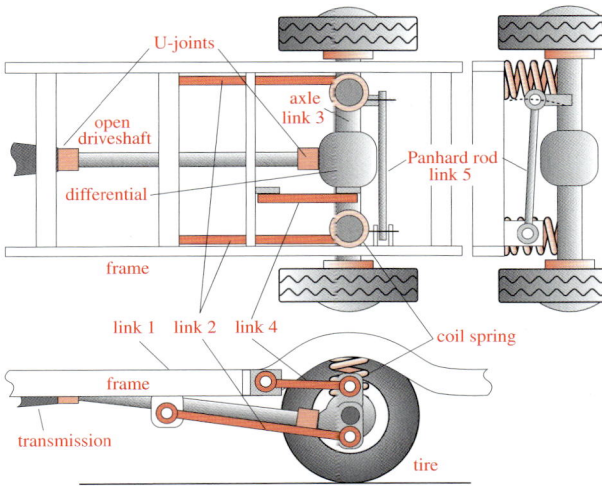

FIGURE 6-20   Multi-ink suspension

cars use a Hotchkiss-like arrangement with an open driveshaft, but with coil springs. These need some sort of linkage to control the fore-aft and side-to-side location of the axle. Several arrangements of links have been used for this situation. We will describe one of these.

One arrangement for an open-driveshaft set-up with coil springs is the five-link arrangement shown in Figure 6-20. Two lower links are pivoted to the frame some distance in front of the axle with their other ends pivoted to the axle housing. One shorter, upper link is also pivoted to the frame and to the top of the differential housing at its other end. Together these links control the vertical axle motion to a curve that approximates a straight, vertical line. They also transmit torque reaction forces to the frame. A Panhard rod is needed to control lateral motion. The fifth link is the frame. A Watt linkage can replace the Panhard rod. Other link arrangements are also used. Some manufacturers use two upper trailing links instead of the one shown in Figure 6-20.

Independent Suspension

As already related, the earliest auto pioneers understood the issues of suspension design and some of them knew that non-independent suspension was not the way to go. But pioneers are often well ahead of their peers and it took a couple of decades for others to get onto the independent-suspension bandwagon. Sizaire-Naudin in France and Christie in the U.S. produced the first automobiles with independent front suspension in 1905. Morgan in England produced the second example with a modified Christie design in 1909 and, amazingly, is still producing cars with the same type of suspension in 2014. Perhaps that says they got it right in the first place.

Nobody else waded into the independent suspension "pool" until the 1920s when several European manufacturers began to experiment with both independent front suspension (IFS) and independent rear suspension (IRS). The first production IRS example appeared in 1921 from Rumpler in Austria, but he had patented his design in 1903. Alvis in England produced a car with both IFS and IRS in 1925. Lancia (Italy) produced its Lambda model with IFS and unit-body construction, a car very much ahead of its time, in 1922. By 1929, there were at least 17 makes that had either IFS or IRS or both. Six had IFS only, two had IRS only, and nine had both. Most significantly, all of these companies were European. The U.S. did not jump into the independent suspension pool until 1934. Chevrolet adopted Andre Dubonnet's 1933 design of IFS in 1934. The same year, Oldsmobile, Buick, and Cadillac introduced an unequal-arm IFS that became one of the two dominant designs today.

Wheel Control

The basic purpose of a suspension system is control of caster, camber and toe while the wheels move through their motions. Discussion of non-independent suspensions in the preceding sections have pointed out some of their deficiencies. The early inventors of various designs

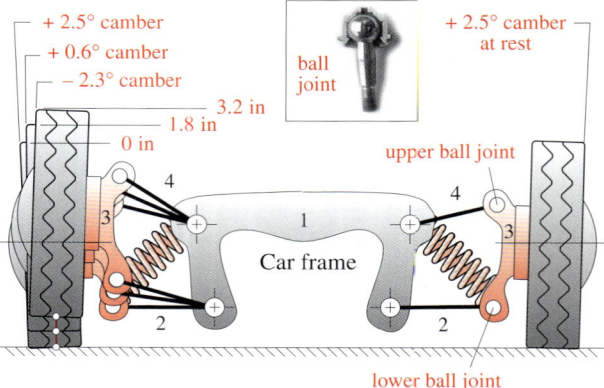

FIGURE 6-21 Fourbar-linkage independent suspension

of independent suspensions appreciated these issues and each made an attempt to improve motion control of the wheels and reduce unsprung weight. Some succeeded better than others and, as with most technological development, the better solutions eventually came to dominate.

Before discussing some of these early designs in detail, it will be useful to first address the theory of linkages in a small way. Most independent suspension designs are variations on one very basic type of mechanism known as the fourbar linkage. This collection of four links pinned together is the basic building block of all machinery. At first glance it appears to have only three links, not four, because the fourth link is the structure, in this case the car frame to which the three moving links are attached.

Figure 6-21 shows a schematic drawing of a generic, fourbar-linkage, independent suspension. The springs are not positioned accurately, but the linkage geometry is accurate. The links are numbered with 1 as the ground (frame), 2 and 4 are the arms or rockers, and link 3 is called the coupler, which is the only link not pivoted directly to ground. Rather, it has a ball joint between it and links 2 and 4 to allow the wheel to turn. The wheel/tire assembly is carried on the coupler—the tire is effectively part of the coupler. It is a property of the fourbar linkage that all points on the coupler describe very high-order curves that can take shapes ranging from circles to teardrops to figure eights to approximate straight lines and everything in between. The suspension designer seeks a linkage geometry that will provide approximate straight line motion to the contact patch of the tire as it moves up and down over bumps. Another desire is that, throughout its motion, the angle of the coupler should change only slightly and in the right direction as this defines the camber angle.

In Figure 6-21, the left wheel is shown going over a bump, and the center of the contact patch of the tire is seen to traverse an approximately vertical straight line, as is desired (red line with white circles). The error in track (the distance between the tire centers) with this linkage is only about 0.3 in per side over the suspension travel shown. The camber is set to an initial value of 2.5° positive at rest. This transitions through zero to about 2.3° negative over the suspension travel shown. Recall from the earlier discussion of the effect of camber on handling that it is desired to transition positive camber to negative on the outside wheel as the car rolls in the turn in order to create a small amount of oversteer at that wheel.

The suspension design shown schematically in Figure 6-21 is variously called **unequal arm** because of the different lengths of links 2 and 4 or **A-arm** suspension. (From a top view, both links 2 and 4 are shaped like the letter A, wide at the frame and narrow at the wheel.) It is also called a **double-wishbone suspension** because the A-shaped arms also resemble a fowl's wishbone. This arrangement is used in both front and rear independent suspensions and has become one of the two most common arrangements in modern cars. However, it was not the first independent suspension arrangement designed. Nevertheless, all its predecessors have a great deal in common with it.

CHAPTER 6   SUSPENSION AND STEERING

FIGURE 6-22   1908 Sizaire Naudin sliding pillar suspension
(Public Domain Courtesy of Buch-t)

Independent Front Suspension (IFS)

Sliding Pillar

From 1905 to 1914, Sizaire-Naudin in France produced automobiles with sliding pillar suspension using a leaf spring; this invention is credited to Decauville of France in 1898. Each wheel is attached to a cylindrical rod (the pillar) which slides in a guide on the frame or axle. Springs of some sort support the axle/frame on the wheels. An example of a 1908 Sizaire-Naudin with Decauville sliding pillars is shown in Figure 6-22. In 1904, J. Walter Christie invented a different form of sliding pillar suspension in the U.S. using coil springs surrounding the pillars. He produced a FWD car using his suspension in 1905. This arrangement is shown on a Morgan in Figure 6-23.

Sliding pillar designs hold camber and caster constant over all positions, by making the wheels slide in straight lines on cylindrical guides in the ends of a sprung dead axle or on the frame. The pillars also serve as the kingpins. Springing on the Sizaire was accomplished by a single, transverse leaf spring double-cantilevered from the frame to support the chassis on the pillars sliding in the axles. While this may not look much like a fourbar linkage, it actually is a degenerate form of one in which two links have been converted from rotating to sliding. This arrangement worked quite well. It was simple and robust. It reduced the unsprung weight at the front to just the wheels and their pillars. Sizaire cars were noted for their good handling compared to the non-independently sprung competition and were made through 1914.

In 1909, Morgan in England also began using sliding pillar suspension on their cyclecars. They chose to use coil springs on the pillar like Christie's design as shown in Figure 6-23 on a modern Morgan cyclecar. But they inverted his arrangement by putting the pillar on the frame and attaching the wheel to a sliding collar around the pillar. Morgan continues to use this type of suspension on its *Plus Four*, four-wheeled automobile to this day.

In 1922, Lancia used a coil-spring, sliding pillar suspension on its *Lambda* model. A 1927 *Lambda* is shown in Figure 6-24. The coil springs are hidden within the sleeves around the pillars. This car received rave reviews on its ride and handling. An interesting, unsolicited trib-

FIGURE 6-23   Morgan sliding pillar suspension
(Courtesy of Dave_7)

181

AUTOMOTIVE MILESTONES

FIGURE 6-24 Lancia Lambda with sliding pillar suspension
(Photo by the author at the Collier Museum)

FIGURE 6-25 Tatra double-spring independent suspension
(Photo by author at the Tampa Bay Auto Museum)

ute to the *Lambda* was found in the discussion of a 1929 paper titled *The Case for the Independently Sprung Wheel* by W. M. Evans in the *Journal of the Institution of Automobile Engineers* (England). The discusser, C. H. Stephenson, reports on his extended experience driving a contemporary *Lancia Lambda* as follows:

"On really bad surfaces this (car) *is altogether out of the ordinary. For example, I can drive across the sort of gullies they cut in mountain roads for drainage purposes, without any discomfort at all. There is a total absence of bouncing or unpleasant shock when crossing those gullies at speeds higher than those at which I should attempt to drive an ordinarily sprung vehicle. It is difficult to exaggerate the superiority of the Lancia springing on these surfaces."* (Author's note: This is a testimonial to the value of low unsprung weight.)

Leaf Spring Linkages

Another variant of IFS spawned a number of designs that used transverse leaf springs alone or in combination with links to make a form of fourbar linkage suspension. Figure 6-25 shows a 1934 Tatra T75 (Czech) RWD car that used two leaf springs as "compliant" links to support and control the front wheel assembly. The top spring can be clearly seen and the bottom spring is identical to it. These cantilevered springs serve as links 2 and 4 in Figure 6-21, but there are some important differences in geometry as compared to that earlier figure. The fact that the springs do not have true pivots at the frame does not make this "not" a linkage. The springs are able to bend through small angles at their roots and this combined with their flexibility allows their outer ends to behave like the ends of rotating links over small distances. In modern parlance, this would be called a compliant linkage, and it is in fact a form of fourbar linkage.

But the significant difference between this "linkage" and that of Figure 6-21 is that it is an equal-arm linkage rather than an unequal arm one. This makes it what is properly called a **parallelogram linkage.** That linkage is unique in that all points on its coupler (the wheel assembly) move in circular arcs rather than approximate straight lines; the coupler does not change angle throughout its motion. Contemporary literature on the subject of independent suspensions indicates that the early designers (excepting the sliding pillar advocates) favored "parallel linkages" of this generic type. Eventually they realized that this was less than ideal, despite its zero camber change, because the arc motion of the wheel during its suspension travel caused significant tire scrubbing. This led eventually to the adoption of unequal arms that can approximate straight-line motion of the contact patch.

CHAPTER 6 SUSPENSION AND STEERING

FIGURE 6-26 Alvis four-spring independent suspension
(Photo by author at the Tampa Bay Automobile Museum)

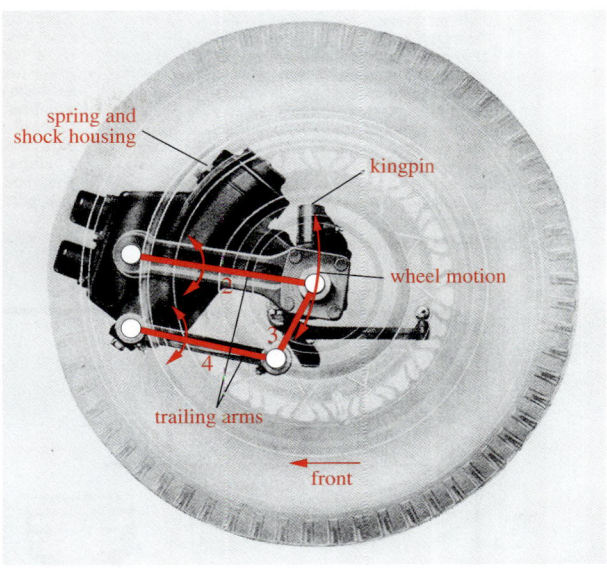

FIGURE 6-28 Dubonnet suspension
(Courtesy of Hemmings Motor News)

The single upper and lower springs of Figure 6-25 have limited resistance to shock loads in the direction of travel. Alvis (England) devised a solution to this in 1928 as shown in Figure 6-26. They used a total of four springs per side, two up and two down angled toward one another in the top view. This begins to look more like the modern A-arm suspension in that it has wide spaced support at the inner end to accommodate fore-and-aft loading. This is a FWD car as can be seen by the half-shaft to the wheel.

Other variants on this theme used one spring either at top or bottom and a pivoted link in the other position. Such an arrangement is shown in Figure 6-27 on a 1957 Talbot Lago (France) that used short, pivoted upper links and a single leaf spring on the bottom. Others inverted this arrangement and put the spring on top and the links on the bottom, but this would not have given as good a straight-line motion to the tire contact patch.

Trailing Arm Suspension

A fourbar linkage can be placed in a longitudinal rather than a lateral plane to suspend the wheels. Dubonnet (France) invented such an arrangement in 1933. Like the sliding pillar, it uses a dead axle with kingpins and attaches tilted cylinders to, and in front of, each kingpin. These serve to house springs and shock absorbers. They also provide pivots for two short links that couple to the wheel assembly, one at the wheel pivot and one below the wheel pivot as shown in Figure 6-28.

The spring housing and links turn with the wheels, but only links 2 and 4 plus the wheel and its pivot contribute to unsprung weight. The wheel center has pure arc motion as it moves up and down. But instead of causing tire scrub as with the lateral parallelogram arrangement, this error only results in a slight variation in angular velocity of the wheel. This was dubbed a

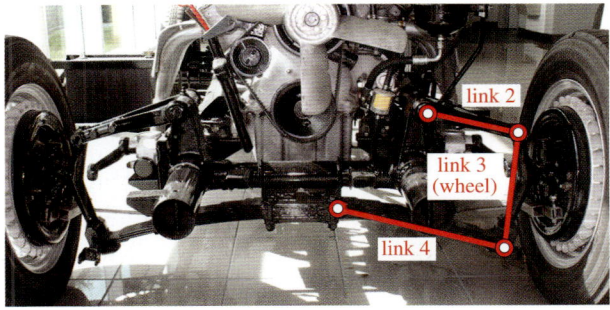

FIGURE 6-27 Talbot Lago spring-link independent suspension
(Photo by author at the Tampa Bay Automobile Museum)

183

AUTOMOTIVE MILESTONES

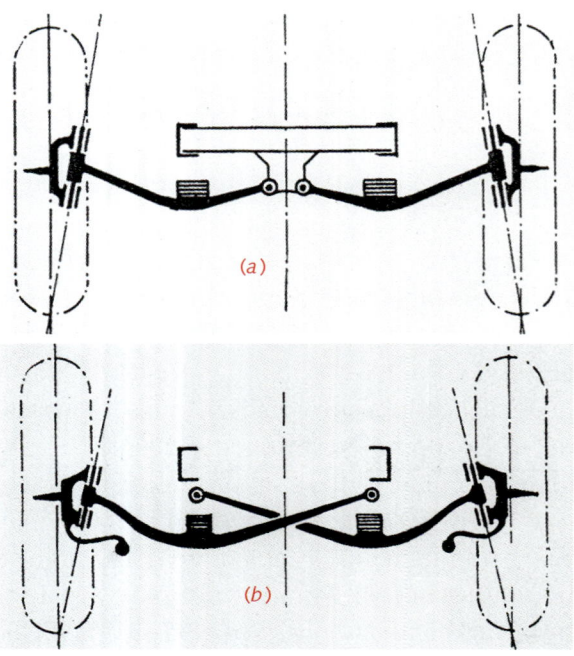

FIGURE 6-29 Swinging axle front suspensions
(Public Domain)

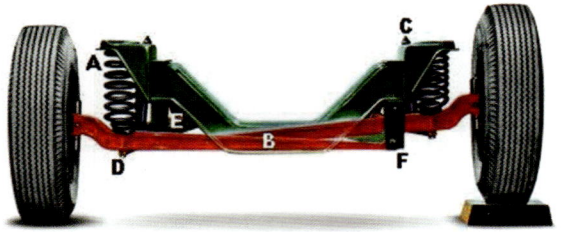

FIGURE 6-30 Ford truck twin-I-beam front suspension
(Courtesy of Hemmings Motor News)

"knee-action" suspension by GM and was very smooth riding. The housing was oil-filled and the oil provided shock absorption. The VW Beetle used a similar, trailing-link, parallelogram linkage in its front suspension but without the spring and shock built in. It used a torsion bar spring and a separate shock absorber.

Swing Axle Suspension

Rumpler invented the first swing axle system in 1921 in Austria and applied it to both his front and rear suspensions. Two variants of this suspension are shown in Figure 6-29. Part (*a*) pivots both half-axles near the center and part (*b*) overlaps the pivots to reduce their swing angle. This system has the disadvantage of variable camber as the wheels move up and down. They can be used with longitudinal leaf springs to control fore and aft movement at the wheels, or either transverse leaf or coil springs when additional control arms are fitted to control longitudinal motion. These were not used very much on front suspensions but were popular at the rear of both RWD and FWD vehicles. Ford 2WD pickup trucks from 1965 to 2002 used the version in part (*b*) of Figure 6-29 on the front wheels and it is shown in Figure 6-30. Ford called it their *Twin I-beam* suspension. Owners complained about the change in camber with varying truck loading and they were difficult to align, requiring the axles be bent to change caster or toe.

Unequal Arm Linkages

Prior to 1930, the Europeans seemed to favor either of two types of IFS: sliding pillar or equal-arm, transverse parallelogram linkages. The issues with transverse parallelogram linkages for suspension, whether using pure rotating links or springs as links, have been pointed out. Mercedes introduced a two-leaf-spring parallelogram suspension like that of Figure 6-25 in 1931 and replaced it in 1933 with a double-wishbone, equal-arm arrangement with coil springs. This may have been the first use of a double-wishbone and coil spring arrangement, but it still had equal-length arms.

In the early 1930s General Motors began work on IFS. The engineer most heavily involved was Maurice Olley, who was well aware of European developments, as he was a regular attendee and presenter at the British society's meetings. Olley and his group at Cadillac developed the first unequal-arm, double-wishbone, fourbar-linkage IFS in 1933. It looked like the generic design of Figure 6-21 except that the

CHAPTER 6 SUSPENSION AND STEERING

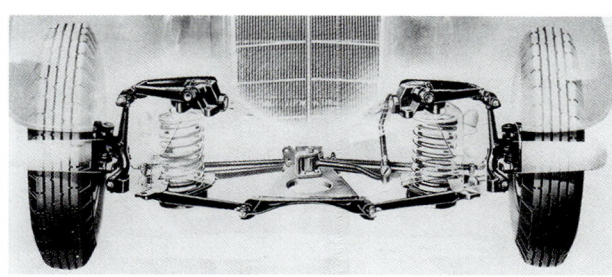

FIGURE 6-31   1934 Cadillac unequal-arm IFS
(Courtesy of Hemmings Motor News)

springs were more vertical and bore against the lower A-arm as shown in Figure 6-31. It also used ball joints to allow the wheels to turn versus the wishbones. They also licensed the Dubonnet trailing arm suspension and worked on its improvement side by side with the work on their own design.

After extensive proving-ground testing of both designs, each GM division was given the choice of the two designs, the Cadillac and the Dubonnet, for their 1934 models. All wanted the Cadillac design. Management overruled Chevrolet and Pontiac on the grounds that if everyone used it they would not be able to make enough of them. So Chevrolet and Pontiac got the consolation prize of the Dubonnet system. While the Dubonnet suspension was well received and worked well, its Achilles heel was the fact that the oil reservoirs had to be topped up periodically. If the owner neglected this and they ran dry, they lost all shock absorption and the car rode poorly. As a result, GM dropped the Dubonnet suspension after 1936 and reverted those cars to beam axles. The Chevrolet and Pontiac finally received GM's own wishbone independent suspension in 1939.

When GM's competitors got wind of the impending release of the first IFS in an American car, they all scrambled to bring out their own. Chrysler was heavily involved in developing their semi-automatic transmission at the time, but nevertheless diverted sufficient engineering effort to bring out a close copy of the GM suspension on most of their own 1934 models.

Studebaker brought out an unequal-arm double wishbone design in 1935, but it used a single, transverse, leaf spring to support the lower wishbones. Packard more or less copied the GM design in 1935 as did Hudson in 1941. Nash used a sliding pillar design in 1940. Ford finally got with the program in 1949 with a design essentially similar to GM's. By that time, there were no American cars left with front axles. Those that had not adopted IFS by that time were out of business. Light trucks continued to use front axles for several more years, but most of them eventually adopted IFS as well. Some Jeeps still use live front axles.

Though IFS had originated in Europe, The U.S. eventually overtook them in its development. When Rolls-Royce decided to use IFS for the first time in 1936, they chose to license the GM design. In a 1936 paper, G.H. Lanchester provides a list of marques then using IFS and IRS in production. He divides IFS into three types, **sliding pillar**, **trailing link**, and **transverse links**. He further subdivides the transverse link category into six subtypes. Five of these are the variants already described that use transverse leaf springs with or without wishbones in various combinations, and the sixth is the **double wishbone** with coil springs like the GM design. He lists two makes with sliding pillars (both European), twelve with trailing arms (all European except for Chevrolet and Pontiac), twenty-two with transverse leaf springs (all European except for Studebaker), and eleven with double-wishbone/coil springs (all American except for Rolls-Royce and Mercedes). Of the last eleven, all but Mercedes were using the GM, unequal-arm design. Mercedes still had equal arms but changed to unequal arms after the war.

Automotive Milestones

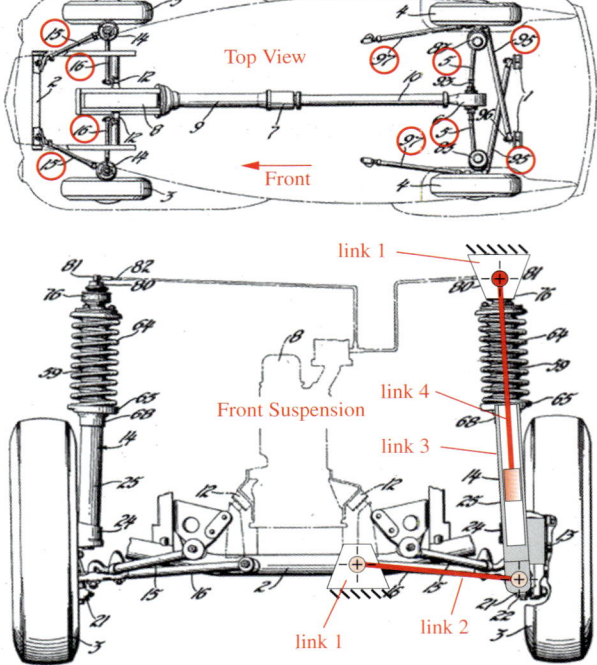

FIGURE 6-32 MacPherson strut 1947 patent drawings

MacPherson Strut Suspension

Earle S. MacPherson was with GM when he invented his eponymous strut suspension. He was assigned the task of designing a completely new, light car for Chevrolet in 1945 to be called the Cadet. He employed ideas that were way ahead of U.S. automakers at the time, but had been used in Europe before the war, such as unit-body construction (no frame) and four-wheel independent suspension. He designed a suspension system to be lightweight, compact, and inexpensive. Drawings from his 1947 patent are shown in Figure 6-32. His suspension uses a **fourbar slider-crank** mechanism with the shock absorber playing the part of the sliding joint. A coil spring surrounds the shock to provide support.

There are two control arms, labeled 15 and 16 in the top view of the patent drawing. Together they form the lower wishbone. By an-gling arm 15 toward the front, it can take the longitudinal loads from the wheel to the chassis. A diagram of a slider-crank mechanism is superposed on the right strut assembly in Figure 6-32. Link 2 of the superposed mechanism represents arms 15 and 16 schematically. The slider assembly (links 3 and 4 of the superposed mechanism) comprise the steering axis. The wheel is carried on the coupler (link 3) of the slider-crank linkage, which gives approximate straight line motion to the contact patch.

The shock-absorber piston rod (link 4) is pivoted to the chassis at the top of the strut in a rubber mount, high inside the fender in what is called the strut tower. This suspension was designed specifically for frameless, unit-body automobiles, and McPherson's intention was that it be used at the rear as well as the front. Note in the top view of Figure 6-32 that the rear wheels are located by two control arms. One (97) is a trailing arm and the other (95) is arranged like a Panhard rod except that it goes to the wheel. Both parts 95 and 97 are pivoted to the bottom of the slider (shock-absorber body), which is link 3 of the superposed mechanism.

Chevrolet cancelled the Cadet project and did not build any. A discouraged MacPherson left to work for Ford where he patented an improved strut design in 1949. The 1951 English Ford Consul and Zephyr became the first production automobiles to use MacPherson strut suspension.

When MacPherson struts are used for the rear suspension, they are usually called **Chapman struts** instead, named after Colin Chapman, the father of Lotus automobiles. Chapman modified MacPherson's design in the rear by using the driveshafts (links 5 in Figure 6-32) also to do the job of the transverse links 95. This reduced weight and cost but meant that the universal joints in the driveshafts had to take suspension forces as well as driving torque.

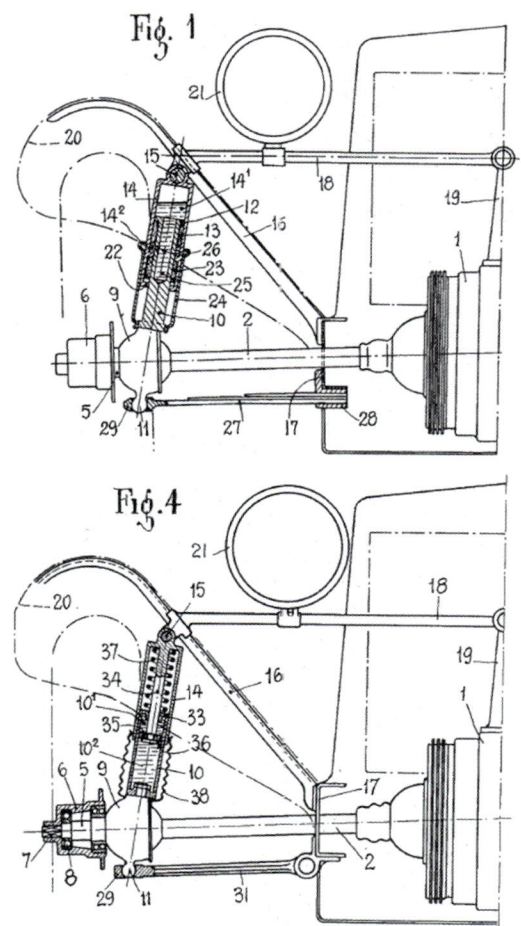

FIGURE 6-33 Fornaca strut 1929 patent drawings

FIGURE 6-34 Morgan cyclecar suspension *(Photo by the author)*

Modern IFS Practice

In 2015, virtually all independent front suspensions worldwide are either the unequal-arm, wishbone type or MacPherson struts, both using coil springs. But, at least one application uses a transverse spring. The Chevrolet Corvette has unequal-arm, upper and lower wishbones loaded by a single, transverse, one-piece, fiberglass, leaf spring instead of coil springs. Figure 6-34 shows a modern, double-wishbone suspension on the new Morgan cyclecar that was redesigned in 2012 to eliminate its original sliding pillar suspension. It also now uses an S&S, V-twin motorcycle engine driving the single rear wheel. The wishbones can be clearly seen and the spring/shock absorber assembly acts between the upper wishbone pivot and the lower wishbone, much like the arrangement depicted in Figure 6-21. The upper wishbones are slightly shorter than the lower ones.

Independent Rear Suspension (IRS)

Rear suspensions were made independent as early as the fronts were. Sizaire used his sliding pillar suspension of Figure 6-22 at all four corners of his 1905 automobile. All the other designs already described for IFS have been used also for IRS systems.

MacPherson was not the first to design a suspension system of this type. Guido Fornaca of Italy patented virtually the same suspension design in the U.S. in 1929. MacPherson cited the Fornaca patent in his second patent application in 1949, but not in his first in 1947, which is the one shown in Figure 6-32. Fornaca showed a strut design with a transverse leaf spring and also an alternate design with a coil spring as seen in Figure 6-33. It appears Fornaca was directing his application to a FWD, frame and body car, as he shows the strut attached to a fender and a headlight bracket. These do not look sufficiently robust to take the suspension loads, however. There is no evidence that the Fornaca design was ever produced.

Swing-Axle IRS

As mentioned earlier, Rumpler invented the swing-axle system in 1921 and by 1936 it had been adopted as IRS by at least eight marques, all German or Austrian. Rumpler used it at both front and rear. These all used the type shown in Figure 6-29a with the half-axles pivoted on either side of the differential and one universal joint in each shaft. Some used transverse leaf springs and trailing arms with swing axles. Others used coil springs with trailing arms.

The swing axle system became quite popular for IRS, especially in Germany, Austria, and Czechoslovakia. Mercedes adopted it on many of their models before WWII and continued using it into the 1960s. Tatra used it on most models before WWII, and the VW Beetle copied it in 1938. VW kept this design throughout the 21-million run of the original Beetle.

The camber change of this design caused some problems. The effect of camber change on steering was described in the section on IFS, and this caused a small amount of rear wheel steering. It was found that setting the initial camber negative (tops of wheels inside the bottoms) gave better handling and tire wear. FWD cars such as the 1935 Audi also used swing axles at the rear.

The bigger problem with swing axles is their tendency to "tuck-under" and try to flip the car over if a driver goes too fast in a turn. As the car rolls toward the outside of the turn, it rotates the outside axle shaft down, increasing positive camber. If the driver does not correct this by slowing down and reducing the steering angle, the outside tire can start sliding sideways and possibly cause a rollover. This tendency is exacerbated in a rear-engined car such as the original VW Beetle or Chevrolet Corvair, which tend to oversteer at the limit because they are tail heavy. It was on this basis that Ralph Nader criticized the Chevrolet Corvair in his book, *Unsafe at Any Speed*. He could have levied the same complaint against the VW Beetle as well.

Other IRS Types

Some prewar marques used IRS arrangements such as those shown in Figures 6-25 and 6-27 and their variants as described in the section on IFS. Since WWII, the transverse-spring types of IRS and IFS have largely disappeared in favor of double wishbone arrangements like that of Figure 6-31, various trailing-arm arrangements, or Chapman struts. Unlike the swing-axle system, these all use four U-joints, two on each half-shaft—one on each side of the differential and one at each wheel.

All IRS systems in RWD cars mount the differential on the frame or unit body, and thus markedly improve the sprung-unsprung weight ratio. Others, such as the Corvette, some Ferraris, and many others use a transaxle, which has the same result. Some also mount the brakes inboard on the axle shafts at the differential. This further reduces unsprung weight. Jaguar was one of the early adopters of this approach on their 1961 E-Type. This car was quite advanced for its time. It had a unit body with a sub-frame to mount the engine, transmission, and suspension at the front and another to mount the differential, inboard brakes, wishbones, springs, shocks, and axle assembly at the rear. Figure 6-35 shows the rear axle assembly from a Jaguar E-Type.

Spatial-Linkage Suspension

We have described how many suspension systems are based on variants of the ubiquitous fourbar linkage. This is a two-dimensional mechanism that is given the third dimension by simply duplicating the links in a parallel plane. Many examples of parallel-planar linkages can

FIGURE 6-35 Jaguar E-Type IRS assembly *(Photo by the author)*

Modern Rear Suspension Practice

Most modern IRS systems in RWD automobiles use a double wishbone arrangement, a trailing-arm arrangement, or a Chapman strut. Some use spatial linkages and these tend to be high-performance, luxury, or sports cars. Some FWD automobiles use MacPherson struts at the rear, but many use a light, beam axle in the rear with coil springs, trailing arms, and a Panhard rod. Light trucks from GM, Ford, Ram, Nissan, and Toyota, most of their SUV derivatives, and Jeeps still use live axles at the rear. Heavy trucks have stuck with axles at both ends.

be found. Think of a common, folding beach chair. Each side of it is a fourbar linkage. The two sides are duplicates separated by a number of tubes that hold them in parallel planes.

There also exist true, three-dimensional, spatial linkages. In the late 1960s, Mercedes developed a seven-link[1] spatial linkage to control the independent rear wheels as shown in Figure 6-36. This gives better control over caster, camber and toe of the rear wheels and can greatly improve handling. Other makes such as the BMW 7-series, Lexus GS/ISF, Cadillac ATS, Ford Expedition, and some Audi AWD models now use similar linkages on their rear suspension systems. The 2001–2009 Lancia Thesis used multi-link suspensions on both front and rear wheels.

Dampers

Engineers use the term damper to refer to any mechanical device whose purpose is to dissipate energy. A damper is the mechanical equivalent of an electrical resistor. It converts mechanical energy to heat. The more colloquial term for the dampers used on an automobile is **shock absorbers**. That name is somewhat unfortunate because their purpose is not so much to absorb shock as to convert energy to heat. But, we will have to accept their colloquial name as it has become a standard term.

The purpose of dampers is to keep the wheels in contact with the road despite encounters with bumps and potholes. When a bump is encountered, the suspension spring is compressed, making the wheel move upward. The purpose of the shock absorber (damper) is to resist that movement and keep the wheel from leaving the pavement. A good damper gives little resistance to a force that is applied slowly. But a suddenly applied force is resisted heartily.

How can this be? The most effective dampers use the resistance to shearing of viscous fluids (like oil) to do their job. As a simple example, imagine the following: In your bathing suit,

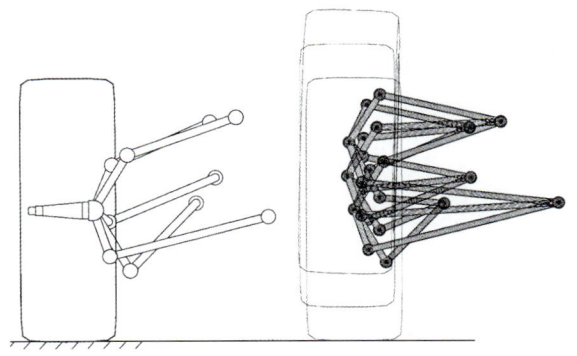

FIGURE 6-36 Mercedes seven-link spatial linkage suspension *(Public Domain Courtesy of Prof. P.A. Simionescu)*

---

[1] Most car publications and manufacturer's descriptions refer to this as a five-link suspension, but they are only counting the five bars that pivot to the chassis and are neglecting the chassis itself (the ground link) and the wheel hub, which is the seventh link. Kinematically, it is a seven-link, spatial linkage.

you slowly lower yourself into the pool. You feel very little resistance from the water. Next you climb to the high diving board and jump off. When you hit the water, it hurts. Why?

When you slowly part the waters with your body you are only gently shearing the liquid. Shearing means separating the water molecules. When you hit the water from the diving board, you asked the water molecules to separate in a hurry, and they resisted that. Any liquid can be pushed aside slowly, but it resists the push in proportion to the **shear rate**—the velocity of the push. This property of a fluid, **resistance to shear rate**, is called **viscosity**. Some liquids have higher viscosity than others. Water has one of the lowest viscosities of any liquid. Alchohol has less, oil has more, and molasses has a lot. The motor oil designations, 10W, 30W, etc., are indications of its viscosity. Higher numbers mean higher viscosity.

A modern, tubular, hydraulic shock absorber can be seen in Figure 6-23. The way a hydraulic shock absorber works is as follows: The tube contains a piston within a cylinder. Seals around the piston prevent leakage past it. The piston has two or more small holes in it that allow a controlled leakage between the oil-filled chambers. Each of these holes has a check valve on it, so that half the holes allow oil to flow from the bottom chamber to the top and the other half allow flow from top to bottom. For each flow direction, the holes are different size so that they present a different resistance to oil flow. The reason for the different size holes and check valves is to allow resistance to bounce (upward motion of the wheels) and rebound (downward motion of the wheel) to be different.

If you have this type of damper in your hands, you can compress it by pushing it together very slowly. A constant force applied over time will make the oil flow slowly through the small orifices from one chamber to the other with moderate force. But, even a large, suddenly applied force will encounter strong resistance due to the viscous oil resisting the rapid shearing required to flow through the small orifices. So a shock absorber cannot hold a car up against its weight, but will resist a sudden force applied to either a wheel or the body.

The primary job of the dampers on a car is to keep the wheels in contact with the ground and to minimize their response to unevenness in the road surface. Without functional shock absorbers, the car body will continue to oscillate up and down well after a wheel hits a bump. Proper dampers dissipate the spring's energy in no more than one bounce and keep the car body from continuing to bounce. There is an optimum amount of damping for any dynamic system and this is called its **critical damping**. If a system has critical damping, when disturbed, it will overshoot slightly and return to its original position in the fastest possible time. More damping than critical will make it sluggish and slower to return. Less damping than critical will cause it to bounce multiple times and also take longer to settle to its original position. If the shocks are worn out, the system will be underdamped and the car will bounce long after it hits a bump. Figure 6-37 shows plots of the response of a system for the three cases of over, under- and critically damped systems. A critically damped system recovers from a disturbance quicker than either the over- or underdamped system.

MagnetoRheological Dampers

Rheology is the study of how fluids flow, and **viscosity** is one **rheological property** that affects their flow. A **magnetorheological (MR) fluid** is one (usually an oil) to which very fine ferrous particles (essentially iron filings) have been added. When a magnetic field is applied to this fluid, the ferrous particles align and clump together. This stiffens the fluid, effectively in-

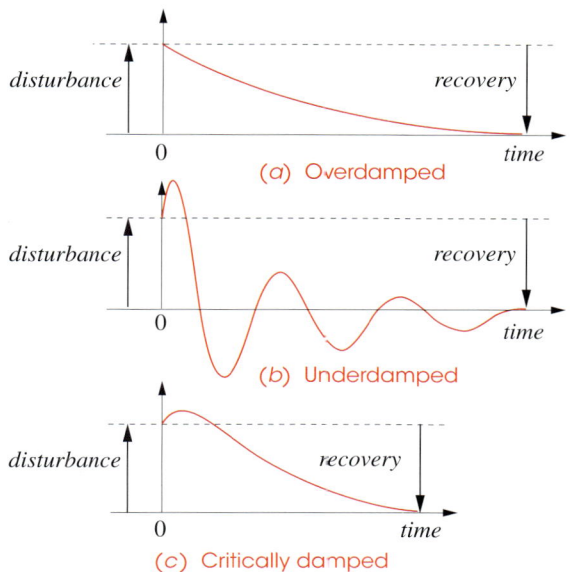

FIGURE 6-37 Effects of different damping levels on response

creasing its viscosity temporarily. The higher viscosity fluid gives higher damping until the magnetic field is removed. Jacob Rainbow introduced MR fluids in the 1940s.

General Motors, through its Delphi division, originally developed **magnetorheological shock absorbers** in 1998. They were first used on Cadillacs and the Chevrolet Corvette, and still are. Electromagnets are built into the shock absorber body. A microcomputer controls the application of electricity to the electromagnets in response to sensors mounted on the suspension. The control system can respond to changes in suspension loading in one millisecond. So this system can change the effective damping rate of the shock absorbers 1000 times a second.

This system is remarkably effective in controlling the automobile over bumps and through corners. The car seems to remain flatter on tight curves, and the ride is excellent. After GM launched this system, other manufacturers tried without success to design their own magnetorheological shocks without violating GM's patents. Ferrari ultimately gave up and licensed the GM design for its 599 GTB Fiorano. BMI Group (Beijing West Industries) bought part of the bankrupt Delphi Corp. in 2009 and now produces *Magneride* shock absorbers to Delphi's design. BMW has developed its own version of an MR system. Ferraris, Maseratis, Audis, Acuras, Camaro, and Buicks are now using GM or BMI **semi-active suspension** systems. Maserati calls it their "Skyhook" suspension, the trope being that the car is hung from a skyhook.

Active Suspension

All the suspensions discussed so far are **reactive** in nature. They can only react to the road contours and imperfections when encountered. This includes those with semi-active suspensions as described above. While the active dampers in these systems can react in milliseconds to road inputs, they are still just reacting.

A true active suspension can anticipate road conditions using various sensors and also actively drive the suspension components up or down rapidly enough to prevent the road contour from altering the vehicle's vertical position versus a stationary frame of reference. The first of these to be developed came from Colin Chapman on his Lotus Formula 1 race cars in the 1980s. This system used hydraulic cylinders to move the wheels, weighed 300 lbs, required 5 HP, and cost thousands of dollars. It was not practical for production cars. It also could only respond fast enough to smooth out large bumps.

Infiniti tried a similar system on its Q45 in 1991. This system's hydraulic cylinders could only drive the wheels downward, but not lift them. It also was slow to respond and only affected gradual suspension motions. It cost $5500, added 200 lbs to the car, and lowered fuel economy by 10%. It was not continued long in production. Mercedes offers Active Body Control (ABC) on some models. But it is not true active suspension because it operates in only one direction to keep the wheels planted on the road and is not fast enough to respond to individual bumps. Mercedes also offers an active system that uses cameras to map the road in front of the vehicle and provide input to the suspension in anticipation of those bumps. It only works in daylight.

Bose Active Suspension System

Dr. Amar G. Bose, founder of the Bose Corporation and an acoustic expert, began working on an active suspension system in 1980 and by 2004 had a prototype system installed in a Lexus LS400. Not surprisingly, given his expertise in speakers, he used an electromagnetic (EM) approach rather than a hydraulic one. An audio speaker is an EM system that consists of a magnet in a coil attached to a diaphragm. Electrical signals vary the field in the coil and cause the magnet to oscillate, driving the cone. The cone motion pressurizes the air, creating sound.

EM speaker systems can respond rapidly enough to generate sounds at 20,000 cycles per second or higher. Such high frequency is not needed for an auto suspension, and it probably could not be achieved because of the high mass of a suspension versus a speaker. The Bose system can respond at 100 Hz and that is fast enough to control the suspension in response to even small bumps. Each wheel is fitted with a Linear Electromagnetic Motor (LEM) with an 8-in stroke and is about the diameter of a suspension spring. The system uses about 2 HP continuously, but each motor is capable of much higher instantaneous power levels. The LEM is capable of generating electricity when it is driven upward by road motion and this is stored in the battery. The LEM can draw on the battery to generate downward forces and velocities sufficient to lift the car in milliseconds. All this is controlled by a microprocessor responding to inputs from sensors on the vehicle.

Bose has a demonstration video at https://www.youtube.com/watch?v=eSi6J-QK1lw in which its Bose-fitted Lexus is driven around a test track along with a standard-suspension Lexus shown on a split screen. The Bose Lexus is dead flat in the turns and over bumps that roll or upset the Lexus with standard suspension. At the end of the video they do a "trick" with the Bose Lexus by driving it at speed toward a 2x6 placed on edge across the road. Before the front wheels reach the board, the motors at the front wheels extend to lift the car up and then retract the wheels back into the fenders as the front wheels sail over the board. This motion is repeated for the rear wheels as they sail over the board as well. For several frames, the car is airborne with all four wheels in the air. At the end of the video, the driver steps out of the car, thanks the audience, bows, and then makes the car bow as well.

This trick is the same one you have done when leaping straight up by extending your legs rapidly from a semi-crouched position, then pulling your legs up to be "airborne" for a few milliseconds before re-extending your legs to land on your feet. The Bose system is powerful and fast enough to lift half the weight of the Lexus (about a ton) plus retract and redeploy the wheels, all in milliseconds without damage to the suspension or car. To date, no manufacturer has put the Bose suspension system into production. But Bose has marketed a custom seat for 18-wheeler trucks that has its active suspension built in. It greatly increases comfort and reduces fatigue for the driver. See: https://www.youtube.com/watch?v=QHiVeuvlnBc

A similar active suspension system was developed at the *University of Texas Center for Electromechanics* in the 1990s. Called the *Electronically Controlled Active Suspension System* (ECASS), it was developed for military vehicles and has been installed on Humvees. A video demonstration of the ECASS system on a Humvee is at https://www.youtube.com/watch?v=-vArnkbQ1ws

Steering

The basic issue in designing a steering system for a four-wheeled vehicle is the fact that the outside wheels must follow an arc of larger radius than the inside wheels. This was recognized well before automobiles came along because all animal-powered carriages had the same problem. Dr. Erasmus Darwin of England is credited with the invention of a linkage steering mechanism to solve the problem in 1758. This is now known as Ackerman steering. It was reinvented in 1817 in Germany by Georg Lankensperger, and he hired Rudolph Ackerman as his agent to obtain a patent in England. Ackerman attached his name to the invention on that patent, and his is the name associated with the mechanism in modern times.

Ackerman Steering

Figure 6-38 shows drawings of this mechanism. The top image shows the trapezoidal fourbar linkage that connects the front wheels and causes the inner wheel to turn through a greater angle than the outer wheel. The steering arms are each part of a wheel assembly, and the tie rod connects them. The mechanism to move the tie rod is not shown. The correct angles for the steering arms are found by extending them to intersect at the center of the rear axle. The velocities of the wheel centers and any point on the vehicle are all perpendicular to their radii to a common center point called the **instant center of rotation**. By definition, every point on the vehicle is rotating about its instant center at all times.

An **instant center**, as its name suggests, might only exist in a particular location for an instant and then move to a new location. But, the instant center of rotation for the vehicle will always lie on the line of the rear axle extended and, of course, will be on a different side of the vehicle for turns of opposite direction. Its po-

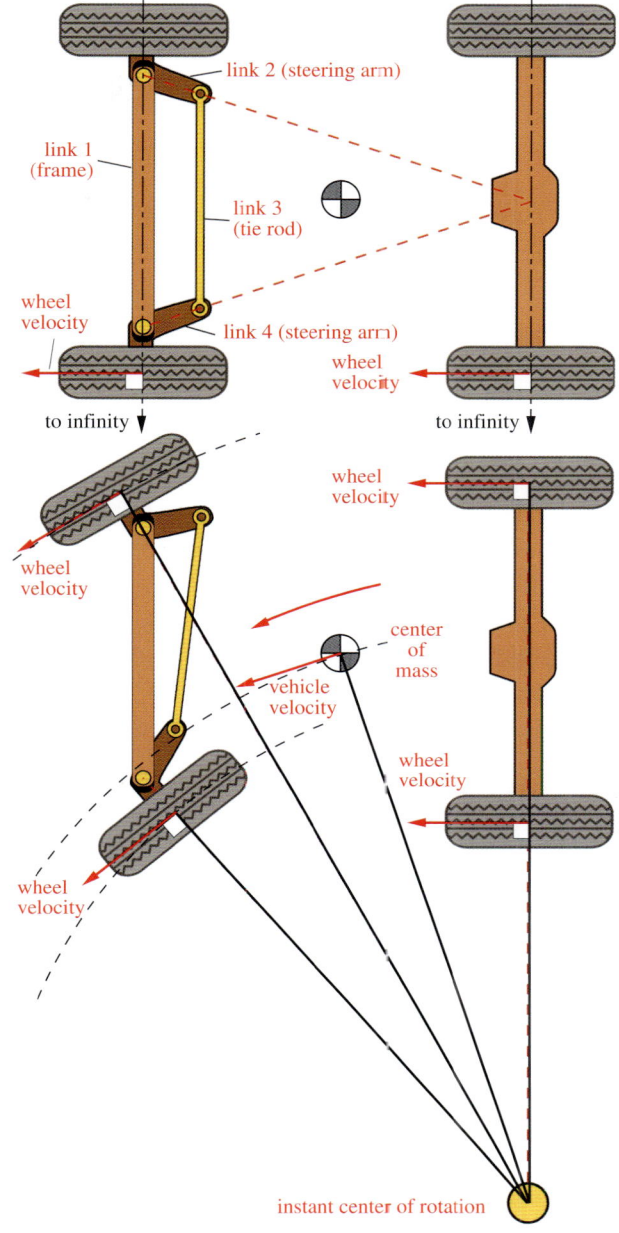

FIGURE 6-38 Ackerman steering
(Courtesy of Andy Dingley)

sition on that line will vary depending on the steering angle of the car. In the top drawing, the car is moving straight ahead and the lines through the centers of the rear wheels and the front wheels are parallel. Then the instant center of rotation is at infinity. If that statement bothers you, consider that straight line motion is no different than motion along an arc of infinite radius. As the driver starts a turn, the instant center moves from infinity toward the car, becoming closer as the steering angle increases. Its position in the bottom drawing represents only the particular steering angle shown.

A fourbar mechanism, such as the trapezoidal linkage shown, cannot give "perfect" Ackerman steering geometry. It can be exact at two points but will deviate slightly over the range of steering motion. The tires' flexibility allows this error with no significant detriment. Setting the link angles as shown in the top drawing gives acceptable Ackerman error. In fact, Ackerman steering only really works at slow speeds. As speed increases, the dynamic loads on the suspension and tires cause further deviation from the ideal and can move the true center of rotation of the vehicle away from its theoretical location as shown in the figure. These factors are quite complex, and suspension designers must consider them to find the best compromise to achieve desired handling characteristics. This issue will be addressed in more detail in the upcoming section on handling.

Steering Mechanisms

Many variations of steering mechanism have been designed for vehicles. We will discuss only the two types most commonly on modern passenger cars and light trucks, the so-called parallelogram steering linkage and the **rack and pinion steering** linkage. Both of them use the basic Ackerman steering-arm layout of Figure 6-38 but differ in the way they convert steering-wheel input to motion of the steering arms.

Parallelogram Steering Mechanisms

Figure 6-39a shows a schematic drawing of a **parallelogram** steering mechanism. The parallelogram portion comprises the **Pitman arm**, which is turned by the steering wheel, the **drag link**, the **idler arm**, and the **frame** between the

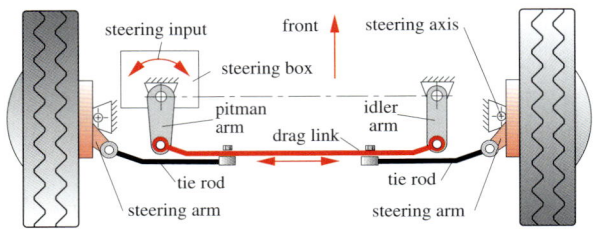

(a) Parallelogram steering

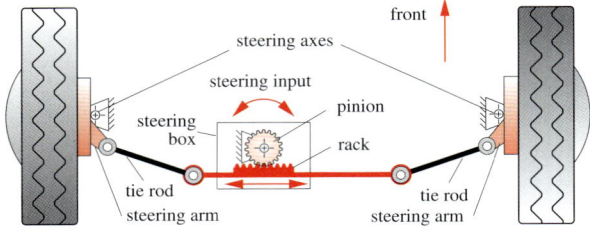

(b) Rack and pinion steering

FIGURE 6-39 Common steering mechanisms

pivots of the Pitman and idler arms. A property of a parallelogram linkage is that its coupler, here the drag link, moves in **curvilinear translation**. This means that it does not change its angle throughout its motion. It always remains parallel to the line between the two fixed pivots of the Pitman arm and idler arm. All points on the drag link describe circles of the same radius as the Pitman arm, so the drag link moves backwards as it oscillates left and right from the position shown. The tie rods accommodate this curvilinear translation motion and transfer it to the steering arms to turn the wheels. There is a mechanism within the steering box to convert rotation of the steering column to rotation of the Pitman arm. This most common arrangement uses a worm on the steering column and a recirculating ball-bearing nut around the worm. The nut has a short rack on it that rotates a pinion segment attached to the Pitman arm. This type of steering was common on vehicles before the 1990s but has largely been replaced by the rack and pinion type in passenger cars. Recirculating ball steering is capable of large mechanical advantage, and that makes it more useful for heavier vehicles such as trucks. But it has more backlash and less feedback of road irregularities and tire performance than does rack and pinion steering.

Rack and Pinion Steering

Figure 6-39b shows a schematic drawing of a rack and pinion steering system. For simplicity, the three-dimensional geometry of the real system has been "flattened" in the drawing. The pinion is usually above the rack. The steering column turns a pinion that engages a rack that moves transverse to the car. Each end of the rack attaches to a tie rod that connects it to the steering arm on either side.

The steering column sometimes uses universal joints to angle it toward the center of the car so the rack can be more symmetrically located. This type of steering is considered to give more direct and better feedback to the driver as to what the wheels are doing. It can have less backlash than the recirculating ball mechanism; it is also less expensive to manufacture. Most modern cars have adopted rack and pinion steering. Light trucks and their SUV derivatives more often use recirculating ball steering with the parallelogram linkage arrangement, as do heavy trucks.

Interestingly, the first self-propelled road vehicle used rack and pinion steering. This was the 1769, steam-powered Fardier de Cugnot described in Chapter 1 and shown in Figure 1-5. Its rack and pinion steering is shown in Figure 6-40.

Power Steering

Patents for power steering systems go back to 1906. The modern form of hydraulic power steering was invented by Francis W. Davis in Waltham, Mass in 1926. In recent years, hydraulic power steering has been replaced by electric power steering. In a 1945 SAE paper on Power Steering (PS), Mr. Davis describes three types of PS then in use: vacuum, compressed air, and hydraulic. The hydraulic type took over from the other two by the 1950s. Chrysler introduced the first power steering on a production automobile in 1951, but it had been used on trucks and heavy vehicles since before WWII.

Hydraulic Power Assist

Two variants on this hydraulic power-assist theme were developed. One was intended to be an after-market retrofit to existing vehicles, and the initial market was trucks, which had heavy steering. The other was developed to be fitted as original equipment on new vehicles. The latter has come to dominate today, because most automobiles and trucks offer power steering either as standard equipment or as an option.

FIGURE 6-40 Fardier de Cugnot rack and pinion steering
(Courtesy of John Perodeau of the Tampa Bay Automobile Museum)

The add-on variety left the manual steering gear in place. If the power system failed, the driver still had complete control. Developers of the original-equipment systems all agreed that a manual connection should be left intact as a safety measure. It would actually have been easier to make a fully-automated, drive-by-wire power steering system that could conceivably be controlled with a joystick rather than a steering wheel, as is the case with all large aircraft. Retaining the mechanical steering system meant that the PS system had to be designed to actually be a **power-assist system**, and this is its proper name.

Various schemes were devised to couple a power assist to the existing steering system. Some fitted a hydraulic cylinder to push on the tie rod. Others added a hydraulic cylinder to the mechanism that drove the Pitman arm. Whatever the details, the trickiest part of the problem was to devise a control system that still allowed the driver some feel of the road. It was easy to create a system that would allow the steering wheel to be turned with the "pinkie finger," but that would not satisfy most drivers who want some feedback as to what the tires are doing.

The most common solution for the parallelogram steering systems of the time was to provide a two-way valve coupled to some part of the mechanical system (Pitman arm, steering column, or other) with light springs between the valve and the mechanical part driving it. One spring was for the left- and one for the right-turn valve motion. This allowed a small amount of steering wheel rotation around the center or straight-ahead position without applying any power assist. During this range of steering-wheel motion, the driver had full road feel. Turning the wheel further compressed a spring sufficiently to exceed the valve's motion threshold and started to feed hydraulic fluid from the engine-driven pump to the cylinder. Thus the boost came on in proportion to how hard and how far the driver turned the wheel and dropped to zero when going straight ahead.

With rack and pinion steering, the hydraulic cylinder acts directly on the end of the rack. The fluid pressure supplied to either side of its two-way piston is controlled by a valve with a dead spot at center, as in the older systems. This approach gives zero boost on center for good road feel with boost coming on as the wheel is turned through a larger angle.

More boost is needed at low speeds and in parking than when the car is at highway speeds. Modern PS systems in high-end and performance automobiles vary the boost with road speed. BMW refers to this as "degressively linked power steering." At highway speeds the amount of boost "degresses" to zero and increases nonlinearly as the car slows. This greatly improves road feel at speed but allows easy parking. Some early cars with PS in the 1960s overdid the boost levels to the point that there was little-to-no road feel at any speed. Ford was the worst offender in this regard, followed by Chrysler, with GM having the best "feel" of cars with PS in those years. Today, there is less difference between makes in this regard and all are better than in the "old days."

Hydraulic power steering (HPS) and electro-hydraulic power steering (EHPS), in which the pump is driven by an electric motor rather than the engine, have essentially been perfected to a high degree and are very satisfactory in all regards. However, as with any hydraulic system, they are not very efficient. To maintain pressure, an HPS pump runs all the time when the engine is on, whether the steering wheel is being turned or not, and this continuously absorbs engine power. Further losses occur from friction in the hydraulic lines and cylinder. This is the main reason that these systems are being replaced by electric power steering (EPS) in many new cars.

CHAPTER 6 SUSPENSION AND STEERING

Electric Power Assist

The cost of electric servomotors has come down in recent years and using an electric motor to directly power the steering is less expensive than using an hydraulic system. Servo motors are very controllable with microprocessors and they draw current only when the steering wheel is being turned. EPS systems increase economy by about 1 mpg over hydraulic systems.

One form of EPS uses a servomotor sitting piggyback on the rack case and driving a ball screw to convert the motor's rotary motion to linear. The ball screw is connected to the rack. Ironically, this is essentially a return to the recirculating ball steering of decades ago because a recirculating ball-nut is the same animal as a ball screw. Another design uses the motor to drive a second pinion in mesh with the rack. With small, light cars the motor can drive the same pinion that is connected to the steering wheel.

Many performance-oriented drivers complain that EPS does not give as good road feel as an HPS system. Car and Driver magazine did a driving test of both types with mixed opinions as to which was better.[1] But, the trend is clear. In 2005, most new automobiles had HPS or EHPS and only a fourth had EPS. In 2011 the shares had switched to 58% EPS and the rest HPS, EHPS, or manual. Cost and economy of operation will eventually drive HPS and EHPS away. Another factor is the advent of driverless cars on the horizon. These may eliminate the steering wheel entirely and have a true drive-by-wire system. Electric motors are the natural choice for that application. Citroën developed a drive-by-wire steering system in 2005 but has yet to put it on a production automobile. Nissan currently offers a drive-by-wire steering system in its Q37 and Q50 models. A clutch in the steering column of the Nissan is supposed to engage to return manual steering control if the electronic system fails.

Handling

Handling is a very complicated topic. The handling characteristics of an automobile depend on many factors including the weight distribution between front and rear axles, height of the mass center above the ground, polar moment of inertia, sprung-unsprung weight ratio, spring rates, shock absorber damping ratios, locations of the roll centers and of the roll axis, suspension travel, suspension control of caster, camber and toe, and wheels and tires. Some of these factors such as polar moment of inertia, spring rates, sprung-unsprung weight ratio, effects of caster, camber, and toe, suspension design, and dampers have already been discussed in earlier sections of this and other chapters. We will address the other factors here.

Weight Distribution

Though there are some dissenters, many experts agree that an equal distribution of vehicle weight on both axles, referred to as a 50/50 distribution, is optimal for good handling. A nose-heavy car will tend to understeer and a rear heavy car to oversteer. These tendencies can be overcome, to a large degree, by suspension design and other factors such as using different size tires front and rear. Perhaps the best example of uneven weight distribution being largely overcome by other means is the Porsche 911 series. This rear-engined, rear-drive car's weight distribution is about 40/60 front to rear, yet is considered to be one of the best handling automobiles. Nevertheless, the Porsche Cayman/Boxster, with mid-engine RWD, has about 45/55 front/rear balance and most reviewers credit it with better handling than the 911. This author has driven both and agrees with the above assessment.

---

1 "Are We Losing Touch? A Comprehensive Comparison Test of Electric and Hydraulic Steering Assist," Car and Driver, January 2012.

BMW has long promoted their 50/50 weight distribution in all their passenger car offerings and they all handle very well. On the other hand, many Ferrari models are mid-engine RWD and are back heavy, but are considered among the best handling cars. The Ferrari 599 GTB Fiorano and the Maserati Quattroporte, which are both front-engined-RWD, have their engines so far back in their chassis that they have a 47/53 F/R weight distribution, and both have excellent handling. Ferrari and Maserati claim that a small rearward weight bias improves handling. Unequal weight distribution can be overcome by other factors.

Front-engined, FWD cars are nose heavy and tend toward understeer—as do most front-engined, RWD cars. With the exception of cars designed for the enthusiast (such as Audi, BMW, Corvette, Ferrari, Jaguar, Maserati, Porsche, and others), most ordinary, front-engined automobiles tend to be nose heavy, regardless of which end has the drive wheels. RWD and AWD cars are generally considered to handle better than FWD cars, though there can be exceptions.

Mass-Center Height

It is logical that the lower the mass center is to the ground, the less the car will want to roll in turns. Tall sedans, SUVs, and trucks roll much more than a low sports car. Most cars designed for handling performance will keep their mass center low.

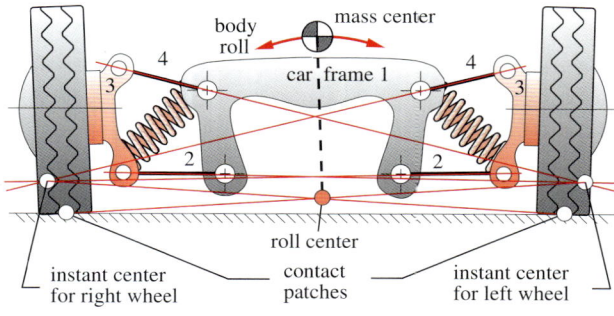

FIGURE 6-41 Suspension instant centers and roll center

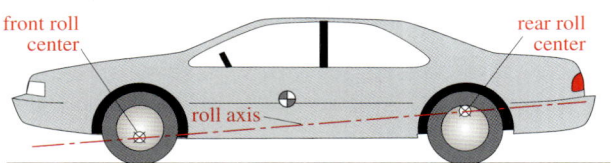

FIGURE 6-42 Roll axis

Roll Center and Roll Axis

A front and rear suspension each have a roll center about which the chassis rolls at that location. The roll center is depicted for one position of a front suspension in Figure 6-41, which shows a generic, unequal-arm, independent, front suspension. Each of the fourbar linkages that comprises the suspension on either side of the car has what is called an **instant center of rotation** for its coupler, link 3, which is the wheel hub. An **instant center** is a point in space about which its related link, here the wheel hub, is in pure rotation at this instant. Its location is found by extending the lines of the upper and lower wishbones (labeled 2 and 4) until they intersect. That intersection is the instant center for that wheel.

The roll center is then found by drawing lines from each wheel's instant center back to the contact patch of that wheel's tire with the road. Where these two lines cross is the instantaneous location of the roll center of the suspension. The mass center of the body will rotate about this roll center in the plane of the front wheels. Other designs of suspensions will have their instant centers and roll center in different locations than that shown here.

The rear suspension will also have its own roll center that can be above or below the front roll center. The rear suspension is usually designed to place its roll center higher than that of the front so that the roll axis, which connects the front and rear roll centers, is tilted downward toward the front of the car as shown in Figure

6-42. This reduces roll toward the rear of the car as the mass center is closer to the roll axis there.

Slip Angle

The handling "feel" of a vehicle comes from interaction between inertial forces generated by lateral acceleration of the mass center induced by turning and the dynamic cornering forces developed by the tires on the pavement to resist those inertial forces. As lateral loads on the tire increase due to the square of the cornering speed, the tires creep to the outside of the turn and move in a direction different than the direction in which the wheel is pointed by the steering input. The driver has to turn the tires for a tighter circle than the tires actually describe on the ground. The difference between the tires' actual direction and the requested direction is called the **slip angle**. The tires are now not in a pure rolling condition. They are sliding sideways as well as rolling forward, called **tire scrub**. Except at very slow speeds, the tires' slip angles will never be zero. In fact, at any speed above a walking pace, there can be no cornering force without a slip angle.

Understeer and Oversteer

The phenomena called **understeer** and **oversteer** have been mentioned several times in the preceding sections and chapters. These phenomena can be more accurately defined in terms of the relative slip angles between front and rear tires. If the slip angles of both front and rear tires are equal, the car has neutral steer. If the front slip angles are greater than the rear, understeer occurs. The driver has to increase the steering angle to compensate for the tendency of the tires to travel at a smaller angle. The driver has to fight the slip angle to force the car to the desired path. If the driver reaches full steering lock and runs out of compensation, then the car will travel its own path, perhaps leaving the road.

If the rear slip angles exceed the front, then the rear tires try to travel a larger radius than the front tires, the turn-rate increases on its own causing oversteer, and the driver has to reduce steering angle to compensate, lest the car spin out backwards. This is why you have to countersteer to react to a car oversteering. If slippery surfaces cause a rear wheel skid, the driver may need to turn the steering wheel to full opposite lock to stop the vehicle's rotation. That will often cause it to begin a skid in the opposite direction with the driver madly wheeling back and forth until it (hopefully) settles down. This is why suspension designers prefer to create an understeering tendency in the suspension setup, as it is somewhat easier to recover from, especially by simply slowing down.

Sway Bars

The tendency for a car to roll in a turn can be reduced by fitting so-called **sway bars** (also called **stabilizer bars**, **anti-sway bars**, or **anti-roll bars**) at the front or rear or both. Invented by a Canadian, Stephen Coleman, in 1919, a sway bar is a torsion bar that is connected across the car between the left and right lower wishbones as shown in Figure 6-43. The long, straight portion is supported on the frame or body in bushings so it can twist about its own axis. The left and right legs of the sway bar connect to the wishbones. When the car is level side to side,

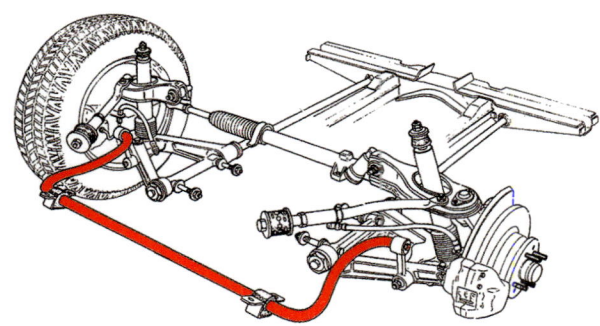

FIGURE 6-43 Sway bar on an Alfetta *(Courtesy of Evan Mason)*

the torsion bar does nothing. As the car leans in a turn, the outside wheels are driven up into the fenders as the suspension on that side compresses, and the inside wheels extend from the fenders as the inside suspension expands.

The compression of the outside suspension lifts that end of the sway bar and that motion puts an upward force on the opposite wishbone. The upward force on the inside wishbone counters the extension of its spring and acts to reduce its motion. At the same time, the outside end of the sway bar is trying to push the outer wishbone down, adding to the spring force on that side. So the sway bar tends to reduce the amount of roll in a turn by increasing the vehicle's roll stiffness. This helps the car remain more nearly level side to side in turns. Sway bars can be added to the front or rear suspensions, or both. If a sway bar is fitted only to the front, it will increase understeer, and if only to the rear will increase oversteer.

Anti-Dive and Anti-Squat

When a car brakes, there is a weight transfer to the front wheels causing the nose of the car to dive. When it accelerates, weight transfers to the rear wheels, causing the back of the car to squat. The front suspension system can be designed to reduce dive, and the rear suspension can be designed to reduce squat, and this is typically done on modern cars. Both anti-dive and anti-squat are expressed as a percent with 100% indicating zero dive or squat. Suspensions are designed to have at most about 50% anti-dive or anti-squat. If the anti-squat were 100%, then the rear wheels would hop and lose traction under hard acceleration. With 100% anti-dive the front wheels would act similarly under hard braking.

Figure 6-44 shows how the percentages of anti-dive and anti-squat are determined. Anti-dive is controlled by angling the wishbones of an unequal-arm front suspension to be closer together at their rear pivots than at their front pivots as shown in part (*a*) of the figure. For anti-squat, the suspension geometry at the rear is angled opposite to the front in a side view. A trailing arm arrangement is shown in part (*b*) of the figure.

In part (*a*) of the figure, the instant center of the wishbones is found by extending their angles until they intersect. A line is then drawn from the instant center to the front tire contact patch. That line crosses a vertical line through the car's center of mass (CG) a distance *a* above the ground. The CG is *b* units above the ground. The anti-dive ratio is *a* divided by *b*. It is about 35% as drawn in the figure.

In part (*b*) of the figure, the instant center of the trailing arms is found by extending their angles until they intersect. A line is then drawn from their instant center to the rear tire contact patch. That line crosses a vertical line through the car's center of mass (CG) a distance *a* above the ground. The anti-squat ratio is *a* divided by *b*. It is about 50% as drawn in the figure.

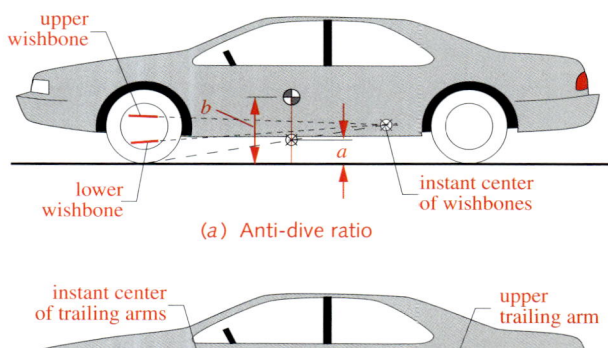

FIGURE 6-44 Anti-dive and anti-squat ratios

CHAPTER 6   SUSPENSION AND STEERING

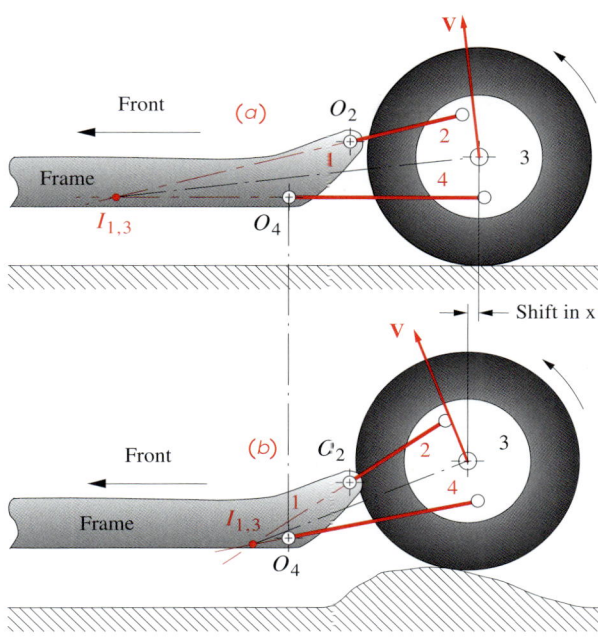

FIGURE 6-45   The "Hopalong Vega"

Bump Steer

Bump steer refers to a car's tendency to spontaneously steer right or left when hitting a bump in the road. Cars with live axle rear suspensions will sometimes jump sideways when one wheel hits a bump severe enough to lift the tire momentarily off the ground while cornering. The lost traction combined with the centrifugal force of the turn may be enough for this to occur. A worse situation is when a suspension design is flawed in such a way that whenever a bump is encountered, the car steers involuntarily.

Figure 6-45a shows a rear suspension design from the Chevrolet Vega of 1970s vintage that was later redesigned because of a disturbing tendency to "bump steer," i.e., turn the rear axle when hitting a bump on one side of the car. The figure is a view looking from the center of the car outward, showing the trailing-arm, fourbar linkage that controls the up and down motion of one side of the live rear axle and one wheel. Links 2 and 4 are pivoted to the frame of the car (which is link 1) at O_2 and O_4. The wheel and axle assembly is attached to the coupler, link 3. The instant center of the suspension linkage at rest is shown at $I_{1,3}$ in the top panel.

Ideally, one would like the wheel to move up and down in a straight vertical line when hitting a bump. Figure 6-45b shows the motion of the wheel and the new instant center location ($I_{1,3}$) for the situation when one wheel has hit a bump. The velocity vector for the center of the wheel in each position is drawn perpendicular to its radius from $I_{1,3}$. You can see that the wheel center has a significant shift in the horizontal component of motion x as it moves up over the bump. This horizontal component causes the wheel center on that side of the car to move forward while it moves upward, thus turning the axle (about a vertical axis) and steering the car with the rear wheels in the same way that you steer a toy wagon. Viewing the path of the instant center over some range of motion gives a clear picture of the behavior of the coupler link because the coupler is always in pure rotation about its instant center. The undesirable behavior of this suspension linkage system could have been predicted from a simple instant center analysis before ever building the mechanism. After several years of production GM changed the Vega's rear suspension to a torque-tube design, and that cured the problem.

Ride Comfort

Maurice Olley of GM, the father of Cadillac's independent front suspension and generally recognized in his time as the ultimate expert on ride and handling said in 1946: *It is generally agreed that for optimum ride, the front end needs to be sprung more softly than the rear, and also that the body-frame structure must be torsionally stiff ... .* This formula only became possible due to the invention of independent suspension with coil springs. Non-independently suspended automobiles rode poorly in part because they were stiffly sprung at front and rear. Their poor sprung-unsprung weight ratio also hurt the ride.

Ride comfort is affected by many factors. Principle among these are the sprung-unsprung weight ratio previously discussed and the natural frequency of the car body on its suspension system. Natural frequency is a physical property of any dynamic system that contains both masses and springs, such as an automobile. The simplest possible dynamic system consists of a single mass suspended on a single spring, and its natural frequency is easily calculated as the square root of the quotient of the stiffness of its spring divided by its mass. The natural frequency of a system as complex as the automobile is calculated the same way using its unsprung mass and the stiffness of its springs.

When a dynamic system is disturbed by a force, it will vibrate at its fundamental natural frequency, measured in cycles per second (Hz) or cycles per minute (CPM). When a car hits a bump, the suspension will vibrate vertically at its natural frequency. The dampers (shock absorbers) damp out this vibration fairly quickly, but the passengers still feel it.

The human body is more sensitive to some frequencies than others, and the sensitivities are different in a head to toe (vertical) direction versus a fore and aft (longitudinal) direction. Very early in automotive development, tests with passengers showed that people prefer a car whose suspension's vertical natural frequency is between 60 and 90 CPM (1 to 1.5 Hz). If the vertical ride frequency reaches 120 CPM (2 Hz), riders report it as harsh. If it drops to 30–50 CPM motion sickness often results.

Passenger cars are typically designed with vertical ride frequencies in the 60–90 CPM range. But high-performance sports cars often have stiffer suspension to reduce roll and improve handling, and these can have vertical ride frequencies in the 120–150 CPM range.

Frequencies in the longitudinal direction at 60–120 CPM also bother people. These result from pitching motions that cause the upper torso to oscillate back and forth. Higher frequencies, in the 1000–2000 CPM range (18–20 Hz) bother the head and neck. So all this presents a complex problem to the designer trying to develop a good riding automobile.

The bump response of the suspension system is not the only source of vibration in an automobile. Tires often generate noise in the 18–20 Hz range. Engines produce vibrations at higher frequencies that can excite natural frequencies within the body structure and create booming. Wind noise is another source of vibration, even with closed windows. Besides the obvious tactics of tuning the suspension frequencies to be appropriate, other techniques used to suppress or hide vibrations to improve ride comfort include rubber motor mounts, rubber bushings in the suspension, and sound insulation.

Chapter 7
Brakes

This chapter will describe the evolution of braking systems and how they work. The earliest automobiles had only rudimentary brakes, inherited from horse-drawn wagons. Wagons traveled slowly so could get away with minimal braking systems. A wagon brake was nothing more than a curved wooden block forced against the iron wheel rim by a long lever. The earliest cars did not move much faster than a horse, so also got away with minimal braking for a while. Wagon brakes were used at first, but after the Michelin brothers invented the inflated rubber tire, wood brakes chewed up tires. As speeds increased and tires replaced iron wheels, brakes began to improve. Daimler was the first to wrap a steel cable around a cast-iron drum on the wheel in 1899. One end of the cable was attached to the frame and the cable made several wraps around the drum with its free end attached to the brake lever. It functioned like a **capstan** when the driver pulled on the lever to tighten the cable on the drum. Unfortunately, a capstan works only in one direction, so this brake could only stop forward motion. If the vehicle stalled on a hill, it would roll backwards down the hill with no braking.

Wilhelm Maybach wrapped a brake shoe lined with friction material around a drum to make the first external-contracting-shoe drum brake in 1901. The brake lever served to squeeze the shoe on the drum to stop it. This brake stopped in both directions. In 1902, Ransom E. Olds used an external-contracting, flexible steel band wrapped around a brake drum to win a race in New York. Instead of requiring the driver to pull on a hand lever, he supplied a brake pedal for that purpose. Other manufacturers were so impressed with the stopping power of Olds' foot-powered design that most had copied him by 1904. Louis Renault is credited with the invention of the internal-expanding-shoe drum brake in 1902. It used two brake shoes inside the drum that were pushed apart to rub on the inside surface of the drum. Putting the shoes inside protected them from the elements. This became the model for drum brakes used to this day.

The disk brake is as old as the drum brake. It uses the same principle as a bicycle's rim brake, squeezing a disk between two pucks. Frederick Lanchester patented a disk brake for the automobile in 1902. He used copper plates to line the pucks, but unfortunately, copper rubbing on cast iron let out a loud screech. Five years later, Herbert Frood replaced the copper with asbestos, which cured the screeching. Asbestos was used as the lining for new brakes and clutches until the 1990s, when it was determined to be a carcinogen. Replacement brake linings still use asbestos. Disk brakes gradually replaced drum brakes in modern automobiles.

The opening photograph is of a 1920 Hispano-Suiza chassis with an early form of mechanical power brakes. It was taken by the author at the Revs Institute Museum in 2014.

Four-Wheel Brakes

The earliest cars had brakes on only the rear wheels. The addition of front brakes was resisted for many years because many thought that braking the wheels used for steering would make the car unstable. Isotta-Fraschini (Italy) was the first to develop four-wheel brakes in 1906 and first offered them in 1909. Eventually, the value of front wheel brakes became accepted and all cars were fitted with them by the 1930s. It is now known that the majority of braking is done by the front wheels because of weight transfer to the front when decelerating. Thus, cars with only rear brakes did not stop well at all.

The fears of instability induced by braking the steering wheels proved to be unfounded. However, in the 1920s, when all U.S. cars used beam front axles and were fighting shimmy, wobble, and bounce (as described in the previous chapter), adding the mass of brakes to the front wheels would have made the shimmy problem worse. This might explain the reluctance of U.S. manufacturers to adopt four-wheel brakes. Most European manufacturers adopted them much earlier, but they also used front independent suspensions decades earlier than the U.S.

Pierce Arrow and Rickenbacker adopted four-wheel mechanical brakes in 1923. In the same year, Bendix developed a mechanical four-wheel braking system and sold it to GM, Willys, and others. Cadillac and Buick used them on their 1924 models, and the other GM marques followed soon after. All of Ford's Model T's had only rear brakes, but their 1928 Model A had them on all four wheels. By the 1930s, all American manufacturers were offering four-wheel brakes.

Drum Brakes

Most cars in the U.S. used only drum brakes until the 1970s. European manufacturers adopted disk brakes much earlier, in the 1930s. The earliest drum brakes used external bands that contracted around the drums. These were soon superseded by internal-expanding-shoe brakes. Putting the friction shoes inside the drum protected them from both dirt and mud and also allowed the brakes to be self-energizing.

Hydraulic Brakes

All brakes were mechanically operated until the first hydraulic brakes were invented in 1918 by Lockheed, primarily for aircraft. Duesenberg was the first to apply four-wheel hydraulic brakes to a production automobile in 1921. They had used them on their race cars since 1914. Walter P. Chrysler was the president of Chalmers in 1923 and put Lockheed hydraulic brakes on its top models that year. The next year, when Chalmers foundered, Chrysler reorganized it and named it after himself. Chrysler redesigned the Lockheed hydraulic brakes by replacing their leaky rawhide cup seals with rubber ones and made them standard equipment in 1924. Lockheed allowed Chrysler to use the improved design royalty free in return for allowing Lockheed's to use the new seals. Bendix bought the Lockheed Hydraulic Brake Company in 1930. By the mid 1930s, most manufacturers had switched to hydraulic brakes. The last to change, as usual, was Ford in 1939.

Hydraulic brakes use a master cylinder, whose piston is connected to the brake pedal, to pressurize hydraulic fluid (oil) in metal tubes that run to all four wheels, where they connect to a wheel cylinder in each brake assembly. The wheel cylinders force the brake shoes against the brake drum to stop the car. A cutaway of a wheel cylinder is shown in Figure 7-1. It contains two pistons with cup seals. A spring keeps the pistons separated. The pressurized fluid is introduced between the pistons, forcing them apart and pushing the cup seals against the cylinder wall. The pistons push the brake shoes

CHAPTER 7 BRAKES

FIGURE 7-1 Drum-brake wheel cylinder
(Courtesy of Nimal Kumar)

through short links. Figure 7-2 shows a schematic drawing of a basic drum-brake assembly.

The wheel cylinder is at the top of Figure 7-2. The brake shoes are pivoted to the backing plate at the bottom. A coil spring just below the wheel cylinder pulls the shoes together to retract them when the brake pedal is released. There is a small clearance between the brake-shoe linings and the brake drum (not shown). The star wheel just below the wheel cylinder is

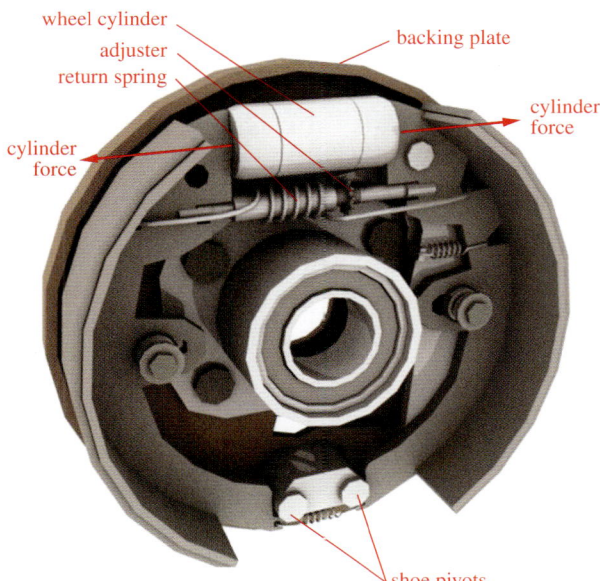

FIGURE 7-2 Drum brake assembly
(Courtesy of Wapcaplet)

the brake adjuster. The adjuster shaft is threaded, and when turned, spreads the brake shoes to make up for lining wear. Early brakes had to be manually adjusted periodically. In 1925 the Cole offered self-adjusting brakes. They were not offered again until 1946 by Studebaker. By the 1950s most cars had them.

The first hydraulic brakes had one circuit to service all four wheels until 1960 when Wagner Electric developed a dual-cylinder brake system. This split the hydraulics into two systems for safety, one servicing the front wheels and the other the rear. Some cars split the brakes diagonally, front right with left rear and vice versa. If one circuit failed, the car could still stop with the other. Cadillac introduced this system in 1962, AMC in 1963, and the Federal Government mandated that all new cars sold in the U.S. be so equipped as of 1967. NHTSA[1] claimed in 1983 that this system had prevented 40,000 accidents.

Self-Energizing Brakes

The relationship between the brake-shoe pivots and the friction surface creates an overturning moment that causes one brake shoe to be pushed against the drum with more force than the wheel cylinder provides, as shown in Figure 7-3. This is called a **self-energizing** force or **servo-effect**. Both shoes in Figure 7-3 are shown in contact with the drum. The shoe on the side of the axle toward the vehicle direction is called the **leading shoe** and the other is the **trailing shoe**. The cylinder's action creates friction forces on both shoes at the interface between the shoe linings and the drum as shown. The reactions to these friction forces occur at the shoe pivots and form friction couples with the friction forces. The moments of these two couples are labeled friction moment. Note that both friction moments are clockwise in this example. A clockwise moment tends to force the leading shoe harder against the drum, but it relieves the force on the trailing shoe.

[1] National Highway Transportation Safety Administration

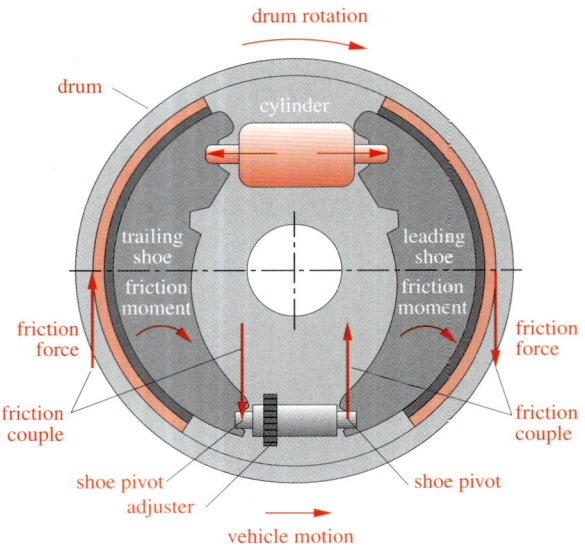

FIGURE 7-3 Self-energizing drum brake

The leading shoe is called **self-energizing** because the car's momentum, converted to friction moment, is creating an additional force on the shoe. When the car reverses direction, the leading and trailing shoes swap sides; the left shoe in the figure becomes self-energizing. It is possible to arrange the brake-shoes' pivots such that both shoes are leading shoes in (say) the forward direction. In that case, the braking power in forward will be enhanced to a greater degree at the expense of reducing it further in reverse. Most modern cars use disk brakes on the front and many also have them at the rear. Others use disk brakes in the front and drums at the rear. Disk brakes cannot be made to be self-energizing. In the now rare case of a vehicle with drums all around, the front drums are often arranged with two leading shoes to increase their stopping power. But the rear brakes use a leading/trailing shoe arrangement to give good braking in reverse.

One disadvantage of self-energizing brakes is that they can be "grabby" and have a nonlinear feel. Light application of the brake gives a somewhat proportional response to pedal force, but stronger application lets the self-energizing take over, which can surprise an inexperienced driver with its vigor. Disk brakes are very linear in their response to pedal pressure.

An advantage of drum brakes is that their shoes do not touch the drum when the brakes are off. There is a small gap between shoe and drum. So application of the brake pedal has a small amount of travel before any braking is felt. But, the gap makes drum brakes efficient as there is no energy-wasting drag when off. Disk brakes operate with zero gap and can have a small drag when off.

The purpose of a braking system is to convert kinetic energy of the vehicle to heat through friction. This raises the temperature of the brake parts, and this heat needs to be dissipated to the environment. Because the brake shoes are contained within the drum, they do not have a good path for heat dissipation. If the brake temperature gets too high, it can cause brake fade, a condition in which the brake loses much of its stopping power.

If the temperature at the wheel cylinder exceeds the boiling point of the brake fluid, the brake fluid will vaporize and then all pressure will be lost. This failure can be catastrophic with the brake pedal going to the floor with no braking power at all. This is why signs on steep downhill roads warn drivers to stop periodically to cool the brakes lest they boil the brake fluid.

Disk Brakes

The earliest examples of disk brakes such as the 1902 Lanchester were manual. By the time disk brakes were beginning to be commonly used in the 1930s in Europe and the 1970s in the U.S., they were only hydraulically energized. Figure 7-4 shows a modern disk brake assembly from a Renault. The caliper contains at least two

CHAPTER 7   BRAKES

FIGURE 7-4   Disk brake assembly
(Courtesy of David Monniaux)

A two-piston, floating-caliper, disk brake is disassembled in Figure 7-5 to show the pistons. It is a floating caliper because, when assembled, the caliper is free to slide on the pad support in a direction parallel to the wheel axis. This allows the caliper to self-center on the rotor as the rotor and pads wear. Some disk brakes have fixed calipers, and these tend to wear the pads unevenly. Some disk brakes in race cars have as many as six pistons, or "pots," per caliper. High performance and sports cars often also have two, four, or six "pots" (pistons) per wheel.

All brakes heat up from friction in use. Disk brakes have inherently better cooling than drum brakes because they are exposed to the airflow. As a result, disk brakes are less prone to brake fade than drum brakes. Some rotors are ventilated with air passages in their center to improve cooling as shown in Figure 7-4. Disk brakes also recover from water exposure faster than drum brakes because their close-fitting pads scrape the water off the rotor.

pistons, one on each side. The pistons push on the back of the brake pads, one on each side of the rotor, which is part of the wheel hub. Some disk brakes have multiple pistons in the caliper to provide more braking force.

The disk rotor is most commonly made of gray cast iron, and if that becomes overheated from hard use, the brakes can fade and the rotor can warp. Race cars and high performance street cars use ceramic- or carbon-composite rotors to obtain higher heat resistance and less brake fade along with lighter unsprung weight. Their brake pads are also made of ceramic- or carbon-composite for lower wear.

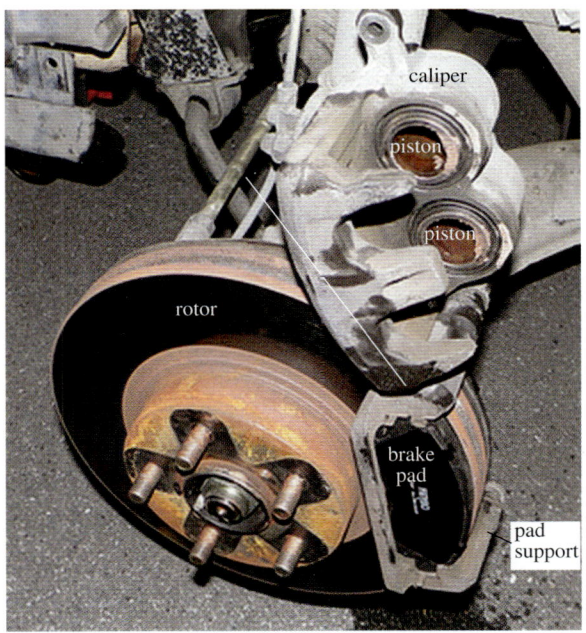

FIGURE 7-5   Two-piston, floating-caliper disk brake
(Courtesy of Awojtas)

Figure 7-6 shows a cross-drilled, ceramic disk rotor on a Porsche, who provide these brakes as standard equipment on their highest performance models and as an option on all other models. The holes lighten the rotor, improve cooling, and vent the outgassing of hot pads. Figure 7-7 shows a carbon-composite disk rotor as used on AMG Mercedes models, also cross-drilled. Carbon disks are used on many race cars and can sometimes be seen glowing red-hot as the cars circle the track at night.

AUTOMOTIVE MILESTONES

FIGURE 7-6 Ceramic disk brake on a Porsche
(Courtesy of Dan Lindsay)

Disk brakes are not self-energizing, which makes them very linear, meaning that the stopping force is proportional to the applied force on the brake pedal over the entire braking range. They give much better braking feel than drum brakes, which are nonlinear due to their self-energizing feature. Disk brakes also stop a car in much shorter distance from the same speed than drum brakes.

FIGURE 7-7 Carbon disk brake on an AMG Mercedes
(Courtesy of Dana60Cummins)

The first automobile to use four-wheel caliper disk brakes in U.S. production was the 1925 Rickenbacker, an expensive car that did not sell well. The Crosley Hotshot was next to offer front disks in 1950. The Crosley brakes proved unreliable and many owners converted their car to drum brakes. Chrysler introduced a different design of disk brake on its Imperial in 1949. This brake used two full disks inside a cast iron drum that were pushed outward against the inside ends of the brake drum. No one else used this very expensive and complicated brake design then or since.

Dunlop developed caliper disk brakes in the U.K. and these were first used with good success on the Jaguar C-Type race car at Le Mans in 1953. Austin Healy put disk brakes on all four wheels of their 100S in 1954, and were the first to do so. The 1953 Triumph used front disk brakes as did the 1955 Citroën DS, which mounted them inboard to reduce unsprung weight. The Jaguar E-Type of 1961 was the street-legal version of their Le Mans-winning D-Type race car and sported four-wheel disk brakes, mounted inboard at the rear as shown in Figure 6-35.

In Germany, Mercedes was first with front disk brakes on its 1961 220SE. In the U.S., Studebaker fitted its avant-garde, 1963 Avanti with a Bendix disk brake system on the front wheels. AMC followed in 1965, making front disk brakes standard on the Marlin and optional on all other models. Ford offered front disks on the Thunderbird and Continental. In 1965, the Chevrolet Corvette was the first U.S. car to make four-wheel disks standard equipment, and has every year since. By the seventies, most manufacturers had switched to disk brakes on the front wheels. Presently, there are no automobiles, except possibly some small econocars, that do not have disk brakes at least on the front wheels. Many have them on all four.

Power Brakes

Power brakes are servo systems. A servo system applies power from some source to a mechanism; the amount of power applied is controlled by a low-level input from an operator. It serves to multiply the control force that the operator applies to a level that the operator is physically incapable of. Power steering is an example of a servo system, as are power brakes.

Mechanical Power Brakes

The earliest power brakes were strictly mechanical. Hispano-Suiza designed a very clever, mechanical-servo-driven, non-hydraulic, four-wheel braking system in 1919. Figure 7-8 shows two views of this brake servo mechanism. A gear-driven *brake drum* is fitted to the *tailshaft* of the transmission and rotates on a horizontal *servo shaft* that runs across the car. This *brake-servo drum* turns whenever the car is moving. When the driver steps on the *brake pedal*, a linkage connecting it both to the *brake-servo output plate* and the brakes applies internal brake shoes within the drum. The drag from the *brake-servo drum* on the brake shoe turns the *servo plate* clockwise from the side shown in part (a). The *servo plate* turns the *servo shaft*, which rotates the *servo output lever* shown in part (b). This *servo shaft* connects to two long rods that run to the brake drums at front and rear to apply the brakes. A duplicate *servo output lever* is at the opposite end of the *servo shaft* to operate the brakes on the other side of the car.

At zero or low speeds the brake pedal directly applies the brakes with little or no assist. But, when the car is moving, the servo system makes all the momentum of this 5000-lb car available to help stop it. Since momentum is mass times velocity, the faster the car goes, the more stopping power it has. And this is all triggered by a very light force from the driver's foot on the brake pedal. That force needs only be large enough to press the servo brake shoe against the servo drum sufficiently to turn the servo plate. The car does the rest and stops itself. Rolls-Royce liked this patented design so

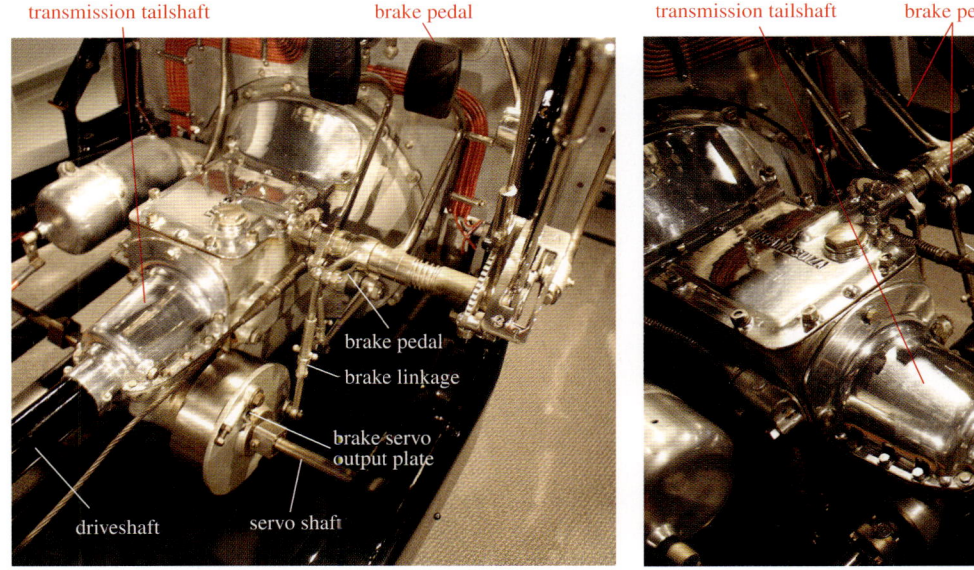
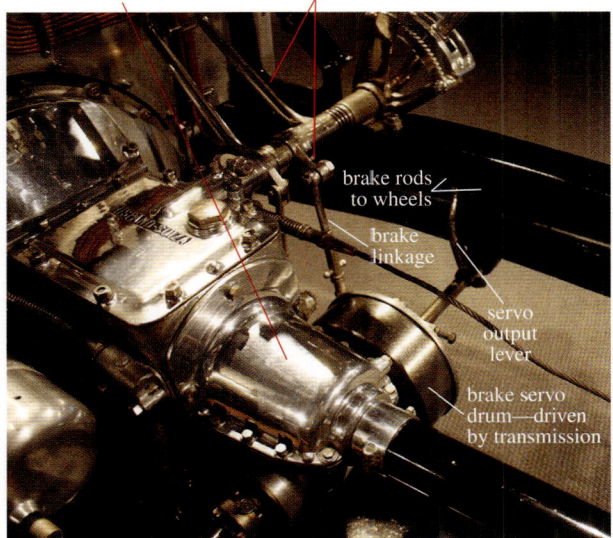

(a)   (b)

FIGURE 7-8 Mechanical servo power brake on a 1919 Hispano-Suiza

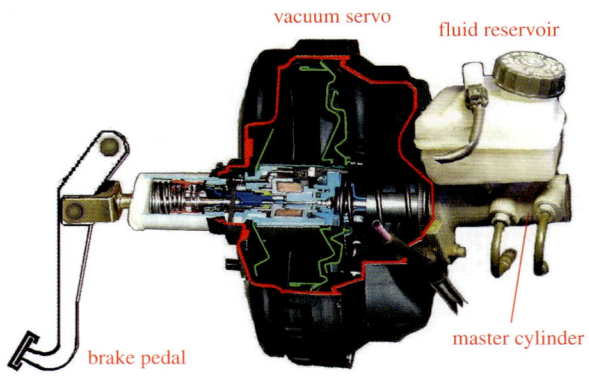

FIGURE 7-9 Power Brake Vacuum Servo
(Courtesy of Paul Day)

well that they licensed it from Hispano-Suiza and used it on all their cars through 1948!

Vacuum Power Brakes

Four-stroke, gasoline engines have vacuum of about 20 in-Hg (inches of mercury) or –10 psig in their intake manifold system and this can be used to do work. Vacuum powered hydraulic brakes work by amplifying the force applied to the master cylinder by the brake pedal. A cross-section of a power-brake vacuum booster for a hydraulic braking system is shown in Figure 7-9. A large-diameter rubber diaphragm is attached to the master cylinder rod. One side of the diaphragm is exposed to manifold vacuum by a valve that is opened when the brakes are applied. Its force adds to that of the force from the brake pedal to increase stopping power.

In 1929, Duesenberg offered the first vacuum-powered, four-wheel, hydraulically actuated power-brake system in the U.S. Pierce-Arrow also offered them at about the same time as Duesenberg. Lanchester (England) followed in 1931 and Chrysler in 1932. Power brakes were initially an option on American automobiles but have now become standard equipment. A power boost is needed particularly with disk brakes because of their lack of self-energizing.

Other Power-Brake Systems

Diesel engines do not generate manifold vacuum because they have no throttle plates or venturi in the intake system to restrict air flow. Their engine speed is regulated by controlling fuel injection. Diesel-engined vehicles typically use a hydraulic pump to boost brake pressure for power brakes, much as is used for hydraulic power steering. Some automobiles with anti-lock brakes (see next section) also use hydraulically boosted power brakes. Some 1985 GM cars used an electrically powered brake booster, which is smaller and lighter than a vacuum booster.

Anti-Lock Braking Systems

Anti-lock braking systems (ABS) have proven to be very effective in reducing accidents, especially under slippery driving conditions. Drivers in cold climates were trained to "pump the brakes" when trying to stop on snowy or icy roads. If one applies the brakes forcefully on snow or ice, the brakes will lock the wheels. Locked wheels will not steer or stop. They will just skid, and the vehicle will proceed in a straight line in whatever direction it was going when the wheels locked up.

The old "cure" was to apply the brakes intermittently, trying to sense incipient wheel lockup (when the vehicle began to deviate from its intended path) and release the brakes for an instant before reapplying them. A skilled driver could keep the wheels on the verge of skidding by this technique and obtain some slowing power from the brakes. To be effective, this "brake pumping" had to be done very rapidly with a frequency of brake application several times a second. All the time the driver's "heart was in his mouth" hoping to avoid a crash. Not fun at all.

ABS has eliminated the need to pump the brakes in slippery conditions because it does so for you and mostly eliminates the "pucker factor." It was first invented for aircraft brakes in 1929 by Gabriel Voisin in France and only found its way into automobiles in significant numbers in the 1970s.

It requires that rotational sensors, called encoders, be fitted to the wheels. These encoders use various technologies such as magnets to sense the rotational position of the wheel within a few degrees. These pulses are monitored by a computer that can detect any differences in wheel speed. If it detects that one wheel has slowed or stopped with respect to the others during braking, it begins to pulse the pressure in the brake line to that wheel, effectively "pumping the brake," as we used to do with the foot. But the computer can release and reapply brake pressure to the wheel much faster that a human foot can do—at 16 or more times a second.

If one learned to pump the brakes before ABS came along, he has to re-learn his braking technique in slippery conditions with an ABS-equipped vehicle. In that case the proper drill is to put maximum pressure on the brake pedal and hold it steady while the ABS does its work. He will feel the pedal pulsating under his foot as the ABS modulates the pressure. The most impressive result is that the vehicle can be steered on a slippery surface with the brake pedal fully depressed when the ABS is operating. Steering was always a dicey proposition when pumping non-ABS brakes on snow. You were happy to get the car to stop and did not much care where it ended up, short of the ditch or another car.

In 1958, a British designed ABS system called *Maxaret* was used on the Jensen FF sports sedan in 1966 (shown in Figure 2-24). In 1969, the Lincoln Continental used a Kelsey-Hayes ABS system on the rear wheels. Chrysler was first to offer a four-wheel ABS system by Bendix on its Imperial in 1971. Ford offered a rear-wheel-only system on its Continental in 1970, and GM introduced a rear-wheel ABS system in 1971 as an option on Cadillacs and the Oldsmobile Toronado. By law, all present day cars must provide ABS as standard equipment on all wheels. It also has become common on motorcycles, beginning with BMW in 1988. The *Insurance Institute for Highway Safety* reported in 2010 that motorcycles with ABS had 37% fewer fatal accidents.

Stability, Cornering, and Traction Control

ABS has also led to the proliferation of stability control systems that use the same sensor systems. Once you have the ability to sense wheel speed at all four corners, much can be done to aid control of the vehicle. **Electronic Stability Control (ESC)** typically adds a **steering angle sensor** and a **gyroscopic sensor**. If the gyroscopic sensor detects that the car's direction does nor agree with the steering angle sensor, it can brake individual wheels to correct the car's trajectory.

The steering wheel sensor also allows the addition of a **Cornering Brake-Control** system **(CBC)** which, if it detects the car is understeering excessively and in danger of a skid, can brake the inside wheels to help turn the car tighter. It can also do the reverse in an oversteering situation, braking the outside wheels.

Traction Control Systems (TCS) also use the wheel sensors to detect when a wheel has lost traction under acceleration. It will then brake that wheel individually to stop its spin, and can also reduce throttle position. No more burnouts! Shucks.

Since 2007, authorities in Europe, have required that all new passenger cars be equipped with ABS. In the U.S., NHTSA has required since 2011 that all new cars be equipped with ABS and Electronic Stability Control. All these systems have brought the possibility of a driverless automobile closer to reality.

Air Brakes

Air brakes are not used on passenger cars, though the 1903 Tincher used them; they are common on large trucks and on trains. To store air pressure, tractor-trailers and other heavy vehicles have engine-driven air compressors and accumulators (air tanks). Their brakes are released by air-driven pistons at each wheel and are spring-loaded on. This is done for safety, so that in the event of a loss of air pressure, the brakes will default to the on position. If you are next to one of these trucks in traffic, you will hear the air exhaust from the brakes as the driver applies them.

Parking Brakes

Parking brakes, also called **emergency brakes** or **hand brakes**, are designed to hold the car stationary when parked. If the service (foot) brake fails, they can be used to stop a moving car, though they have much less stopping power than the service brakes. They also provide a redundant stopping system for safety.

They typically act on the rear wheels only, but some FWD cars, such as the Citroën, fit them to the front wheels. They are mechanical rather than hydraulic and either act on the brake shoes or pads, or use a separate, smaller drum brake. Some older cars put the emergency brake drum on the driveshaft rather than on the axle, but this approach seems to have gone out of favor.

A parking brake is often actuated by a hand lever positioned between the front seats, allowing a passenger to access it if the driver becomes disabled. Some cars, especially sedans, use a foot pedal placed at the far left to actuate the parking brake. Whatever the mechanism, it will always have a latching mechanism so that when set on, it stays on until the driver releases it. A handbrake handle usually has a button at its tip that can be depressed with the thumb to release the latch. A foot-pedal brake will have a lever pull on the dash to release it, or in some cases a second press on the pedal will release it. The brake handle or pedal pulls on steel cables that actuate the brakes.

If the car has drum rear brakes, then the parking brake cables act on the service-brake shoes. Some cars with rear disk brakes will squeeze the pads against the rotor with the parking brake cable motion. Others add small drum brakes to the rear wheels to serve as parking brakes.

Some newer cars have electrically operated parking brakes. Instead of a lever or foot pedal, they have a switch or button to engage/disengage the parking brake. Their operation is usually automatic. The brake is set on when the ignition is switched off and automatically released when the car is put in Drive or Reverse with the foot brake applied. Automatic transmission cars require the foot brake to be engaged before the transmission will come out of Park.

Putting an automatic transmission in Park, or a manual transmission in low or reverse gear when parked, is usually sufficient to prevent the car from rolling. Theoretically, the parking brake alone, if set properly, should hold the vehicle stationary regardless of the transmission's state. However, some safety experts recommend that drivers use both the transmission and the parking brake when leaving a car parked.[1] In Germany this is required by law and an insurance company can refuse to pay for damages that result from a parked car rolling if both systems were not set.

---

[1] Nevertheless, drivers who live where snow, sleet, and rain accompanied by freezing temperatures occur frequently have learned that leaving a car parked with the parking brake on in such conditions can create a problem. If the underside of the car was wet when parked, and the temperature drops before the driver returns to use the car, he may find the parking brake frozen in the on condition. Denizens of these parts of the country (like Boston) quickly learn that it is a bad idea to set the parking brake in winter, and many (this author included) do not.

Chapter 8
Body Design

Body design has changed markedly since the first automobiles and has gone through several phases. Most early car bodies were open, possibly with canvas folding tops, but no windows other than the windshield. If they offered any weather protection at all, it was limited to removable side curtains. The first car bodies were built in much the same way that carriages had been built up to that time. Wood was the principal material for carriage bodies and frames. Automobiles soon had to use steel frames because of the heavy engines and running gear, but early bodies were often made entirely of wood. This soon evolved into wood-framed bodies with thin sheet metal hand-hammered into shape over the wood skeleton. Floorboards and dashboards were wood. The firewall between engine compartment and passenger space had to be metal to live up to its name, though its inside frame was wood. The hood and radiator were made of metal.

Some manufacturers produced only a chassis, i.e., the frame, suspension, steering gear, engine, and transmission, and shipped it to a body builder to add the body to the customer's design. Those manufacturers who were able to sell large numbers of cars gradually took body building in-house and standardized on a few body designs. But the high-end, low-volume manufacturers such as Duesenberg, Bugatti, Delahaye, and Rolls-Royce continued to make only chassis for some years. Cadillacs in the 1910s were bodied exclusively by Fisher Body, later bought by GM. Duesenberg never made any bodies.

In terms of body design, the railroad industry was far ahead of the nascent automobile industry at the turn of the 20th century. Railroad passenger cars were all steel from 1899 on when Edward Budd, a graduate engineer, had designed them for the Pullman Company. The Dodge Brothers contracted with the Budd Company in 1916 to produce all-steel bodies for their line of cars, creating something of a revolution in the auto industry. After Speedwell presented a closed, four-passenger sedan in 1919, their competitors were forced to offer closed cars, whether of steel over wood or all steel. Eventually the all-steel body would take over, but steel over wood frame persisted well into the 1930s. Budd later became expert in working with and welding aluminum for bodies. Though some European auto companies would engage Budd's aluminum expertise for bodies, U.S. manufacturers mostly stuck with steel. That is now changing because of the demands for higher mpg, and current manufacturers are redesigning their bodies and frames of aluminum. The 2014 C7 Corvette's frame is aluminum, as is the cab and body of the 2015 Ford F-150 pickup truck.

The opening photograph is of a 1905 Pierce Great Arrow limousine with aluminum-body by the coachbuilder Bodies of Aluminum Company of America taken by the author at the Seal Cove Auto Museum in 2014.

Body-on-frame cars continued to be the norm in the U.S. until after WWII, but starting in the 1920s, European manufacturers began to explore frameless, unit-bodied cars using the body's sheet metal to provide structural strength. Most passenger automobiles now use unit-body construction. Light trucks, their SUV derivatives, and some Jeeps are still body-on-frame. The Chevrolet Corvette still needs a frame because of its composite plastic body. We will flesh out these different body-design techniques in the following sections.

FIGURE 8-2   1904 Stanley Steamer Runabout
(Photo by author at Boothbay RR Village Museum)

Early Body Designs

At first, all bodies were open because the low power of the early engines could not handle the weight of a closed car. An open two-seater without a top like the 1900, wood-bodied Skene Steamer shown in Figure 8-1 was called a Buggy. Add a top and it was a Runabout such as the 1904 Stanley Steamer with mother-in-law seat in Figure 8-2. If it had only two bucket seats on a sporty chassis, it was a Raceabout like the 1912 Mercer shown in Figure 8-3. In this case the lack of a top was not due to lack of power, as these were essentially race cars for the street.

Soon customers were demanding seating for four, but the short wheelbases of the time left little room for a rear door. Several interesting solutions were offered, one being the *vis-a-vis* (French for face-to-face) like the 1899 De Dion Bouton shown in Figure 8-4. An alternative arrangement was the *dos-a-dos*, or back to back, in which the passenger seat faced rearward behind the driver's seat. Another arrangement was the rear tonneau with a door in the middle of the back to access the rear seat, such as on the 1904 Pope-Hartford in Figure 8-5.

As wheelbases got longer, a rear door could be fitted as on the 1904 Knox Tudor Surrey of Figure 8-6, whose only doors are for the rear seats. Figure 8-7 shows a very expensive 1912

FIGURE 8-1   1900 Skene Steamer Buggy
(Photo by author at Seal Cove Museum)

FIGURE 8-3   1912 Mercer Raceabout
(Photo by author at Heritage Museum)

CHAPTER 8   BODY DESIGN

FIGURE 8-4   1899 De Dion-Bouten vis-a-vis
(Photo by author at Seal Cove Museum)

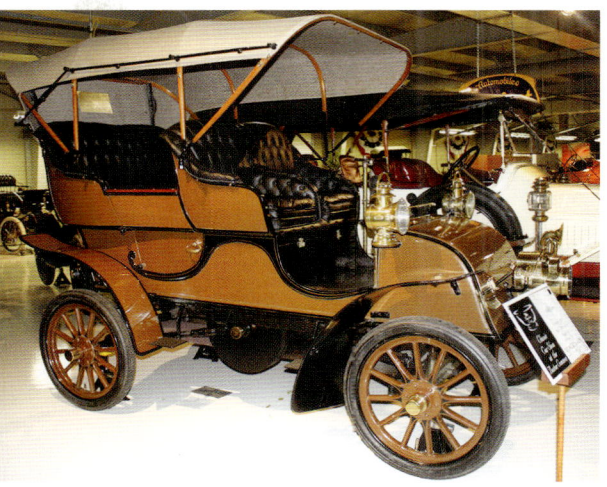

FIGURE 8-6   1904 Knox Tudor Surrey
(Photo by author at Seal Cove Museum)

FIGURE 8-5   1904 Pope-Hartford rear entrance tonneau
(Photo by author at Seal Cove Museum)

This car and chauffer in 1912

FIGURE 8-7   1912 Packard Victoria
(Photo by author at Heritage Museum)

Packard Victoria, which term refers to the fact that the top covered only the rear seat. This style was also called a "To hell with the chauffeur" car, as he sat out in the weather.

All these early cars were body-on-frame with the body made all of wood or a wood frame with a metal skin, hand-formed over it. Figure 8-8 shows an example of an all-wood bodied, 1913 Peugeot, Boat-Tail, Dual-Cowl Phaeton, which **is** actually built like a boat. The term Phaeton refers to an open car with rear-doors. Dual Cowl means that there is a second cowl across the car between the front and back seat where a second windshield is often mounted. Most metal-on-wood-frame bodies

215

AUTOMOTIVE MILESTONES

FIGURE 8-8 1913 Peugeot Dual-Cowl Phaeton
(Photo by author at Seal Cove Museum)

FIGURE 8-9 1913 aluminum Stevens-Duryea Tourer
(Photo by author at Owls Head Transportation Museum)

used steel, but a few used aluminum. Figure 8-9 shows a 1913 Stevens-Duryea tourer with an unpainted aluminum-over-wood body and an aluminum frame. The body and frame were originally painted, but the most recent owner of this car was apparently so impressed with it being aluminum that he had the paint stripped and left it in bare metal. The car is presently in the Owls Head Transportation Museum collection in Maine.

Closed Bodies

Renault was the first to enclose a car body in 1900 when they made a two-passenger, single-cylinder "Doctor's Coupe." It was taller than it was long and was rather ungainly looking. In the U.S., the Speedwell Motor Car Company introduced a two-door, four passenger car that they called a sedan in 1911. This is apparently the first use of "sedan" to describe a body style. In Britain they still refer to a sedan as a "saloon."

As described above, Dodge offered an all-steel touring car and followed it with an all-steel four-door sedan in 1919. Though this body eliminated the wood frame, it was not really all-steel because the roof was open except for the edges, and a canvas-over-wood-and-chicken-wire panel closed it. Despite many early adopters of "all-steel" bodies, this innovation was not complete until the late thirties. In the early thirties, many manufacturers, including GM, Chrysler, and Ford, were still making metal-on-wood bodies. Ford and Chrysler eliminated the wood in 1934. GM pioneered the "turret top" body with an all-steel roof in 1935 but still had wood frame elsewhere in the body until 1938.

In 1919, Hudson's low-priced brand, Essex, offered a Budd-built two-door sedan for only $100 more than their open tourer, and that forced competitors to follow suit. By 1922, 55% of Essex production was closed cars. In 1920, only 17% of all cars had closed bodies. By 1927, that number was 85%. Figure 8-10 shows a 1925, four-door Rickenbacker sedan. This was a very expensive car that had four-wheel disk brakes. But it did not sell well.

Europe was slow to adopt the all-steel body because their production runs were much smaller than in the U.S., and they could not justify the large investment needed for the $60,000 presses to make body panels. So, they stuck with metal-over-wood bodies much longer than the Americans. In fact, Morgan is still making metal-on-wood bodies for its Plus Four and Plus Eight lines of two-seater sports cars in England. A modern Morgan Plus Eight is shown in Figure 8-11. This car also has sliding-pillar suspension, as most Morgans have had since 1909.

CHAPTER 8   BODY DESIGN

FIGURE 8-10   1925 Rickenbacker four-door sedan
(Photo by author at Heritage Museum)

FIGURE 8-11   Late model Morgan Plus-Eight
(Photo by the author)

The first European manufacturer to adopt all-steel bodies was Citroën in 1925. They paid royalties to the Budd Company in Philadelphia to use their patents. Morris in England followed in 1927 by buying the pressed steel panels from the Budd-affiliated Pressed Steel Company in Cowley, GB, who had bought their presses from the U.S.

Fabric Bodies

Metal-on-wood bodies quickly developed squeaks and rattles as they loosened up on rough roads. To stop the squeaks, Charles Weymann patented a different method of making wood-framed bodies by connecting all the wood-frame members with thin metal plates. The wood parts now no longer touched one another, so they wouldn't squeak. He then covered the frame with fabric. This allowed the body to flex with the chassis. It also reduced weight and eliminated the tedious and time-consuming painting process as the fabric was precolored, and also could have a basket-weave or other pattern. This was a big hit at the 1921 Paris Auto Show and many European manufacturers used fabric bodies. Only a few American coach builders used them, even though Weyman had an office in Indianapolis.

Woodies

The station wagon with a wood body on a sedan chassis was first offered as an after-market modification to the Ford Model T in 1919 by the Stoughton Wagon Company in Wisconsin. Early models were sometimes called depot hacks and were used by hotels and resorts to pick up patrons at the railroad station. Figure 8-12 shows a 1923 Model T Ford Depot Hack.

Many farmers also found them useful for their ability to double as a truck and they bought so many Model T Woodies that Henry Ford decided to offer one from the factory in 1929 on the Model A chassis. He outsourced the body

FIGURE 8-12   1923 Model T Depot Hack
(Photo by author at Boothbay RR Village Museum)

to Briggs in Detroit who used wood from Kentucky. The Ford Woodie wagon cost $695 as compared to $525 for a sedan and they accounted for only about 5000 of Ford's production of 1.5 million cars in 1929. The real Woodie station wagons did not survive WWII, but the station wagon continued into the 1980s as a popular family car with wood-grained decals on their all-steel bodies.

Streamlining

The early cars were quite obviously not streamlined as is apparent in all but one of the figures in the preceding sections. Speeds were so low in the beginning that reducing resistance to motion through the air was a low priority for designers. The burgeoning aircraft industry in the 1920s was teaching some lessons about the value of streamlining. Airplanes travel much faster than automobiles, and the negative effects of drag were well understood by aeronautical engineers. They knew that resistance to motion through the air, called drag, is a function of the square of speed. Also larger frontal area increases drag, and the shape of the vehicle is also an important factor.

Work in Germany by aerodynamicist Paul Jaray on the design of lighter-than-air ships called Zeppelins showed that a teardrop shape, with the fat end of the teardrop forward, most effectively reduced drag. This generic shape is used on the fuselage and wings of present day commercial aircraft. A rectangular box or "brick" shape is about the worst possible from an aerodynamic standpoint. As automobile speeds increased from 30–40 mph to 60–70 mph, the importance of streamlining became apparent to automotive engineers.

A 1932 wind-tunnel test using scale models of a car typical of 1932 designs and a proposed

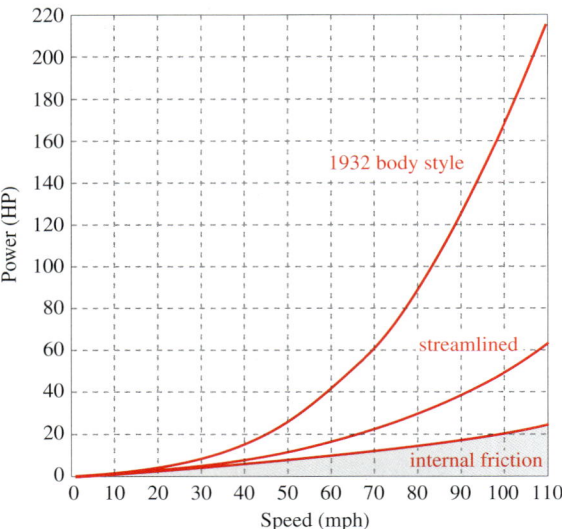

FIGURE 8-13 Power to move a car through the air at speed

teardrop-shaped body showed the effects on power required to move the two shapes through the air.[1] Figure 8-13 shows a plot of their test data. The bottom curve represents the losses due to friction in the engine and drive train. The top curve is the power needed to overcome total losses from internal friction plus air drag for a car of typical, boxy, 1932 style.

The curve labeled 'streamlined' represents the power requirements for a 1932 engineer's concept of what a streamlined car body should look like. These data show the significant economy that streamlining provides. It happens that the streamlined curve is not far from representing a modern auto-body design. A modern passenger car needs about 15–20 HP to push it down a flat road with no headwind at 50–60 mph.

Drag Coefficient (C_d)

The drag coefficient is a number that represents the relative resistance of a body to motion through a fluid, either liquid or air. The very

---
[1] "Economy of Streamlining the Automobile," *SAE Journal*, Vol. 30, No. 3, March, 1932, pp. 150-152.

complicated formulas to calculate drag all contain this coefficient of drag, usually denoted by C_d. If you read road tests of vehicles in enthusiast magazines like *Road and Track*, you will often see the C_d factor listed for a tested vehicle. It is a useful metric to compare how "slippery" a given vehicle is through the air compared to another. The lower the C_d number the less drag it has.

A list of the C_d values for a host of vehicles can be found at http://en.wikipedia.org/wiki/Automobile_drag_coefficient. They range from 0.60 for a typical pickup truck to 0.48 for the original VW Beetle or Rabbit, 0.42 for a Lamborghini Countach, 0.39 for a Chevrolet Tahoe, 0.35 for a BMW M3, 0.29 for a C6 Corvette, and 0.24 for a Tesla Model S.

The Chrysler Airflow

In 1934, Chrysler brought out a radically redesigned and restyled automobile called the *Airflow*. Its shape was based on extensive wind-tunnel testing begun before 1930. Carl Breer, chief engineer at Chrysler, watched military aircraft landing at a base on his commute and wondered if their sleek shape could be applied to the automobile to improve its performance. He assigned an engineer, Bill Henshaw, to do some experiments.

Henshaw knew Orville Wright and enlisted his help. Wright supervised the construction of a wind tunnel at Chrysler's Highland Park facility, and they began testing models of car bodies. They knew the theoretical value of the teardrop shape, but they were not designing an airship, so had to compromise the ideal shape in order to house engine and passengers in a reasonable length vehicle. At one point, Breer suggested that they turn their model of the current Chrysler body backwards in the wind tunnel. They discovered that running the car backwards had 30% less resistance to airflow than running in a forward direction. After that test, Breer looked out a window at the cars in the parking lot and said "Just imagine, all those cars have been running in the wrong direction all this time."

All this work led to a radical new design for a car body to be called the Airflow. Cars of the time had about 65% of their weight on the rear wheels. With three passengers in the back seat, this increased to 75%. Breer moved the rear seat forward from its standard position on top of the rear axle to just forward of the axle. This required that the engine be moved forward from its standard position behind the front axle to be on top of the front axle in order to make room for the front passengers. This resulted in close to a 50/50 weight distribution between the wheels.

When all this was done, they discovered that they could use softer springs all around that gave a much better ride than on earlier cars. Because of all the weight over the rear axle, the rear springs on previous cars had to be stiff, and this gave the passengers sitting above it a rough ride.

The shape of the all-steel Airflow's body as shown in Figures 8-14 and 8-15 was designed to be more aerodynamic than the previous 'two-box' designs. Chrysler paid a royalty to Paul Jaray to use some of the ideas in his patents.

FIGURE 8-14   1934 Chrysler Airflow front view
(Courtesy of Randy Stern)

FIGURE 8-15 1934 Chrysler Airflow rear view
(Courtesy of Trekphiler)

FIGURE 8-16 1937 Chrysler Airflow after a facelift
(Courtesy of dave_7)

The Airflow design did away with the vertical radiator at the front and made the nose of the car curve smoothly downward from hood to bumper, like a waterfall. The headlights were integrated into the body instead of being perched on the fenders. Moving the rear passengers forward allowed the tail of the body to be sloped to come to somewhat of a point at the rear bumper. The windshield, previously a flat, vertical sheet, became a V, slanted in both horizontal and vertical planes to part the wind better.

Chrysler revealed its radical new Airflow at the January, 1934 Detroit Auto Show. It generated much excitement and they took many orders. Unfortunately, due to production delays, the car was not ready for delivery to dealers until May, and the early models had problems. This resulted in many recalls for issues as serious as engines coming loose from their mountings at 80 mph. Some customers had canceled orders and chosen another car before the Airflow became available. Given its teething problems, many customers were dissatisfied with their purchase. All the bad press suppressed sales.

The result was that the car was a colossal flop in the market, and Chrysler dropped it after 1937. They lost a lot of money on it. It seemed that the public was not ready for such a radical restyling of a product that they had become used to. The 'waterfall' grill engendered the strongest negative reaction. For 1935–37 they redesigned the nose to look more conventional, as shown in Figure 8-16.

Despite its problems, the Airflow has to be considered an engineering success and is a true milestone in the development of the automobile. Once the bugs were worked out, it became a very reliable and pleasant car to own and drive.

Its ideas were widely copied by every manufacturer in the years that followed. No sedan puts its rear seat passengers above the rear axle anymore. Fifty-fifty weight distribution is now considered a goal, and modern cars are much more aerodynamic than even the Chrysler Airflow was. Moreover, the Airflow was the first American car to attempt a form of unit-body construction—the subject of a later section of this chapter. Most modern cars use unit-body construction rather than body-on-frame. Chrysler was an early advocate of this method in the U.S., though the Airflow did have a frame and non-independent suspension. They were the first in the U.S. to weld all body panels together and this resulted in a much stiffer body bolted to the frame, which reduced squeaks and rattles. It can be considered a stepping stone on the way to a true unit-body.

CHAPTER 8 BODY DESIGN

FIGURE 8-17 Paul Jaray's model of the Tatra T77
(Courtesy of Cimmerian Praetor)

FIGURE 8-18 A production Tatra T77 on the test track
(Courtesy of AlfvanBeem)

European Streamliners

All good ideas in design, engineering, and science seem to spring up in multiple locations around the world at about the same time. Streamlining is one example. Probably as a result of research done by the young aircraft industry, the value of streamlining became well known around the world in the 1920s. European automobile manufacturers were ahead of the U.S. in this as well as in many other automotive technologies.

Tatra in Czechoslovakia was a leader in streamlining its automobiles. They had hired Paul Jaray, the aerodynamicist responsible for the later Zeppelin designs, to help them design their car bodies. An early result was the 1934 Tatra T77. This was, like most Tatras, a rear-engined, air-cooled, rear drive car, and that made streamlining it easier. It did not have a bulky engine and radiator in its nose, and the passengers could be moved forward to allow the body to approach the ideal teardrop shape. Figure 8-17 shows Paul Jaray's 1933 model of the Tatra T77 and Figure 8-18 shows a production version rounding a test track.

This was a very advanced automobile for its time. It had a central-tube steel chassis, a rear-mounted, 75-HP, V8 engine with overhead valves, hemispherical combustion chambers, and four-wheel, fully independent suspension using swing axles at the rear. It used extensive magnesium alloy in the engine, transmission, suspension, and body to reduce weight. And, it had a drag coefficient of only 0.24, making it more slippery than a C6 Corvette and the equal of a Tesla S. All this in 1934! No wonder Ferdinand Porsche was ordered to copy Tatra!

Figure 8-19 shows the evolution of automobile body shapes from the early 1900s to WWII. They went from squared-off boxes to a more rounded contour. As the reader knows, modern body styles have become much more aerodynamic since 1939. Many present-day sedans have C_d numbers of 0.30 or less. The cars of 1939 were probably closer to a C_d of 0.50.

FIGURE 8-19 Evolution of automotive body shape
(Public Domain)

Unit-Bodies

As has been described, the first automobiles used a frame as the main structural member and their bodies were fastened to the frame at a number of points. A body-on-frame does not add any significant stiffness to the assembly. As early as 1915, engineers began experimenting with a different approach, which came to be called the **unit-body**, or **unibody.** This concept dispenses with the frame entirely or reduces it to vestigial forms to carry the suspension components at each end of the car. Instead of a frame, the sheet metal of the body is reinforced with box- or tubular-sections throughout to provide the needed strength and stiffness. When properly done, this results in a much lighter and stiffer container for the passengers and drive train than does body-on-frame construction.

A unit-bodied car is also sometimes referred to as a **monocoque**, French for "single shell." A monocoque structure has a "stressed skin," meaning that loads on the body are distributed throughout the entire skin, or shell, of the structure. This construction is common in aircraft and boat hulls, but some unit-bodied automobiles take the loads through the reinforced box- or tubular-sections that are part of the body. Technically, these are not true monocoques because much of the skin is not highly stressed. Modern race cars, supercars, and some production cars are designed as true monocoques. A few older production automobiles such as the Jaguar E-Types were true monocoques. It was based on the monocoque, Jaguar D-Type racer that won Le Mans several times. Some older Alfa models also have monocoque construction.

Ruler Frameless—1915

In 1915, while presenting a paper on the virtues of unit-body construction, H. Jay Hayes announced to an SAE audience that the Ruler Auto Company would launch such a car called the *Ruler Frameless* that week. Ruler apparently made about 3000 cars and then faded into history.

Lancia Lambda—1922–1931

The first successful unit-bodied car[1] was the Lancia Lambda made in 1922 through 1931. They produced over 11,000 in that time. Vincenzo Lancia was reported to have conceived the idea of a monocoque car body while on an Atlantic voyage. He studied the ship's construction and decided to apply its lessons to automobile bodies. He received a patent for a monocoque auto-body design in 1919, which was the basis for the 1922 Lambda. All of its inner-body panels are fully stressed and have 10 times the torsional stiffness of a body-on-frame car. The inner monocoque structure is shown in Figure 8-20 along with a complete 1927 Lambda Torpedo. The outer body panels are carried on the inner body and are unstressed.

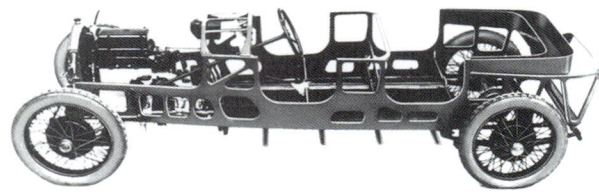

FIGURE 8-20 Lancia Lambda Torpedo and monocoque body
(Photo by the author at the Revs Institute Museum)

---

[1] Vauxhall in England produced a prototype, unit-bodied car in 1903, and Lagonda (England) also did so in 1913, but its production was interrupted by WWI.

CHAPTER 8 BODY DESIGN

FIGURE 8-21 1933 Adler Trumpf
(Photo by author at Tampa Bay Automobile Museum)

FIGURE 8-22 Citroen Traction Avant *(Courtesy of Traction.fr)*

FIGURE 8-23 Citroen Traction Avant body
(Courtesy of Andy Dingley)

The Lancia Lambda also had sliding-pillar, independent front suspension, which was described in Chapter 6. In that section, a contemporary testimonial to this car's handling is reproduced, and the praise it received was as much due to its stiff, unit-body construction as to its independent suspension.

Adler Trumpf—1932–1939

The Adler Trumpf, from Germany, was an early unit-bodied car. It had a 4-cylinder, flathead engine with a four-speed, non-synchromesh transmission. Its all-steel body was supplied by Budd's German subsidiary, Ambi-Budd of Berlin. It was front-wheel drive. They made 50,000 of this model over its run. A Cabriolet Limousine is shown in Figure 8-21.

Citroën Traction Avant—1934–1957

The Citroën Traction Avant was described in Chapter 2 for its pioneering front-wheel drive and also in Chapter 6 for its air/oil independent suspension on the 15CV. These cars also used unit-body construction. They were made from 1934 to 1957 as the 7CV, 11CV, and 15CV models. A 7CV is shown in Figure 8-22 and its bare unit body is shown in Figure 8-23. Note that the roof and sides are integral with the floor pan. It is made from multiple steel panels welded together.

The stiffness of a structure in bending is a function of its depth (or height) raised to the third power. Consider the difference in height of this unit body compared to that of a frame rail and you can begin to appreciate why a unit body is so much stiffer in bending than a car frame. Stiffness in torsion (twisting) is a function of the combination of width and height to the fourth power and that makes the unit body much more resistant to twisting over bumps than a frame-and-body car.

Cord 810/812—1934–1937

Cord brought out its L-29 FWD car in 1929, but succumbed to the depression shortly thereafter.

AUTOMOTIVE MILESTONES

FIGURE 8-24 Cord 812 *(Courtesy of Mr.choppers)*

FIGURE 8-25 Lincoln Zephyr *(Courtesy of Walter Vermeir)*

They went out of business in 1931, but emerged from the ashes in 1934 with a new and radically designed FWD car, the Cord 810/812. The "coffin-nose" Cord, shown in Figure 8-24, is probably the most recognizable American auto to this day. The 810 and its successor, the 812, were made until 1937.

The Budd Company of Philadelphia had filed a patent for a unit-body automobile in 1927 and built an experimental car to their design soon after. Cord worked with Budd to make its 810 a unit-body car, the first to be manufactured in the U.S. It had a sub-frame forward of the firewall to carry its engine and FWD drivetrain, but its body from the cowl back was otherwise frameless. Not a monocoque, it relied on sheet-metal box-sections to take the stresses. The relatively new technique of spot-welding, developed earlier and jointly by Budd and Dodge, was used to join the body panels. The first use of spot welding for car bodies had been on the 1927 Dodge Victory, and it soon became the standard method. It still is.

Lincoln Zephyr—1936–1941

Henry Ford's son Edsel conceived a luxury car to sit below the big Lincolns in their lineup and called it the Zephyr. He had John Tjaarda of Briggs design it as a unit-body car. But Edsel did not like the nose John had styled, so he had his chief designer Eugene T. Gregorie restyle the front end. It was an attractive automobile as can be seen in Figure 8-25, and it also had a lower coefficient of drag ($C_d = 0.45$) than the Chrysler Airflow. It used a small, 75° V12, flathead engine based on their V8. It was made from 1936 to 1941 and was the first Ford to have an all-steel roof. Earlier sedan bodies of most makes typically used a non-removable canvas panel to fill a large opening in the roof to reduce weight and lower the car's center of gravity. Unfortunately, the V12 engine was underpowered and proved unreliable due to some design flaws. The Zephyr was dropped in 1942.

Tatra T87—1937–1948

Tatra was a revolutionary automobile company in Czechoslovakia that introduced many innovations to automotive design. Their influence on the VW Beetle has already been reported. They were one of the first companies to embrace Paul Jaray's theories on aerodynamic design as described earlier. Their T77 described in the section on Streamlining was streamlined, but it was not unit-bodied. Its successor, the T87, added unit-body construction. The T87 also was designed by Paul Jaray with a teardrop shape to be very aerodynamic. It used the Budd patents for all-steel, unit-body construction with a central, tubular-steel backbone. An example of a 1942 model in the collection of the Tampa Bay Automobile Museum is shown in Figure 8-26. Note the dorsal fin.

CHAPTER 8 BODY DESIGN

FIGURE 8-26   1942 Tatra T87 *(Courtesy of John Perodeau)*

Its rear-mounted, air-cooled, V8-engine was also quite advanced with an aluminum block and overhead cams. The car had independent suspension on all wheels and was capable of 100 mph. Because of the rear engine, it had a tendency to oversteer. It was the favorite car of Nazi officers of the Wermacht during WWII. But so many of them were killed in accidents in it that the German High Command ordered their officers not to drive it.

Volkswagen Beetle—1938–2011

The original VW Beetle was not a true unit-bodied car, but it came close. Along with the Citroën Deux Chevaux, the Borgward, and Mercedes 180 and 220 models of 1950s vintage, the Beetle, while lacking a conventional frame, is properly called a "platform-framed" car. These cars typically use a sheet metal platform with a tubular backbone running front to rear. The engine, transmission, suspension, cowl, and wheelhouses are attached to the platform. The body is welded to the platform, which provides overall stiffness similar to a true unit-bodied car. A true unit-bodied car has, in contrast, a three-dimensional "space frame" of sheet metal. The platform frame is considered to be a less-expensive way to achieve the benefits of a unit-body.

Nash Ambassador 600—1941–1949

Sold from 1941 through 1949, the number in its name came from the claim that it could go 600 miles on a tank of gasoline. It is considered to be the first mass-produced, unit-body automobile made in the U.S. It was 500-lb lighter than the previous year's model, which had used body-on-frame construction. Budd helped in its design. It was offered in both fastback and notchback styles, the latter having a trunk bulge. In the interest of aerodynamics, it had no protruding lights, running boards or door hinges. Figure 8-27 shows the car, and Figure 8-28 shows an "X-ray" depiction of its unit-body structure. Nash carried the design over to 1945 essentially unchanged.

Nash continued to build unit-body cars through a merger with Hudson in 1954 to form

FIGURE 8-27   Nash Ambassador 600 *(Courtesy of CZmarlin)*

225

AUTOMOTIVE MILESTONES

FIGURE 8-28 Nash Ambassador 600 *(Courtesy of CZmarlin)*

AMC and the introduction of the Rambler in 1950, the first post-war compact car. Rambler even made a unit-bodied convertible with full window frames to enhance body stiffness.

The Unit-Body Revolution—1960–on

The swing to unit-bodies began in earnest in 1960–61 when Chrysler switched all models but its Imperial to this construction. Imperial converted in 1967. In 1960, the "Big Three" American auto companies (GM, Ford, and Chrysler) were facing serious competition from compact, economical imports coming from Europe and Japan. They scrambled to bring out smaller, lighter, more economical cars and some of these such as the Chevrolet Corvair, Plymouth Valiant, and Oldsmobile F85 were unit-bodied.

By the 1990s, there were few frame-on-body passenger cars being made in the U.S. The exceptions were the largest sedans from Ford and GM. Eventually even these were replaced with unit-body designs. One of the last to give up in 2012 was Ford with its long-running Crown Victoria, a favorite of police departments and taxi companies. The Lincoln variant of the "Crown Vic," the Lincoln Town Car, also hung on to the last, kept in production by demand from limousine companies.

By 2014 the only American frame-on-body vehicles left were trucks and their SUV derivatives like the Chevrolet Suburban/Tahoe, and their GMC and Cadillac Escalade twins. The Ford Expedition/Lincoln Navigator and some Jeep models are still body-on-frame. The Dodge Durango, which originally was built on their small-truck, Dakota chassis, was converted, during Chrysler's Daimler era, to a unit-body with four-wheel independent suspension based on a Mercedes-Benz GL design. The Ford Explorer, originally based on a truck chassis, was also converted to unit-body in 2010.

Aluminum Bodies

The 1913 aluminum-bodied Stevens-Duryea in Figure 8-8 shows that aluminum was used for bodies from the early days of the automobile. All Pierce-Arrow models from 1904–1928 had aluminum bodies. A number of other manufacturers also made aluminum bodies back then, but aluminum fell out of favor, possibly due to its cost. Aluminum-bodied cars began to reappear in Europe in the 1930s with France in the lead. Some of these were body-on-frame, and some of those used aluminum castings for the cowl, which greatly stiffened the body. But, many had unit-bodies made of aluminum. Many automobile, trucks, and SUV manufacturers have used some panels made of aluminum, such as hoods, trunk lids, and tailgates. This section will not deal with those, but will deal only with manufacturers who made essentially the entire body of aluminum.

Pierce-Arrow—1904–1928

Packard, Peerless, and Pierce-Arrow were called the "Three P's," and were considered the best automobiles of their era. Pierce-Arrow was not satisfied with the quality of the typical, metal-over-wood-frame bodies of the time because of their tendency to loosen up and squeak. In 1904, they hired James R. Way from Manhattan's top coach builder, Brewster & Co., to help them de-

CHAPTER 8   BODY DESIGN

FIGURE 8-29   1919 Pierce-Arrow Touring
(Photo by the author)

sign a new way to build car bodies. Way's idea was to cold-rivet cast-aluminum panels together making a more rigid body structure that would not loosen up over time. Pierce worked with the Aluminum Company of America (Alcoa), which was in Buffalo, NY near their factory, to develop the molding process for the thin-walled body parts needed. Alcoa went on to supply the raw castings for production.

The aluminum panels were sand-cast using wood or metal patterns to impress the shape into the sand mold. This process, used for many centuries before, is capable of creating complex curved shapes in metal. Sand-casting is still used to make engine blocks and other complex parts from many metals. The cast-aluminum body succeeded remarkably well and enhanced Pierce's reputation for building a quality product. It also meant that their customers bought a complete car rather than a chassis to be fitted with a custom body, as was the case with some of its competitors.

Pierce's trademark became the graceful tulip-shaped headlights **faired** into the fenders from their introduction in 1913 to the end. These made a Pierce instantly recognizable and they were made possible by casting. While the Pierce aluminum bodies were much superior to a wood-framed one, they were expensive to make. Pierce's new 1906 factory devoted more space to body production than to any other process. They built cast-aluminum bodies until 1920 and then switched to sheet aluminum until 1928 when they went to steel. Figure 8-29 shows a 1919 Pierce-Arrow Touring car in the Heritage Museum collection.

Peugeot Darl'Mart—1930–1953

Emile Darl'Mart was a Peugeot distributor in Paris who also produced low-volume, sports cars based on Peugeot chassis in his own factory. They proved popular enough that Peugeot took their production into its plant at Sochaux in 1936.

The best known of the line is a sports car based on the Peugeot 302, four-cylinder chassis with Cotal transmission. Its aluminum body was designed by Paulin and built by Pourtot, a well-known French coach builder. This model raced at Le Mans in '37 and '38 with some success. Two 2-liter cars finished in fifth and eighth place. It had an aluminum body-on-frame made in roadster, convertible, coupe, and race-car models. Only the fenders are steel. The example shown in Figure 8-30 is a convertible in the Tampa Bay Automobile Museum collection. Its windshield can be cranked down flush with the cowl to enjoy the breeze. Gorgeous car, and aerodynamic too!

FIGURE 8-30   1937 Peugot Darle'Mart sports car
(Courtesy of John Percdeau)

FIGURE 8-31 1938 Amilcar Cabriolet
(Photo by author at Tampa Bay Automobile Museum)

FIGURE 8-32 1953 Hotchkiss-Gregoire
(Courtesy of John Perodeau, Tampa Bay Automobile Museum)

Amilcar—1938–1941

The French Amilcar, in Figure 8-31, is a Cabriolet Limousine (a convertible with window frames). It used an aluminum frame and a cast-aluminum body. It was front-wheel-drive, and had all-independent suspension, rack-and-pinion steering, and torsion bars. It was quite advanced for its time.

Hotchkiss-Gregoire—1950–1954

Benjamin B. Hotchkiss was an American arms manufacturer who supplied weapons and ammunition to the Union in the Civil War. In 1867, Napoleon II asked him to establish an arms factory in France, which he did. Hotchkiss expanded into automobiles in 1903, making his first car in 1904 in France. The Hotchkiss driveline was his invention.

In 1948, Hotchkiss bought the rights to a luxury FWD car from its designer J.A. Gregoire and produced the Hotchkiss-Gregoire in the early 1950s. This was a very advanced and expensive car for the time. It used a streamlined, all-aluminum body with cast aluminum frame and firewall.

Gregoire had a long association with Aluminum Francaise (the French equivalent of Alcoa), and they worked together to explore aluminum as a material for car bodies. The 1953 model shown in Figure 8-32 is the first of only seven, two-door models built. It was presented at the 1953 New York Auto Show and purchased by Ed Cole, who was then the President of the Chevrolet Motor Division of GM.

Jaguar D-Type—1954–1957

Though the D-Type Jaguar was designed to race at Le Mans, it was also sold in small numbers as a road car. Total production was 87 cars, of which 18 were factory team cars, 53 were sold to customers, and the remaining 16 were fitted with additional road-going equipment such as a passenger seat, passenger door, side windows and bumpers to qualify them for American production sports car races. These were re-badged as XKSS versions and sold to the public. The D-Type Jaguar is a famous marque because of its wins at Le Mans in 1955, 1956, and 1957, not to mention its timeless beauty.

It had an all-aluminum monocoque body with an aluminum subframe to support the engine and front suspension. Its suspension wishbones were magnesium. It had disk brakes all around, mounted inboard at the rear. Its 3.8-liter, dual-overhead cam, hemi, inline six was canted at 8.5° to lower the hood.

CHAPTER 8   BODY DESIGN

FIGURE 8-33   Jaguar D-Type *(Courtesy of Sfoskett)*

The dorsal fin integrated in the driver's headrest in Figure 8-33 was added for stability on the Mulsanne straight at Le Mans and was not on the street cars. The E-Type was based on the D-Type but was made more civilized for public use, and it used a steel body. It was produced from 1961–1975 in six- and twelve-cylinder versions.

Mercedes 300SL—1955–1957

Based on the W194 300SL race car, the Gullwing in Figure 8-34 set the automotive world on its heels in 1955. It was the first production, gasoline-powered automobile to use direct (mechanical) fuel injection, which doubled the power of the 3-liter, inline six taken from the Mercedes 300 sedan. Its gull-wing doors were a first as well. With a top speed of 161 mph,

FIGURE 8-34   Mercedes 300SL Gullwing *(Courtesy of Sfoskett)*

FIGURE 8-35   Mercedes 300SL Roadster *(Photo by the author)*

it was the fastest production car to that time. The "base model" had a steel body with aluminum doors, hood, and trunk lid. But an all-aluminum-skinned body was available at additional cost. It was followed by the 300SL Roadster from 1957 to 1963 as shown in Figure 8-35.

Acura NSX—1990–2005 and 2015–?

The NSX from Honda's Acura division was a revolutionary development in 1990. It was targeted to compete with the Ferrari 328 in terms of performance at about 1/4 the cost. Like the Ferrari, it was a two-seat, RWD car with a transverse-mounted-mid-engine, aluminum V6 with titanium connecting rods. It had all the benefits in handling that a mid-engine arrangement confers. It also had an all-aluminum monocoque body by famous Italian designer Pininfarina with aluminum-alloy subframes and suspension

FIGURE 8-36   Original Acura NSX *(Courtesy of ed g2s)*

FIGURE 8-37   2015 Acura NSX *(Courtesy of Latvian98)*

components. This kept its weight to about 3000 lb. It was the first volume-production automobile with an aluminum monocoque body. A photo is shown in Figure 8-36.

At the 2015 North American International Auto Show in Detroit, Honda unveiled the eponymous successor to the original NSX as shown in Figure 8-37. This car is also a mid-engined, RWD two-seater coupe, but it has a hybrid gasoline-electric drive train with all-wheel drive. The longitudinally mounted, twin-turbo V6 drives through a 9-speed, dual-clutch transmission and is augmented by three electric motors, two on the front wheels and one in the rear.

Audi A8—1994–Present

In 1982, Audi entered an agreement with the Alcoa company in the U.S. to develop an all-aluminum car that would be lighter than any others in its class. They developed an aluminum-monocoque, space-frame body that became the new Audi A8, shown in Figure 8-38.

FIGURE 8-38   Audi A8 *(Courtesy of My Hobo Soul)*

FIGURE 8-39   Audi A2 *(Courtesy of Rudolf Stricker)*

Available in either FWD or Quattro AWD models, it uses a V6, V8, or W12 longitudinal, front engine driving the front wheels, with a Torsen differential feeding the rear wheels in the Quattro.

Audi A2—1999–2005

The Audi A2 was a so-called supermini car, designed for an entry-level-market customer. It was a five-door hatchback with transverse 1.2L to 1.6L engines driving the front wheels. It was built of aluminum as a unit-bodied, space-frame car along the same lines and in the same factory as its big brother, the A8. The outer body panels are unstressed. Its light weight provided very good fuel economy, up to 75 mpg with a diesel engine. Its production costs proved too high to make it profitable and it was dropped after a few years. A photo is shown in Figure 8-39.

Morgan Aero 8—2001–2010

The Aero was the first new design of a car from the venerable Morgan Motor Company since 1948. Morgan has never been known to rush into anything new. But when they finally do, the results are usually interesting. The Morgan Aero 8, shown in Figure 8-40 in its AeroMax coupe body, has an adhesive-bonded aluminum frame and aluminum body panels.

FIGURE 8-40 Morgan AeroMax Coupe *(Courtesy of Ed Callow)*

FIGURE 8-41 Jaguar XJ8 *(Courtesy of IFCAR)*

This was the first deviation from Morgan's wood-framed, aluminum-covered bodies since they began production in 1909! Change happens slowly in Malvern Link, England. The Aero is offered as a convertible, coupe, and Targa-roofed coupe. It is front-engined, RWD, and is powered by a 4.8-liter, 360-HP, BMW V8 engine. It weighs only 2600 lb, so has a weight to power ratio of 7.1 lb/HP. For comparison, the 650-HP, 3536-lb, 2015 Corvette Z06 has 5.4 lb/HP.

Jaguar XJ (X350)—2004–2007

The Jaguar XJ sedan has had a long production run from 1968 to the present as the flagship of the Jaguar line. The XJ model made from 2004–2007 had an all-aluminum body and came with a V6 or V8 gasoline, or a V6 diesel engine. Depending on engine, it was badged as an XJ6 or XJ8. A photo is shown in Figure 8-41.

Ford F150—2015–?

The U.S. government's mandate to raise CAFE economy standards to a fleet average of 54.5 mph by 2025 has driven the manufacturer of the Ford F150 pickup truck, the best-selling vehicle in the U.S., to give it an all-aluminum body for 2015. This major redesign reduced its weight by 600 lb and increased its economy figures by a few mpg. This improvement in economy may have had as much to do with the change from a V8 to a turbocharged V6 engine as with the reduction in weight, because its rival, the all-steel Chevrolet Silverado 6.2L V8, weighs about the same and gets similar mpg. It remains to be seen whether other truck manufacturers will follow suit by adopting aluminum for their bodies. It is likely that legislative pressure to reduce fossil-fuel usage may drive more manufacturers of all types of vehicles to move to lighter materials such as aluminum or plastic in the future.

Plastic Bodies

Many modern automobiles use a lot of plastic in them. Intake manifolds, headlight surrounds, grilles, entire nose and rear fascias, dashboards, and many other parts are made of various molded plastics to reduce vehicle weight. Steel is over five times as heavy, and aluminum is nearly twice, as the heaviest plastic. An increasing number of cars are being made with bodies entirely of plastic, and others have used plastic panels for the outer skin with a metal space-frame beneath. This section will discuss only the latter two categories, because it is hard today to find a metal car that does not have some plastic parts.

Fiberglass and Carbon Fiber

Fiberglass refers to glass fibers combined with a liquid polymer (plastic) that hardens to effectively glue the glass fibers together into a strong, solid material. It is commonly referred to as GRP (glass-reinforced-plastic). GRP consists of thin fibers of glass about the thickness of a human hair that are either woven into cloth or matted together like felt to make a non-woven blanket.

In the fiberglass hand-layup process, the mats or cloth are laid over a male "buck" or into a female mold of the shape desired. Then a polymer resin is applied to soak the glass fibers and adhere them to one another. The polymer resin can be any one of a number of thermosetting plastics, such as polyester or epoxy, to which a chemical hardener has been added. An exothermic (heat-generating) chemical reaction ensues that hardens the polymer.

GRP is very strong but, unlike steel and aluminum, it is not ductile. This makes it dent resistant, but, if hit hard enough it will crack or shatter because it is brittle. It is nevertheless repairable. GRP is properly termed a **composite material**, composed of multiple components each of which brings a particular attribute to the mix. The fibers give it tensile strength far above that of the polymer and the polymer provides compressive strength. GRP is the most common material used for small-boat construction. It is best suited to relatively low-volume production as hand-layup is labor intensive.

Other fibers than glass can be used. Carbon fibers are stronger and lighter than glass and so allow thinner, lighter sections to be produced. Carbon fiber has become the high-tech solution to making the lightest possible structure. However, carbon fiber (CF) is more expensive than glass fiber and is about 20 times the price of steel per pound, so CF reinforced plastic (CFRP) has so far been limited mainly to very high-priced, exotic automobiles like Lamborghinis, McLarens, and Bugattis. Formula 1 car bodies are now mostly made with carbon fiber.

Chevrolet Corvette—1953–Present

The Corvette was the first production automobile to use a body made entirely of GRP. Its body sat on a steel frame. Chevrolet has made significant improvements to speed the GRP manufacturing process since the Corvette was launched over fifty years ago and no longer uses the slow, hand-layup method. Rather, it uses matched dies to mold the body panels more efficiently.

The original 1953 Corvette, shown in Figure 8-42, was made in fiberglass just to create a concept car to show at the 1953 GM Motorama. Chevrolet had not intended to produce it right away, but the demand created from its showing was so strong that, to avoid the delay of tooling a steel body, they put it into production in fiberglass that same year, making 300 by hand-layup in 1953. Mechanically this first Corvette was not very sophisticated. It was not yet a true sports car in terms of performance or handling. It used a shortened frame and the inline six from a sedan, a Powerglide automatic transmission, a live axle at the rear, and drum brakes. It was essentially a shortened sedan in a very sexy body.

When Chevrolet's new V8 became available, it was put into the 1955 Corvette, but the suspension was still from a sedan. Other than body re-styling, no significant mechanical changes occurred until 1963 when the model C2 Sting Ray, shown in Figure 8-43, was introduced. From its beginning, this car had four-

FIGURE 8-42 1953 Corvette *(Courtesy of Chevrolet)*

CHAPTER 8　BODY DESIGN

FIGURE 8-43　1963 Corvette *(Courtesy of Chevrolet)*

FIGURE 8-44　2014 Corvette *(Courtesy of Sarah Larson)*

wheel independent suspension and a powerful V8 engine. By the end of its run in 1967, it had become a true sports car, with four-wheel disk brakes, and engines of up to 427 cu. in displacement available. The Corvette has remained in continuous production through its current seventh generation, the C7, shown in Figure 8-44. This car still uses GRP for most of its body panels with the exception of the hood and removable top, which are now carbon fiber. It also now has an aluminum frame. Early-on, Chevrolet developed more efficient methods to produce its plastic components that have eliminated the costly hand-layup method used on the C1 model.

Lotus Elite—1957–1963

Lotus has always been a low-volume producer of small, lightweight, underpowered, two-seat, good-handling, somewhat spartan sports cars. The Elite coupe was an ultra-light model due to its stressed-skin, monocoque, fiberglass

FIGURE 8-45　Lotus Elite *(Courtesy of Tom at Picassa)*

body—a first for any automobile manufacturer. Lotus used a steel sub-frame at the front to support the independent suspension and the engine. In 1957, fiberglass was just starting to be used for monocoque boat hulls and other applications and its strength properties were not yet well known. Probably as a result of this, the Elite's suspension attachment points to the monocoque body sometimes pulled out.

This car weighed only 1100 lb, barely over half a ton, and was powered by a tiny 1.2L inline, 4-cylinder, all-aluminum front-mounted engine of 75 HP driving the rear wheels. It did not have much power, but it was light and had a drag coefficient of only $C_d = 0.29$, so its performance was adequate. A photo of an Elite in racing trim is shown in Figure 8-45. Lotus reused the Elite name on a larger, 2-door hatchback coupe from 1974-82, but that car used a steel backbone with an unstressed fiberglass body shell like the Lotus Elan described below.

FIGURE 8-46　Lotus Elan backbone *(Courtesy of Joc281)*

AUTOMOTIVE MILESTONES

FIGURE 8-47 Lotus Elan coupe *(Courtesy of Fridongis)*

Lotus Elan—1962–1975 and 1989–1995

Lotus replaced the Elite with the Elan and abandoned the monocoque, fiberglass body for a steel backbone structure to support engine and suspension as shown in Figure 8-46. An unstressed fiberglass body was mounted on the backbone. It had fully independent suspension, a 1.6L, DOHC inline four, and was front-engine RWD. It was made in a roadster with optional hardtop and a coupe body style. The coupe is shown in Figure 8-47.

Pontiac Fiero—1984–1988

The Fiero was the first mid-engined car produced by a U.S. manufacturer. So far it is also the only one unless and until Chevrolet makes good on its long-rumored mid-engined Corvette. The Fiero's transverse-mounted, 2.5L "Iron Duke" IL4 used in other GM economy cars did not give the Fiero much of a sports-car performance. The car used a steel space-frame for its inner body. This was covered with unstressed, molded plastic panels (not fiberglass) for the outer body. The body was styled after the Ferrari 308GTB and was attractive. But most of its components came from the GM parts bin, so it was no Ferrari. A photo is shown in Figure 8-48.

Saturn—1985–2010

Saturn was a wholly owned subsidiary of General Motors who created it to manufacture and market a new small car to compete with the Japanese and European brands that were taking sales from GM. The Saturn was built in a new factory in Tennessee. GM negotiated a more favorable contract with the United Auto Workers union than those that were then in place at their northern factories. The car was sold through an independent dealer network with a policy of no hassles and no negotiations that was popular with customers.

The car used a space-frame, unit body of steel to which molded plastic panels were attached to make the outer body, much like the Fiero. This manufacturing approach was used for all Saturn models until 2008, when falling sales forced them largely to abandon their own designs. For the 2009–10 model years, most of the Saturn line consisted of re-badged Opels from GM's factories in Europe. By 2011, Saturn was dead, a victim of the stock-market crash in 2008, the resultant Great Recession, and the bankruptcy of GM.

Dodge Viper—1992–Present

The Viper was originally produced as a one-off concept car and shown at the 1989 North American International Auto Show in Detroit. Public reaction was so strong and positive that Chrysler decided to put it into production. The first generation Viper debuted in 1992 with a fiberglass roadster body over a steel-tube space frame.

FIGURE 8-48 Pontiac Fiero *(Courtesy of Jonrev)*

It was long on power and short on creature comforts. It had an 8-liter, V10, aluminum engine designed by Lamborgini, which was then owned by Chrysler. The engine was based on Chrysler's V8 with two cylinders added, and it retained the 90° V-angle of the V8, which made it uneven firing.

It was spartan with no exterior door handles, no side windows, and no roof, save for a vestigial soft top "kit" that could be assembled to cover the cockpit in a pinch. Its top speed was 150 mph, and it could circle the skid pad with 1-g lateral acceleration, a very impressive number both then and now. Subsequent generations added a coupe body of the same construction as the roadster with twin bubbles in the roof to accommodate racing helmets. It was more track car than street car.

Later models gradually added creature comforts such as air conditioning and safety items like anti-lock brakes. By the fourth generation in 2008, power was up to 600 HP from an 8.4L V10 engine with variable valve timing. This model did 0–60 in 3.6 seconds, with a claimed top speed of 202 mph. In late 2009, with the economy down and automakers in serious financial trouble, Dodge announced Viper production would cease in mid-2010. It went back into production for 2013 with an extensive redesign shown in Figure 8-49. The steel spaceframe is now wrapped with carbon fiber for all body panels except for its aluminum doors.

FIGURE 8-50 Bugatti Veyron *(Courtesy of M 93)*

Bugatti Veyron—2005–2014

One thousand horsepower, 268-mph top speed, 16-cylinder mid-engine, and a two million dollar price tag perhaps define the word *supercar*. The Bugatti Veyron holds the Guinness Book's record as the fastest production automobile in the world. Its carbon-fiber, monocoque body, reinforced with stainless steel and aluminum, weighs only 110 lb. That is only 3% of the car's curb weight of 4162 lb. The rest of the weight comes in part from its 8-liter, W16, 64-valve engine, ten radiators, four turbochargers, and a seven-speed, dual-clutch, automated manual transmission. It is built by VW who bought the rights to the Bugatti name to revive the marque. A photo is shown in Figure 8-50.

Lexus LFA—2010–2012

Lexus began development if its supercar, which went through several generations of designs before debuting as the LFA in 2010 (Figure 8-51). Earlier prototypes were built on an aluminum frame but that was changed to a carbon fiber monocoque body in 2007, which is what the pro-

FIGURE 8-49 2013 SRT Viper *(Courtesy of IFCAR)*

FIGURE 8-51 Lexus LFA *(Courtesy of Motohide Miwa)*

duction LFA has. It also has a front-mounted, 72°, 4.8L V10 engine (72° is the correct angle to make a V10 even-fire) driving the rear wheels through a six-speed SMG transaxle. Only 500 were made and delivered to selected customers.

Lamborgini Aventador—2011–Present

The Lamborgini Aventador, is a mid-engined supercar with a carbon-fiber monocoque body as shown in Figure 8-52. The monocoque comprises the entire passenger compartment from firewall back, as is typical. The engine and front suspension are carried on a space-frame of rectangular tubing made of carbon fiber. Note the front-suspension springs, which are horizontal and act on the wheels through linkages. A similar sub-frame structure supports the engine and rear suspension, which also sports horizontal suspension springs.

Power is from a 6.5L, 60° V12, mounted longitudinally in the body in front of the rear axle and driving the rear wheels through an SMG, seven-speed transmission. The car is made in both coupe and roadster versions, the latter of which has two carbon-fiber, removable roof panels that weigh only 13 lb each. Lamborgini claims a top speed of 217 mph and a 0–60 time of 2.9 sec. A photo of an Aventador coupe is shown in Figure 8-53. Only 4000 will be built and for about $400,000 in 2015 dollars, one can be yours.

FIGURE 8-53 Lamborgini Aventador coupe *(Courtesy of Arnaud Clerget)*

McLaren MP4-12C—2011–2014 and McLaren P1—2014–Present

The McLaren supercars are the closest thing a person can buy to a Formula 1 race car. Their 1992 F1 was the first production automobile with a carbon-fibre, monocoque body. Both of the latest models, the MP4 and the P1, are two-seat, mid-engined-RWD cars and have carbon-fiber monocoque construction for chassis and body. The newer P1 is a hybrid with a similar 3.8L, twin-turbocharged, V8 engine of 727 HP plus a 176 HP electric motor giving a combined 903 HP. The older MP4's V8 engine makes "only" 592 HP in comparison. The P1 accelerates from 0 to 62 mph (100 kph) in 2.8 seconds. A photo of the MP4 is shown in Figure 8-54.

The carbon fiber "tub" for both these cars, which they call the "Carbon MonoCell," is produced in a much more efficient manner than those of its competitors. The carbon fiber cloth has fibers woven in two and three directions to

FIGURE 8-52 Lamborgini Aventador carbon fiber body *(Courtesy of J.Smith831)*

FIGURE 8-54 McClaren MP4-12c *(Courtesy of M 93)*

provide the required strength and stiffness for the body. The MonoCell is made in a single pressing between matched metal dies in a set of patented processes. The resulting body weighs only 176 lb, which is 6% of the 2868-lb weight of the finished MP4. This manufacturing process has reduced the time needed to build the body from 3000 hours for the previous generation McLaren F1 to 4 hours for the MP4. The P1 also uses this technology to make its body. For comparison, the Mercedes-Benz SLR McLaren took 500 hours to make its body.

Mercedes-Benz SLR McLaren—2003–2010

McLaren also had a hand in the design and production of the Mercedes-Benz SLR. They produced its carbon fiber body and assembled the car at their Woking, Surrey factory in England. At the time, Mercedes-Benz owned 40% of McLaren Group.

The SLR is a "front-mid-engined" car, which means its supercharged, 5.4L-V8 SOHC engine is positioned well back of the front wheels (by 20 inches) as is done in the Maserati Quattroporte. The car's design was inspired by the original 1955 Mercedes 300 SLR shown in Figure 8-35, including its gull-wing doors. A photo of the new SLR is shown in Figure 8-55.

FIGURE 8-55 Mercedes-Benz SLR McClaren (Courtesy of Cars en travel)

FIGURE 8-56 BMW i3 (Courtesy of TTTNIS)

BMW i3—2014–Present

The BMW i3 is a small, economy car offered as a pure electric or as a hybrid when the optional "range-extender" two-cylinder gasoline engine is specified. The electric-only i3 has a claimed 81-mile range; the range-extender engine raises that to 150 miles. It is the first mass-produced automobile to use a carbon-fiber reinforced plastic body. A photo is shown in Figure 8-56.

Though the i3 cars are assembled in Leipzig, Germany, the carbon fibers are made in a new factory in Moses Lake, WA, with materials imported from Japan. One of the reasons for this choice of location is that carbon-fiber production is very energy intensive and this plant is near abundant hydroelectric power. Electricity in Germany costs seven times its price in Washington State. The carbon fiber is shipped to Landshut Germany to be made into bodies.

Crushable Bodies

Older, frame-on-body cars were designed with relatively rigid frames and bumpers in the mistaken thought that this would mitigate damage and injury in a crash. After unit-bodies were adopted, it became possible to design them to make the core passenger compartment stronger and more crush resistant in a front or rear collision than the portions of the car forward and back of the passenger compartment. In the

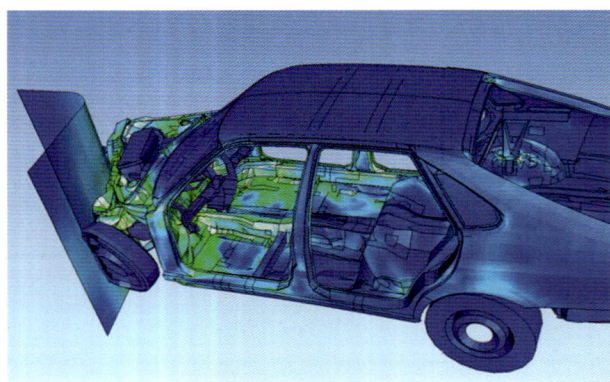

FIGURE 8-57 Computer simulation of a crash *(Public Domain)*

1980s, German car makers began to use computer simulations to study the way car bodies of various designs respond to a collision. This led to what has now become standard practice to provide "crumple zones" outside the core passenger compartment. Figure 8-57 shows a computer simulation of a car body in a front-corner collision with a stationary object like a concrete wall. Note that all the damage is to the nose of the car and the passenger compartment is intact.

An automobile has significant mass, and when a mass is given a velocity, it gains kinetic energy. This energy is proportional to the body's mass times its velocity squared. So a crash at 60 mph is four times worse than one at 30 mph, all other factors equal.

When an impact between two masses occurs, the impact forces are greater the faster the event takes place. Stopping a mass with a given velocity in 1 second will require four times the force that stopping it in 2 seconds requires. If the stopping event can be extended to 3 seconds from 1 second, the force will be only 1/9 as large as the 1-second case (1 over 3 squared). So the key to reducing the forces to which passengers are exposed in a collision is to make the event last longer. This is accomplished by designing the nose and tail of the car to crush in a controlled fashion when hit. The firewall is also reinforced and sloped to force the engine down and under the passenger compartment rather than into it.

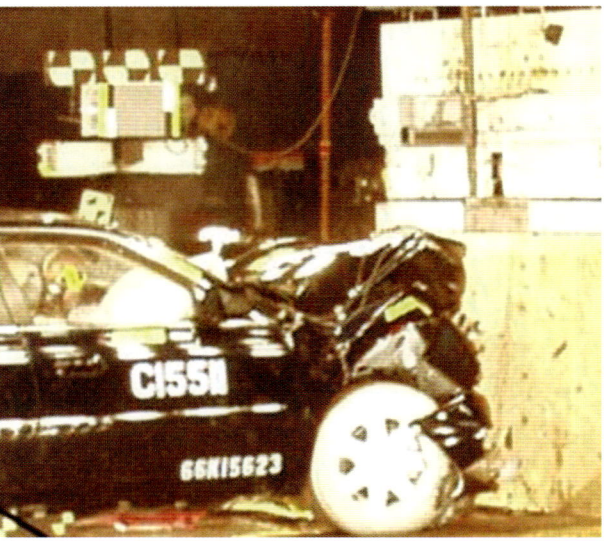

FIGURE 8-58 Crash test at General Motors *(Public Domain)*

All present-day automobiles are designed this way so that they can pass government-mandated crash tests. Before any new car model can be sold in the U.S., the manufacturer must supply several complete examples to the National Highway Transportation Safety Administration (NHTSA), who then crashes them into barriers at controlled speeds with crash-test dummies inside. NHTSA then gives each car a crash-safety rating and publishes the information. Most car manufacturers also have crash-test laboratories where they replicate the tests before submitting a car to NHTSA for testing. Figure 8-58 shows a crash test done at the GM Vehicle Safety and Crash-Worthiness Laboratory.

Pedestrian-Friendly Cars

Most fatalities from car accidents happen to pedestrians. Europe and Asia have adopted regulations that require new automobiles sold there to be less dangerous to pedestrians in collisions. When a car hits a pedestrian, the most common scenario is that the bumper catches the person in the legs, then flips them onto the hood and into the windshield as shown in Figure 8-59.

FIGURE 8-59 Pedestrian crash simulation *(Public Domain)*

Most fatal injuries of pedestrians result from the collision with the car body rather than from hitting the ground. A head hitting the hood or windshield causes traumatic brain injury. These regulations require higher hood-heights, more compliant hoods, and softer front ends. A 20-mm (0.8 in) clearance between the hood and the highest engine part is now required to allow a crush zone before hitting the non-compliant engine. NHTSA is currently reviewing accident data and preparing to issue similar regulations for U.S. cars to reduce pedestrian-crash injuries.

NVH

NVH stands for *Noise, Vibration, and Harshness*, a term that has come to define the most subjective aspect of a human's experience with an automobile. Modern cars are much quieter and smoother than previous generations, and this is a direct result of manufacturers' attention to reducing NVH.

There are many sources of NVH in an automobile. Engine, tires, and the aerodynamics of the body all contribute. Engine noise can be quite effectively controlled with mufflers and insulation around the engine compartment. Tire tread design is a major factor in road noise and tire companies do research on tread design to control noise. Wind noise is caused by disruptions in the smooth flow of air past the body. Simple things like the shape of outside mirrors and window moldings can have significant effect on wind noise. External radio antennas can cause whistling if not properly designed.

All these factors are extensively studied in elaborate NVH laboratories at all major automobile manufacturers. These labs contain anechoic chambers equipped with multiple microphones that can listen to an automobile while it "runs" in place with the wheels on rotating drums. Wind noise is studied in wind-tunnel tests. Adding a smoke-like substance to the air in the wind tunnel makes the turbulent, noise-producing eddies around body contours visible. Microphones inside the car measure interior noise while driving on a test track over various road surfaces. Manufacturers routinely test their competitors' vehicles in their NVH labs and strive to outdo them in order to create the quietest vehicle.

Nevertheless, makers of high-performance cars are now using artificial means to add "pleasant" sounds to their cars to give the impression of power. Cars that have had their natural engine sounds muffled to near silence (and even electric cars) are now using computer technology to pump artificial engine "growl" on acceleration into the cabin through the radio speakers to make the car sound more "aggressive" and "powerful." This allows the crop of recent V6 engines to sound more like the rumbling V8s of old.

Summary

Even though it has essentially no moving parts, the automobile body turns out to be a quite complicated and high-tech part of the automobile system. This chapter has described a remarkable evolution from the very crude forms of auto-body design inherited from the horse-drawn era to an extremely sophisticated state. Modern automobiles use concepts from aircraft technology to fashion automobiles of impressive performance and appearance.

Chapter 9
Summary

I hope that the reader has enjoyed reading the preceding chapters as much as I have enjoyed writing them. Though I felt that I knew quite a bit about the history of automotive technology before beginning the task, I learned a great deal more in the process of researching all the topics. Many reference books were consulted, and several hundred technical papers, articles, and other sources were used, as are listed in the bibliography.

In the book's introduction, the saying *There is nothing new under the sun* was quoted as being very applicable to automotive technology. The previous chapters have revealed that hybrid cars and planetary transmissions go back to 1900, air suspensions to 1901, disk brakes and the V8 engine to 1902, the torque converter, automatic transmissions, and cylinder deactivation to 1905, and the list goes on. In fact, lists of automotive milestones, sorted both by date, and separately by date within categories, are provided in the appendices.

Modern automobiles are better than any of their predecessors by far, and the list of improvements will go on for the foreseeable future. I well remember the dark days of the late 1970s and 1980s when emission regulations after the gas crisis of 1973 had essentially crippled engine performance, and rollover regulations seemed to have doomed all convertibles. But, here in the adolescence of the 21st century, we are witnessing the largest resurgence of performance cars with disappearing roofs and incredible statistics, like under 3-second 0–60 times, in the history of the automobile. And because of hybrid technology being put to the task of improving performance as well as lowering fuel usage, we are seeing cars with both better performance and economy than ever before.

One may not look forward to the advent of driverless cars, as that could take much of the fun out of the experience, but the advance of technology is essentially unstoppable. Whatever the future brings, the optimistic view has to be that, all things considered, it will be better overall. Eliminating traffic jams on interstate-highway commutes would be a good thing. Hopefully there will still be back roads without built-in sensors to control speed and direction where enthusiasts can explore the performance of their newest supercar.

The opening photograph is of a 1912 Pierce Arrow Runabout with the author and his friend, John Lothrop, to whom this book is dedicated. John, who owned and restored the car, is driving.

Appendix A

Glossary of Terms

Terms in **bold** within a definition are also entries within this glossary

TERM	DEFINITION
acceleration	the time rate of change of **velocity**—units are inches per second per second or meters per second per second
amphibious	able to move on either ground or water
angular velocity	the time rate of change of the angle of a rotating body—units are **radians/sec**
atmospheric engine	an external-combustion engine that expands its charge (steam) by cooling it in the cylinder or an internal-combustion engine that does not compress the fuel-air charge prior to burning it
BDC	bottom dead center—the lowest point of a **crank**'s motion
bore	the internal diameter of a hollow **cylinder**
BTU	British Thermal Unit—a measure of heat energy; it is the amount of heat needed to raise the temperature of one lb of water by one degree Fahrenheit
cam	a shaped element (usually metal) that drives a follower as the cam turns or translates
camshaft	a shaft containing a number of **cam**s—often used to actuate engine **valves**
capstan	a drum that turns with rope or a cable wound around it to move or lift a heavy weight
closed-circuit system	a circuit (fluid or electrical) that is self-contained with no connection to the outside environment
clutch	a device to mechanically connect or disconnect two rotating parts to transmit **torque**
complex motion	a combination of simultaneous **rotation** and **translation**
condenser	a device that allows steam to transfer its heat to the environment to cool it (also called a radiator)
connecting rod	a **link** that connects a **crank** to a **piston** (see Fig 1-12)
conrod	a **connecting rod**
couple	a pair of forces acting on a body; the forces are equal in magnitude but opposite in direction; they are also separated by a distance called the moment arm; their effect on the body is to create a turning force, also called a **moment** which tends to rotate the body
crank	a link that pivots about a ground plane and rotates fully in a circle (see Fig 1-12)
crankshaft	a shaft containing one or more **crank**s
crosshead	a guide to control the **linear translating** motion of a piston rod (see Fig 1-4)
cylinder	in an engine, the circular **bore** that contains the **piston** (see Fig 1-12)
desmodromic	a valve train in which the cams both open and close the valves, eliminating the need for valve springs; desmodromic engines have higher maximum speed, which gives more power, all else equal

die-casting	*a process to cast molten metals into shapes, using a steel die in which an impression the shape of the part has been machined; the molten metal is pumped into the closed cavity of the die and allowed to cool and solidify; the die is opened and the solid part removed from the die; the process is highly automated*
differential	*a device that connects the left and right driving wheels of a vehicle and allows them to rotate at different speeds while turning a corner, and simultaneously passes* **power** *to one or both of the wheels*
driveshaft	*a shaft typically connecting the tailshaft of a* **transmission** *to the rear axle through the* **differential;** *it can also connect the engine directly to a rear-drive* **transaxle**
dog clutch	*a* **clutch** *that uses "dogs", or interlocking lugs to lock two shafts together and transmit torque; the shafts have to be stopped to engage the dogs, unlike a friction clutch*
dynamometer	*a machine designed to measure the power of an engine; most are a large electric motor/generator that serves to load the engine by making it generate electricity into a bank of resistors; the voltage and current of the electricity are measured; their product is power*
eccentric	*a form of* **crank** *in the shape of a circle with the crankpin placed at a distance from its pivot*
ECE	*External Combustion Engine—one in which the burning of fuel occurs outside the engine*
electric motor	*a device that converts electrical voltage and current to mechanical* **torque** *and* **velocity**
energy	*see* **work**
epitrochoid	*a geometric curve traced by a fixed point on one circle which rotates around the perimeter of another circle*
exhaust port	*a port or hole, typically in the side or top of a* **cylinder**, *to allow burnt gases to escape the cylinder*
faired	*to make the connection or junction of surfaces smooth and even*
fluid	*matter in liquid or gaseous form that takes the shape of its container and is easily deformed*
fluid coupling	*a coupling between two rotating elements ("fan blades"), immersed in a liquid (oil) that uses the pumping of liquid from one "fan blade" to another to transmit* **torque** *between the elements; (picture two window fans, face to face, one powered one not—the airflow from the powered one rotates the other)*
flywheel	*a circular element that stores* **kinetic energy** *when its rotation is sped up and returns that energy to the system by slowing down when needed; it smooths out variations in speed due to* **torque** *fluctuation*
force	*an effort imposed on a body that tends to move or deform it according to various physical laws*
gear	*a machine element on which shaped teeth are cut to mate with teeth on another similar element to make a gear pair or gearset—can be round, straight, or any shape as long as the mating gear can accommodate its shape*
gear ratio	*the ratio formed by the quotient of the number of teeth on the output gear divided by the number of teeth on the input gear—the output torque will equal the gear ratio times the input torque and the output speed will be the inverse of the gear ratio times the input speed*
gear train	*a combination of two or more* **gearset***s arranged in parallel to obtain a larger or smaller overall* **gear ratio**, *which will be the product of all the individual gearset ratios*
gearset	*a pair of circular gears in mesh, the smaller of which is called the pinion and the larger the gear—used to change the speed and torque between the two shafts*

APPENDIX A GLOSSARY OF TERMS

harmonic	refers to the vibration of a body such as a guitar string; the string will have a theoretically infinite number of harmonics, each of which is an integer multiple of the frequency of a lower frequency one; the lowest frequency is called the fundamental or first harmonic and the others are numbered second, third, etc., with increased frequency; in an engine, the fundamental frequency or first harmonic is equal to the rotational frequency of the crankshaft (rpm) and the higher harmonics are equal to two, three, four, etc. times the rotational frequency of the crankshaft; unitless
heat pump	a device that "pumps" heat from one space to another—can be used to either heat or cool a space because it is reversible
hybrid	composed of elements of different kinds—in this context it refers to a vehicle with more than one type of motive power—typically **ICE** plus electric or **ICE** plus hydraulic
hydraulic motor	a device that converts fluid pressure and flow to mechanical **torque** and **velocity**
hydraulic pump	a device that converts mechanical **torque** and **velocity** to fluid pressure and flow
IC	Internal Combustion
ICE	Internal Combustion Engine—one in which the burning of fuel occurs inside the engine
impeller	a rotating component of a centrifugal pump, which transfers energy from the motor that drives the pump to the fluid being pumped by accelerating the fluid outwards from the center of rotation
inertial force	a force due to the acceleration of a mass (from Newton's second law: force = mass times acceleration); used in respect to an engine, it describes the force felt at the engine mounts as a reaction to the motions of the pistons and other moving parts within the engine; also referred to as a **shaking force**
intake port	a port or hole, typically in the side or top of a **cylinder**, to allow the air/fuel charge to enter the cylinder
jackshaft	an auxiliary shaft typically used to transfer **torque** from one device or system to another
joint	An element that connects two or more **link**s together to allow motion between the links
kinetic energy	mathematically equals **mass** times **velocity** squared; a moving body possesses kinetic energy; units are in-lb or N-m (joules)
linear translation	**translation** in a straight line
link	an element containing holes or surfaces to allow it to connect to and move versus another link
linkage	a collection of **links** and **joints** that allows controlled motion between the links
linkwork	another word for a **linkage**
mains power	electrical power supplied over "main" lines of wire to houses and businesses
mass	a property of a body that defines its resistance to acceleration by a force according to Newton's second law, which states that **force** equals **mass** times **acceleration**—units are lb-seconds squared per inch (called blobs) in the U.S. or kilograms in the SI system
moment	a turning **force** that results from the action of a force on a body, when the force is applied away from the center of gravity (CG) or rotational center of the body; the moment's magnitude is the product of the **force** times the distance it is displaced from the CG or center of rotation; its effect is to try to rotate the body; when you grab the steering wheel with both hands at opposite sides and turn it, you are applying a **couple** with your two hands (force up on one side and down on the other) and the **moment of the couple** turns the steering wheel; its units are lb-in, lb-ft, or N-m
moment of a couple	the product of the magnitude of either of the forces of the **couple** and the distance between the forces; units are lb-in, lb-ft, or N-m

momentum	*mathematically equals the product of* **mass** *and* **velocity**
natural frequency	*a property of a physical system that contains energy storage elements such as mass (which stores* **kinetic energy***) and spring (which stores* **potential energy***); natural frequency is that frequency (cycles per second) at which the system naturally vibrates like a bell or tuning fork; the strings of a musical instrument are tuned so that their natural frequencies correspond to the notes of a musical scale; a body, assembly, or system with more complicated geometry than a string will have multiple natural frequencies at which it readily vibrates*
over-running clutch	*also called a one-way clutch; friction devices internal to the clutch allow it to transmit torque and motion in one direction, while free-spinning in the other direction; so if the output side tries to run faster than the input side, it "over-runs" the input and spins freely; if the output side is turning slower than the input side, the input will drive it*
oversteer	*the tendency of a car to overrespond to steering inputs and want to turn tighter than planned*
parallelogram	*a four-sided figure or linkage in which opposite sides are of equal length; opposite links of a parallelogram linkage remain parallel to one another when moved*
pawl	*a tooth-shaped object used to impel or stop the motion of a* **ratchet** *by engaging a tooth on the ratchet*
phase	*the angular relationship of two rotating elements*
phase angle	*the angle between the first crank (cylinder #1) on a crankshaft and any other crank on the shaft*
pi (π)	*a mathematical constant that appears in many equations; pronounced "pie," it is universally denoted by the Greek letter π (also pronounced "pie"); its value is approximately 3.14159*
pinion	*a circular toothed gear that rotates and can engage a* **rack** *or a larger circular* **gear**
piston	*a cylindrical element fitted into another hollow cylinder to provide motion in response to change in pressure in the cylinder on one side of the piston (see Fig 1-12)*
potential energy	*mathematically equals force times distance; a stationary body positioned above a surface in a gravitational field has potential energy equal to its weight times the height above the surface; if dropped, an instant before it hits, its potential energy will be zero, having been converted completely to* **kinetic energy***; fluid under pressure also contains potential energy as does a mechanical spring when deflected*
power	*the time rate of doing* **work** *or expending* **energy** *per unit time—units are in-lb/sec or N-m/sec (watts); one horsepower (HP) equals 746 watts or 6600 in-lb/sec; in a translating system, power equals force times velocity; in a rotating system, power equals torque times angular velocity; in a fluid system, power equals pressure times velocity; in an electrical system power equals voltage times current*
primary moment	*refers to the first* **harmonic** *of the* **shaking moment** *of an engine*
pumping losses	*this term refers to the energy losses in an IC engine due to the resistance to airflow in the induction system and also to the resistance to airflow in expelling exhaust gases through the exhaust system*
rack	*a straight, toothed gear that* **translates**
rack & pinion steering	*a* **pinion** *on the end of the steering column is turned by the steering wheel and engages with a* **rack** *that translates left and right across the width of the car; each end of the rack connects to the steering linkage to cause the wheels to turn in response to the steering wheel input*

APPENDIX A GLOSSARY OF TERMS

radian	a unit of angular measure; one radian is approximately 57.3 deg. there are 2π radians in 360 deg
radian per second	a unit of angular **velocity**, usually abbreviated as rad/sec or rad/s (but never r/s or rps which mean revolutions per second); one rad/sec is (very) approximately 1/10 **rpm**
ratchet and pawl	a ratchet is a toothed element, often arranged on a circle, whose motion is controlled by a **pawl** that stops or causes motion
ratchet wrench	a tool to tighten or loosen bolts that uses a ratchet wheel and pawl to allow torque to be applied in one direction to turn the bolt and allows the tool handle to reverse without turning the bolt
rectilinear translation	translation in a straight line (see also **translation**)
red-line speed	maximum permitted speed for an IC engine—denoted by a red line on its tachometer
regenerator	a collection of heat absorbing elements that can store heat energy and release it based on temperature differentials
resonance	a condition in which a dynamic system that contains both masses and springs vibrates excessively because the exitation frequency is the same as or close to the system's **natural frequency**
rotation	motion of an element that allows no **translation** (see also transation and **complex motion**)
rpm	revolutions per minute—a measure of angular **velocity**
scow	a flat-bottomed boat with a blunt bow, often used to haul freight; also called a barge
secondary moment	refers to the second **harmonic** of the **shaking moment** of an engine
shaking force	a force due to the acceleration of a mass (from Newton's second law: Force = mass times acceleration); used in respect to an engine, it describes the force felt at the engine mounts as a reaction to the motions of the pistons and other moving parts within the engine; also referred to as an **inertial force**; its units are lb or N
shaking moment	a moment due to the accelerations of a pair of masses in opposite directions that create a **couple**; used in respect to an engine, it describes the moment or turning force felt at the engine mounts as a reaction to the motions of the pistons and other moving parts within the engine; it tends to rock the engine back and forth in its mounts; its units are lb-in, lb-ft, or N-m
slide valves	valves that slide in **rectilinear translation** to expose or cover an intake or exhaust port
slipper piston	a piston longer on either side and relieved under the wrist pin to wrap around the crankshaft counterweight and allow a shorter conrod and shorter block; the longer skirts on either side provide the needed guiding to resist the side force on the piston from the angle of the conrod
spool valve	a valve in the shape of a spool inside a cylinder, which, when translated along its axis, alternately connects or blocks fluid ports on opposite sides of the cylinder (see Fig 1-4)
spring rate	a property of any spring that defines its stiffness; it is the amount of force required to deflect the spring one unit of length; it is measured as lb/in or N/m; metal springs typically have a constant value of rate over most of their range; air springs, on the other hand, have a nonlinear spring rate that increases exponentially with deflection; (see Fig 6-8)
sprung/unsprung ratio	the ratio of the weight of a vehicle that is suspended on the springs to that which is in contact with the road directly; more unsprung weight (a lower ratio) makes the vehicle ride harsher as the larger unsprung mass reacts with larger forces into the chassis when encountering a bump or pothole

stall	*refers to the condition of an engine or motor whose **rotation** is stopped temporarily*
stroke	*the distance through which a **piston** moves from TDC to BDC—a full revolution of the crank contains two strokes*
synchromesh trans	*a geared **transmission** that contains clutches at each gear to alternately lock them to the transmission shaft or let them spin freely; this allows all gear pairs in the transmission to remain in constant mesh; shifting is achieved by engaging and disengaging the synchromesh clutches on each gear in turn. this eliminates the "crashing" of gear teeth when a conventional transmission shifts the gears into and out of mesh to change speeds; Chapter 5 describes this in detail with figures*
TDC	*top dead center—the highest point of a crank's or piston's motion*
torque	*a turning force—mathematically force times distance—units are lb-in or N-m; do not confuse torque with **work** and **energy,** which appear to have the same units; they are not the same thing*
torque converter	*a fluid coupling to which a stationary blade has been added to provide a boost in torque*
transaxle	*a transmission combined with the final drive and differential in either front- or rear-wheel drive cars*
transfer port	*a passage in a two-stroke engine between the crankcase and the intake port in the cylinder above the piston to allow the fuel-air mixture to be pushed from the crankcase into the combustion chamber by the descending piston*
translates	*moves without rotating, always maintaining its angle constant*
translation	*motion of an element that allows no rotation (see also **rotation** and **complex motion**)*
transmission	*a collection of rotating elements (typically gears) to allow different and selectable ratios between its input and its output shaft; it trades **velocity** for **torque**, the product of which is **power**; may be manually or automatically shifted between gears*
turbine	*a rotating device containing multiple "fan" blades that are turned by a jet of fluid (steam or gas)*
understeer	*the tendency of a car to resist steering inputs and try to go straight rather than turn*
valve	*a device that alternately allows or blocks the flow of a liquid or gas (a kitchen faucet is a valve)*
velocity	*the time rate of change of distance—units are in/sec, ft/sec, or m/sec*
volatility	*a property of a liquid that allows it to readily vaporize and combine with air or other gas*
weight	*the force on a body having mass due to its presence in a gravitational field; mass is a physical property of a body but its weight is not; weight of a mass depends on the gravitational field in which it finds itself; weight equals mass divided by the local gravitational acceleration ($W = m/g$); on earth at sea level, gravitational accceleration (g) is 32.2 ft/sec/sec, or 386 in/sec/sec, or 9.8 m/sec/sec; on the moon it is about 1/6 as much and in outer space it is essentially zero; so you would weigh 1/6 as much on the moon as on earth and weigh essentially zero in space, but your mass is the same in every location (unless you go on a diet in the meantime)*
work	*mathematically equals the product of a force times a distance and is the same as mechanical **energy**—units are in-lb, ft-lb, or N-m (joules)*

Appendix B

Milestones by Year

YEAR	MILESTONE
1050 BC	The differential mechanism is invented to guide caravans across the desert (China)
1490	Leonardo da Vinci draws a stepless, continuously variable transmission (Italy)
1758	Erasmus Darwin invents what we now call Ackerman steering (England)
1769	Cugnot builds the first self-propelled road-going vehicle—steam powered (France)
1802	R. Trevithick builds the first passenger-carrying steam carriage (England)
1805	Oliver Evans builds the first amphibious vehicle—steam powered (U.S.)
1816	Robert Stirling invents the Stirling external combustion engine (Scotland)
1817	Georg Lankensperger reinvents Erasmus Darwin's steering idea (Germany)
	Rudolph Ackerman patents Lankensberger's steering in his own name (England)
1837	Robert Anderson builds the first electric carriage (Scotland)
1860	Roots invents their eponymous air pump later to be used as a supercharger (U.S.)
1872	George Brayton invents the Brayton cycle engine (U.S.)
1873	Benz invents double-pivot steering (Germany)
1876	Luigi De Cristoforis invents the fixed venturi carburetor (Italy)
	James Starley invents the chain-drive differential (England)
1877	Nikolaus Otto patents the first practical internal combustion engine (Germany)
1878	Dugald Clerk invents the two-stroke engine (Scotland)
1879	Milton Reeves invents a variable speed transmission (U.S.)
1885	Benz invents the first motorcycle (Germany)
1886	Benz patents the first gasoline-powered automobile—a three-wheeler (Germany)
	Daimler introduces the first gasoline-powered four-wheel automobile (Germany)
1889	Daimler builds the first V-twin engine (Germany)
1891	Panhard creates the *Systeme Panhard* front-engined, RWD arrangement (France)
1892	James Atkinson invents the Atkinson cycle engine (U.S.)
1893	Bramah Diplock patents three-differential-all-wheel-drive (AWD) (England)
	Rudolf Diesel invents the diesel cycle engine (Germany)
	Duryea forms the first American automobile company (U.S.)
1895	Daimler introduces the first inline, two-cylinder engine (Germany)
	Duryea wins first American race—Chicago to Evanston; a Benz is second (U.S.)
	Panhard-Levaseur develops the first sliding-gear transmission (France)
1897	Benz builds the first horizontally opposed (boxer) twin (Germany)
	First use of a differential in a (steam) car by David Shearer (Australia)
	Felix Millet builds the first rotary engine (France)
	The Stanley brothers produce the first practical steam car (U.S.)
1898	First patented front-wheel-drive car—the Societe Parisienne (France)
	F. W. Lanchester patents first compound, 3-speed, epicyclic transmission (England)
1899	Mors invents full-pressure lubrication through crankshaft and conrods (France)
1900	F. W. Lanchester builds the first perfectly balanced engine (England)
	Wilson-Pilcher builds the first boxer four-cylinder engine (England)
	Wilson-Pilcher builds the first boxer six-cylinder engine (England)
	Ferdinand Porsche designs the first hybrid gasoline/electric car (Germany)
	F. W. Lanchester uses the first planetary transmission in an automobile (England)
1901	William Humphreys patents an air-spring system (U.S.)
	John Wilkinson designs the first air-cooled four-cylinder inline engine (U.S.)

1901	Oldsmobile makes first mass-produced car on an assembly line (U.S.)
	C. W. Manly builds the first radial engine (England)
	White makes the first steam car with a condenser and a flash boiler (U.S.)
	Chadwick experimented with supercharging (U.S.)
	Oldsmobile uses the first 2-speed planetary transmission in an automobile (U.S.)
1902	Buffum builds an opposed, four-cylinder engine (U.S.)
	Frederick W. Lanchester patents an automotive disk brake (England)
	Louis Renault invents the internal-expanding drum brake (France)
	Lambert patents friction drive (England)
	Locomobile produces the first water-cooled inline four cylinder engine (U.S.)
	Henry M. Leland introduces interchangeable parts in production (U.S.)
	Alex Craig designs the first SOHC engines for Maudslay Motors (England)
	Levavasseur patents a V8 engine—builds for speedboats and aircraft (France)
1903	Spyker builds the first four-wheel drive automobile (Netherlands)
	Edmund Rumpler patents the swing axle (Austria)
	Spyker builds the first inline, six-cylinder engine (Netherlands)
	Marr Auto Car builds the first U.S. SOHC engine (U.S.)
	Adler builds the first V8 engine for a race car (France)
1904	Pierce-Arrow makes the first cast-aluminum automobile body (U.S.)
	J. Walter Christie patents a FWD car and builds a prototype (U.S.)
	J. Walter Christie invents the sliding pillar IFS (U.S.)
	Buick builds the first overhead-valve (OHV) engine (U.S.)
	The first V12 engine is produced by Putney Motor Works (England)
1905	The Sturtevant brothers patent a mechanical automatic transmission (U.S.)
	Skinners Union (SU) invents the variable venturi carburetor (England)
	Sturtevant offers a car with cylinder deactivation (U.S.)
	Hermann Föttinger patents the fluid coupling (Germany)
	Leon Levasasseur developed a working mechanical fuel injection system (France)
	Sizaire-Naudin pioneers parallel arm independent front suspension (France)
	J. Walter Christie builds a FWD car with sliding pillar suspension (U.S.)
	Wolseley builds the first 12-cylinder, inline engine (England)
	Knight invents the dual-sleeve-valve "Silent Knight" engine (U.S.)
	Hermann Föttinger patents the torque converter (Germany)
	Alfred J. Buchi invents the turbocharger (Switzerland)
	Marmon builds the first, low-production V6 engine (U.S.)
1906	Isotta-Fraschini is first to develop front wheel mechanical brakes (Italy)
	F. W. Lanchester invents the crankshaft torsional damper (England)
	Anzani builds the first W3 engine (France)
1907	Bollee pioneers hydraulic tappets (France)
	Chadwick makes the first race car with a supercharged engine (U.S.)
1908	Daimler makes the first production engines with Knight sleeve valves (England)
1909	Morgan uses sliding pillar suspension and continues to use it to this day. (England)
	Ford builds the Model T, a car that every worker can afford (U.S.)
1910	Isotta-Fraschini is first to offer four-wheel mechanical brakes (Italy)
	De Dion builds the first V8 engines for production (France)
1911	Bendix invents the automatic releasing starter drive (U.S.)
	Oldsmobile markets a compressed air engine starter (U.S.)
	Burt and McCollum patent a single-sleeve-valve engine (Scotland)
1912	Peugeot designs the first dual overhead cam engine (France)
	Cadillac introduces the electric self-starter (U.S.)
1913	F.W. Lanchester invents the harmonic balancer (England)
	Dodge introduces the first cars with all-steel bodies (U.S.)
	Chevrolet is the first to use the Bendix starter (U.S.)
	Cadillac produces the first American production V8 engine (U.S.)

APPENDIX B MILESTONES BY YEAR

1915	Auguste Rateau developed a practical turbocharger on an aero engine (France)
1916	Packard makes the first V12 engine (U.S.)
	Hudson makes the first counterweighted crankshaft (U.S.)
1917	Chevrolet introduces the first OHV V8 engine—produced only through 1918 (U.S.)
	Napier builds the first W12 engine (France)
1918	Lockheed develops hydraulic brakes (U.S.)
1919	Isotta-Fraschini offered the first inline, eight-cylinder automobile engine (Italy)
	Hispano-Suiza introduces mechanical servo-powered 4-wheel drum brakes (Spain)
1920	George Messier invents a system using air bladders at each wheel (France)
	Duesenberg produced the first American inline, eight-cylinder engine (U.S.)
	Ricardo invents the slipper piston (England)
	Lancia makes the first V4 engines for automobiles (Italy)
1921	Duesenberg markets the first four-wheel hydraulic brakes on an automobile (U.S.)
	Edmund Rumpler makes the first car with swing axle IRS (Austria)
	Edmund Rumpler designs an aerodynamic car body (Austria)
1922	Lancia makes the first monocoque automobile body on its Lambda
1923	Sampson made the first series production dual overhead cam engine in the world (France)
	Chalmers-Maxwell offers Lockheed 4-wheel hydraulic brakes on top models (U.S.)
	Packard makes the first straight-eight engine in a mass produced vehicle (U.S.)
	Mercedes offers the first mechanically supercharged automobile (Germany)
1924	Chrysler produces the first high-compression engine (U.S.)
	Chrysler fits the first mass-produced cars with standard four-wheel hydraulic brakes (U.S.)
1925	Alvis enters a FWD car in a hill climb (England)
	Harry Miller runs the first front-wheel drive car in the Indianapolis 500 (U.S.)
	Alvis pioneers a leading arm IRS with four-leaf-spring IFS (England)
	Chrysler fits rubber engine mounts to control vibration (U.S.)
1926	Tracta invents the first constant-velocity universal joint (France)
1927	Packard invents and first uses a hypoid rear axle (U.S.)
	Ford sells the 15-millionth Model T (U.S.)
1928	Francis W. Davis patents the first practical hydraulic power steering system (U.S.)
	Earl Thompson invents synchromesh and sells it to GM (U.S.)
	Cadillac offers the first synchromesh transmission (U.S.)
	Chrysler adds crankshaft impulse neutralizer and vibration damper (U.S.)
	Isotta-Fraschini builds the first W18 engine (Italy)
1929	First production of front-wheel drive cars in the U.S.
	Chevrolet introduces the first mass-produced OHV inline-six (U.S.)
	Duesenberg has first vacuum-powered mechanical brakes (U.S.)
	Felix Wankel invents the Wankel engine (Austria)
1930	Cadillac makes the first V16 engine and is first to use hydraulic tappets (U.S.)
1931	Mercedes offers IFS with two transverse leaf springs as parallel arms (Germany)
	Plymouth introduces the "floating power" method of engine mounting (U.S.)
1932	Chrysler introduces vacuum power hydraulic brakes (U.S.)
	Ford introduces the first V8 engine in a low-priced car (U.S.)
1933	Mercedes switched to equal length A-arms and coils springs (Germany)
	Andre Dubonnet designs a trailing-arm independent front suspension (France)
1934	Citroën introduced its Traction Avant FWD car (France)
	Oldsmobile, Buick and Cadillac offer first, U.S. independent front suspension (U.S.)
	Chevrolet and Pontiac offer Dubonnet independent front suspension (U.S.)
	W. Barnes invents the overdrive transmission; Borg-Warner licenses it (U.S.)
	Chrysler offers the first Borg-Warner automatic overdrive transmission (U.S.)
	Chrysler introduces the streamlined Airflow models (U.S.)
1935	Alf Lysholm invents the screw compressor for a supercharger (U.S.)
1937	Chrysler adopts the hypoid rear axle (U.S.)
1938	Volkswagen starts producing the Beetle "People's Car" (Germany)

AUTOMOTIVE MILESTONES

1939	Oldsmobile makes the first hydraulic-automatic transmission (Hydramatic) (U.S.)
	Minerva makes the last car with a Silent Knight engine (Belgium)
1946	Most U.S. automakers adopt the hypoid rear axle (U.S.)
	Studebaker introduces the first postwar-(Loewy)-designed automobile (U.S.)
1948	Buick introduces the first automatic transmission with a torque converter (U.S.)
	Jaguar introduces a mass-produced, dual overhead cam, hemi-head, inline-six engine (England)
	Citroën introduces the Deux Cheveaux automobile (France)
	Hudson introduces its step-down model (U.S.)
	Dudley invents polydyne cam functions to control vibration in valve trains (U.S.)
1949	Packard makes the first automatic transmission with a lockup torque converter (U.S.)
	Cadillac and Oldsmobile introduce the first high-compression, OHV V8 (U.S.)
	Cadillac designs the first oversquare engine—bore larger than stroke (U.S.)
	Ford makes new line of automobiles, longer, lower, wider, with modern styling (U.S.)
	Cadillac is the first in U.S. to use the slipper piston (U.S.)
1950	Bendix develops a closed-loop electronic fuel injection system (U.S.)
	Flywheel/diesel hybrid buses are put into service (Switzerland)
	The first series production V6 engine is made by Lancia (Italy)
1951	Chrysler introduces full-time power steering (U.S.)
	Chrysler introduces the V8, OHV, hemi engine (U.S.)
1952	Porsche designs an alternative synchromesh mechanism to Thompson's (Germany)
1953	Chevrolet Corvette is the first fiberglass-bodied production automobile (U.S.)
1954	Citroën introduces the first air/oil hydro-pneumatic suspension system (France)
	The toothed timing belt is invented (U.S.)
	Daimler introduces a single-clutch AMT (SMG) transmission—(England)
1955	Flathead (L-head) engines production ceases in favor of OHV engines (U.S.)
	The last U.S. straight-8 engine is replaced by Packard with a new OHV V8 (U.S.)
	Chevrolet introduces its small-block OHV V8 based on the 1949 Cadillac engine (U.S.)
1957	Cadillac offers an air suspension system with self-leveling on the Eldorado (U.S.)
	Ralph Miller invents the Miller cycle engine (U.S.)
	AMC is the first company to offer Bendix fuel injection in its Rambler model (U.S.)
1958	Jensen introduces anti-lock brakes on its limited-production FF model (England)
1959	The Morris Mini-Minor starts a revolution in transverse-engine FWD cars (England)
1962	Buick produces the first uneven-firing, V6 engine with a 90° V-angle (U.S.)
1966	First postwar-production FWD car made in the U.S.—the Oldsmobile Toronado (U.S.)
1967	American Motors pioneers diagonal split-braking systems (U.S.)
1969	Fiat improves on the Mini's FWD design with the Fiat 128, and it is still widely copied (Italy)
1970	Toyota offers an automatic start-stop engine in its Crown model (Japan)
	Saab builds a two-stroke, inline, three-cylinder engine (Sweden)
	Viscous coupling is invented by Tony Rolt (U.S.)
1971	Chrysler introduces four-wheel antilock brakes on a production automobile (U.S.)
1972	Alpha Romeo makes the first engine with variable valve timing (VVT) (Italy)
1974	Mercedes introduces the first inline, five-cylinder engine—a diesel (Germany)
1977	Buick modifies the crankshaft to make its 90° V6 even-fire (U.S.)
1981	Cadillac offers the first cylinder deactivation system in its V8-6-4 engine (U.S.)
	Harry Webster develops a dual-clutch, automated manual transmission for Porsche (England)
1984	Izuzu introduces a single-clutch automated (SMG) manual transmission (Japan)
1989	Honda makes the first VVT system that varies valve lift and duration (Japan)
1991	Dodge makes the first production V10 engine (U.S.)
1992	McLaren produces the first carbon-fiber monocoque automobile (England)
2001	BMW has first VVT system to vary valve lift and duration continuously (Germany)
2003	Volkswagen ends production of the Beetle after 21 million produced (Germany)
2005	Volkswagen introduces the Bugatti W16 engine (Germany)
2009	UPS puts the first hydraulic/diesel hybrid trucks into service (U.S.)
2015	Ford produces the first pickup truck with an all-aluminum body (U.S.)

Appendix C

Milestones by Category

CATEGORY	YEAR	MILESTONE
air suspension	1901	first air-spring system—Humphreys (U.S.)
air suspension	1920	first use of air bladders—Messier (France)
air suspension	1954	first air/oil suspension—Citroën (France)
air suspension	1957	first self-leveling air suspension—Cadillac (U.S.)
all steel body	1914	first automobiles with all-steel bodies (except roof)—Dodge (U.S.)
all wheel drive	1893	first patent for an AWD car with three differentials—Diplock (England)
all wheel drive	1903	first all-wheel-drive automobile produced—Stryker (Netherlands)
antilock brakes	1958	first anti-lock brakes on an automobile—Jensen (England)
antilock brakes	1971	first production 4-wheel antilock brakes—Chrysler (U.S.)
auto start-stop	1970	first automatic start-stop engine—Toyota (Japan)
automatic transmission	1905	first mechanical automatic transmission—Sturtevant brothers (U.S.)
automatic transmission	1939	first hydraulic automatic transmission—Oldsmobile (U.S.)
automatic transmission	1948	first torque-converter automatic transmission—Buick (U.S.)
automatic transmission	1949	first lockup torque converter—Packard (U.S.)
balancing	1900	first perfectly balanced engine—F.W. Lanchester (England)
balancing	1913	the harmonic balancer is invented— F.W. Lanchester (England)
balancing	1916	first balanced, counterweighted crankshaft—Hudson (U.S.)
body	1904	first cast-aluminum car body—Pierce-Arrow (U.S.)
body	1921	first aerodynamic car body—Rumpler (Austria)
body	1922	first monocoque car body—Lancia Lambda (Italy)
body	1934	Chrysler introduces the streamlined Airflow models (U.S.)
body	1953	first fiberglass-bodied automobile—Corvette (U.S.)
body	1992	first carbon-fiber monocoque body—McLaren (England)
boxer engines	1897	first horizontally opposed (boxer) twin—Benz (Germany)
boxer engines	1900	first boxer 4-cylinder engine—Wilson-Pilcher (England)
boxer engines	1900	first boxer 6-cylinder engine—Wilson-Pilcher (England)
carburetor	1876	first fixed venturi carburetor—Cristoforis (Italy)
carburetor	1905	first variable venturi carburetor—Skinners Union (SU) (England)
chassis	1891	*Systeme Panhard* front-engined, RWD arrangement (France)
combustion cycle	1816	first Stirling external combustion engine—Robert Stirling (Scotland)
combustion cycle	1872	first Brayton cycle engine—George Brayton (U.S.)
combustion cycle	1877	first practical IC engine—Nicolas Otto (Germany)
combustion cycle	1878	first two-stroke engine—Sir Dugald Clerk (Scotland)
combustion cycle	1892	first Atkinson cycle engine—James Atkinson (U.S.)
combustion cycle	1893	first diesel cycle engine—Rudolph Diesel (Germany)
combustion cycle	1957	Miller cycle engine invented—Ralph Miller (U.S.)
companies	1893	first American automobile company—Duryea (U.S.)
constant velocity joint	1926	first constant velocity (CV) universal joint—Tracta (France)
CVT	1490	drawing of a stepless CVT—Leonardo da Vinci (Italy)
CVT	1896	first continuously variable transmission (CVT)—Daimler (Germany)
cylinder deactivation	1905	first car with cylinder deactivation—Sturtevant (U.S.)
cylinder deactivation	1981	first modern cylinder deactivation system—Cadillac (U.S.)
differential	1050 BC	geared differential mechanism is invented (China)
differential	1876	first chain-drive differential—Starley (England)
differential	1897	first differential in a steam car—David Shearer (Australia)

AUTOMOTIVE MILESTONES

disk brake	1902	first patented automotive disk brake—Lanchester (England)
dual overhead cam	1912	first DOHC engine—Peugeot (France)
dual overhead cam	1923	first production DOHC engine—Sampson (France)
dual overhead cam	1948	DOHC, hemi-head, inline-6 engine—Jaguar (England)
drum brake	1902	first internal-expanding drum brake—Renault (France)
electric vehicle	1837	first electric carriage—Anderson (Scotland)
engine starter	1911	first automatically releasing starter drive—Bendix (U.S.)
engine starter	1911	first compressed-air engine starter—Oldsmobile (U.S.)
engine starter	1912	first car with electric self-starter—Cadillac (U.S.)
engine starter	1914	first car to use the Bendix starter—Chevrolet (U.S.)
flathead engines	1955	flathead engine production ceases (U.S.)
fluid coupling	1905	fluid coupling patented—Föttinger (Germany)
four-wheel brakes	1910	first 4-wheel mechanical brakes—Isotta-Fraschini (Italy)
friction drive	1902	first patented friction drive—Lambert (England)
front-wheel brakes	1906	first front wheel mechanical brakes—Isotta-Fraschini (Italy)
front-wheel drive	1898	first front-wheel-drive car—Societe Parisienne (France)
front-wheel drive	1904	first to patent and build a FWD car in U.S.—Christie (U.S.)
front-wheel drive	1925	first FWD car to compete in a hill climb—Alvis (England)
front-wheel drive	1925	first FWD car to compete in the Indianapolis 500— Miller (U.S.)
front-wheel drive	1929	first production of a FWD car in the U.S.—Cord (U.S.)
front-wheel drive	1934	Traction Avant FWD car—Citroën (France)
front-wheel drive	1966	first postwar U.S. FWD car—Oldsmobile Toronado (U.S.)
fuel injection	1905	first mechanical fuel injection—Levasasseur (France)
fuel injection	1950	first electronic fuel injection system—Bendix (U.S.)
fuel injection	1957	first Bendix fuel injection used in a car—AMC (U.S.)
high compression	1924	first high-compression engine—Chrysler (U.S.)
hybrid vehicles	1900	first hybrid gasoline/electric car—Lohner-Porsche (Germany)
hybrid vehicles	1950	Flywheel/diesel hybrid buses first used (Switzerland)
hybrid vehicles	2009	first hydraulic/diesel hybrid trucks—UPS (U.S.)
hydraulic brakes	1918	hydraulic brakes invented—Lockheed (U.S.)
hydraulic brakes	1921	first 4-wheel hydraulic brakes—Duesenberg (U.S.)
hydraulic brakes	1924	first standard 4-wheel hydraulic brakes—Chrysler (U.S.)
hydraulic tappets	1907	hydraulic tappets invented—Bollee (France)
hydraulic tappets	1930	first hydraulic tappets in series production—Cadillac V16 (U.S.)
hypoid gearing	1927	invention and first use of a hypoid rear axle—Packard (U.S.)
hypoid gearing	1937	Chrysler adopts the hypoid rear axle (U.S.)
hypoid gearing	1946	Most U.S. automakers adopt the hypoid rear axle (U.S.)
independent suspension	1903	the swing axle is patented—Rumpler (Austria)
independent suspension	1904	sliding pillar IFS invented—Christie (U.S.)
independent suspension	1905	parallel-arm IFS -Sizaire-Naudin (France)
independent suspension	1905	first FWD car with sliding-pillar suspension—Christie (U.S.)
independent suspension	1921	first car with swing axle IRS—Rumpler (Austria)
independent suspension	1931	two-transverse-leaf-spring IFS—Mercedes (Germany)
independent suspension	1933	trailing-arm IFS invented—Dubonnet (France)
independent suspension	1934	first U.S. independent front suspension—GM (U.S.)
independent suspension	1934	Chevrolet and Pontiac offer Dubonnet IFS (U.S.)
inline engines	1895	first two-cylinder inline engine—Daimler (Germany)
inline engines	1901	first air-cooled 4-cyl inline engine—Wilkinson (U.S.)
inline engines	1902	first water-cooled 4-cyl inline engine—Locomobile (U.S.)
inline engines	1903	first 6-cyl inline engine—Stryker (Netherlands)
inline engines	1905	first 12-cyl inline engine— Wolseley (England)
inline engines	1919	first 8-cyl inline engine—Isotta-Fraschini (Italy)
inline engines	1920	first U.S. 8-cyl inline engine—Duesenberg (U.S.)

APPENDIX C MILESTONES BY CATEGORY

inline engines	1923	first mass-produced 8-cyl inline engine—Packard (U.S.)
inline engines	1955	last 8-cyl inline engine replaced by V8—Packard (U.S.)
inline engines	1970	two-stroke, 3-cyl inline engine—Saab (Sweden)
inline engines	1974	first inline, 5-cyl engine—Mercedes (Germany)
interchangeable parts	1902	first interchangeable parts used— Henry Leland (U.S.)
lubrication	1899	first full-pressure engine lubrication—Mors (France)
mass production	1901	first car mass-produced on an assembly line—Oldsmobile (U.S.)
milestone vehicles	1769	first self-propelled steam vehicle —Cugnot (France)
milestone vehicles	1802	first passenger-carrying steam carriage —Trevithick (England)
milestone vehicles	1805	first amphibious vehicle (steam)—Evans (U.S.)
milestone vehicles	1886	first gasoline-powered 3-wheel automobile—Benz (Germany)
milestone vehicles	1886	first gasoline-powered 4-wheel automobile—Daimler (Germany)
milestone vehicles	1909	first widely affordable automobile—Ford Model T (U.S.)
milestone vehicles	1922	Lancia produces first production, monocoque, unit-body vehicle (Italy)
milestone vehicles	1927	Ford sells the 15-millionth Model T (U.S.)
milestone vehicles	1938	Volkswagen starts making the Beetle (Germany)
milestone vehicles	1948	Deux Cheveaux automobile—Citroën (France)
milestone vehicles	1959	transverse-engined FWD car—Morris Mini-Minor (England)
milestone vehicles	1969	Fiat improved Mini's FWD design with the Fiat 128 (Italy)
milestone vehicles	2003	VW Beetle production ends after 21 million made (Germany/Mexico)
motorcycle	1885	first motorcycle—Benz (Germany)
overdrive	1934	overdrive transmission invented—Barnes (U.S.)
overdrive	1934	Borg-Warner licenses Barnes' overdrive (U.S.)
overdrive	1934	first Borg-Warner overdrive offered in a car—Chrysler (U.S.)
overhead valve	1904	first overhead-valve (OHV) engine—Buick (U.S.)
overhead valve	1929	first mass-produced OHV inline-six—Chevrolet (U.S.)
overhead valve	1949	first high-compression, OHV V8—Oldsmobile (U.S.)
oversquare engine	1949	first oversquare engine—Cadillac (U.S.)
post-war designs	1946	first postwar-designed automobile—Studebaker/Loewy (U.S.)
post-war designs	1948	Hudson introduces its step-down model (U.S.)
post-war designs	1949	Ford introduces an all new line of automobiles (U.S.)
power brakes	1919	first 4-wheel mechanical power brakes—Hispano-Suiza (Spain)
power brakes	1929	first vacuum power mechanical brakes—Duesenberg(U.S.)
power brakes	1932	first vacuum power hydraulic brakes—Chrysler (U.S.)
power steering	1928	first hydraulic power steering system—Davis (U.S.)
power steering	1951	first full-time power steering—Chrysler (U.S.)
racing	1895	first American race—Duryea wins, Benz is 2nd (U.S.)
radial engines	1901	first radial engine—Manly (England)
rotary engines	1897	first rotary engine—Millet (France)
single overhead cam	1902	first SOHC engine offered—Craig (England)
single overhead cam	1903	first U.S. SOHC engine offered—Marr Auto Car (U.S.)
sleeve valves	1905	dual-sleeve-valve engine invented—Knight (U.S.)
sleeve valves	1908	first production of Silent Knight engines—Daimler (England)
sleeve valves	1911	single-sleeve-valve engine invented—Burt & McCollum (Scotland)
sleeve valves	1939	last car made with a Silent Knight engine—Minerva (Belgium)
slipper piston	1920	slipper piston invented—Ricardo (England)
slipper piston	1949	first to use slipper pistons in a U.S. engine—Cadillac (U.S.)
split braking system	1967	first diagonal split braking system offered—AMC (U.S.)
steam vehicle	1897	first practical steam car made—Stanley brothers (U.S.)
steam vehicle	1901	first steam car with condenser and a flash boiler—White (U.S.)
steering	1758	"Ackerman" steering first invented—Darwin (England)
steering	1873	double-pivot steering invented—Benz (Germany)

supercharging	1860	air pump for supercharger invented—Roots (U.S.)
supercharging	1901	first experiments with supercharging—Chadwick (U.S.)
supercharging	1907	first car with supercharged engine—Chadwick (U.S.)
supercharging	1923	first supercharged production automobile—Mercedes (Germany)
supercharging	1935	screw-compressor supercharger invented—Lysholm (U.S.)
synchromesh	1928	Thompson invents synchromesh and sells it to GM (U.S.)
synchromesh	1952	Porsche designs a new synchromesh mechanism (Germany)
timing belt	1954	toothed timing belt invented (U.S.)
torque converter	1905	torque converter patented—Föttinger (Germany)
transmission	1879	variable speed transmission invented—Reeves (U.S.)
transmission	1895	first sliding-gear transmission—Panhard-Levaseur (France)
transmission	1898	first 3-speed epicyclic transmission—Lanchester (England)
transmission	1900	first use of a planetary transmission—Lanchester (England)
transmission	1928	first synchromesh transmission—Cadillac (U.S.)
transmission	1954	first single-clutch AMT (SMG) transmission—Daimler (England)
transmission	1981	first dual-clutch AMT invented—Webster (England)
turbocharging	1905	turbocharger invented—Buchi (Switzerland)
turbocharging	1915	first turbocharger on an engine—Rateau (France)
V-engines	1889	first V-twin engine—Daimler (Germany)
V-engines	1902	first aircraft/boat V8 engine—Levavasseur (France)
V-engines	1903	first V8 engine for a race car—Adler (France)
V-engines	1904	first V12 engine—Putney Motor Works (England)
V-engines	1905	first, low-volume V6 engine—Marmon (U.S.)
V-engines	1910	first V8 engines for production—De Dion (France)
V-engines	1914	first U.S. production V8 engine—Cadillac (U.S.)
V-engines	1916	first production V12 engine—Packard (U.S.)
V-engines	1917	first OHV V8 engine—Chevrolet (U.S.)
V-engines	1920	first V4 automotive engine—Lancia (Italy)
V-engines	1930	first V16 engine—Cadillac (U.S.)
V-engines	1932	first V8 engine in a low-priced car—Ford (U.S.)
V-engines	1950	first series production V6 engine—Lancia (Italy)
V-engines	1951	first OHV Hemi V8 engine—Chrysler (U.S.)
V-engines	1955	Chevrolet introduces its small-block OHV V8 (U.S.)
V-engines	1962	first 90°, uneven-firing, V6 engine—Buick (U.S.)
V-engines	1991	first production V10 engine—Dodge (U.S.)
variable valve timing	1972	first engine with VVT—Alpha Romeo (Italy)
variable valve timing	1989	first VVT system to vary lift and duration—Honda (Japan)
variable valve timing	2001	first continuously variable VVT system—BMW (Germany)
vibration control	1906	crankshaft torsional damper invented—Lanchester (England)
vibration control	1913	harmonic balancer invented—Lanchester (England)
vibration control	1916	first counterweighted crankshaft—Hudson (U.S.)
vibration control	1925	first rubber engine mounts—Chrysler (U.S.)
vibration control	1928	crankshaft impulse neutralizer and vibration damper—Chrysler (U.S.)
vibration control	1931	"floating power" engine mounting—Plymouth (U.S.)
vibration control	1948	polydyne cam functions invented—Dudley (U.S.)
viscous coupling	1970	viscous coupling invented—Rolt (U.S.)
W-engines	1906	first W3 engine—Anzani (France)
W-engines	1917	first W12 engine—Napier (France)
W-engines	1928	first W18 engine—Isotta-Fraschini (Italy)
W-engines	2005	first W16 engine—Bugatti (Germany)
Wankel engine	1929	Wankel engine invented—Felix Wankel (Austria)

Appendix D

List of Figures

LICENSE KEY:

AC - Author Original
CC - Creative Commons Attribution-Share Alike - Free use with attribution
PD - Public Domain
PM - Permissioned

CHAPTER 1   Motive Power

Fig	Subject	Source	License
1	Watt sun and planet drive	author original	AO
2	Watt straight-line linkage	author original	AO
3	Watt steam engine	author original	AO
4	locomotive steam engine	author original	AO
5	Fardier de Cugnot	Tampa Bay Automobile Museum	PM
6	ratchet/pawl mechanism	author original	AO
7	Stanley Steamer	author photo	AO
8	Locomobile Steamer	author photo	AO
9a	White Steamer	author photo	AO
9b	White Steamer engine	Dr. Robert Dyke	PM
10	Doble Steamer	author photo	AO
11	Stirling engine	author original	AO
12	slider crank linkage	author original	AO
13	four-stroke cycle	author original	AO
14	two-stroke cycle	author original	AO
15	Atkinson patent drawing	public domain	PD
16	Miller engine	public domain	PD
17	Wankel engine	http://en.wikipedia.org/wiki/Wankel_engine	CC
18	Milburn Electric	author photo	AO
19	Lohner-Porsche	http://en.wikipedia.org/wiki/Lohner-Porsche_Mixte_Hybrid	CC
20	Prius engine/transmission	http://en.wikipedia.org/wiki/Hybrid_Synergy_Drive	CC

CHAPTER 2   Chassis Layouts and Drivelines

Fig	Subject	Source	License
1	Benz 1886	author photo	AO
2	Daimler 1885 motorcycle	http://www.daimler.com/dccom/.html	PM
3	Daimler 1886	http://www.daimler.com/dccom/.html	PM
4	Flocken Elektrowagen	http://en.wikipedia.org/wiki/File:1888_Flocken_Elektrowagen.jpg	CC
5	Panhard 1898	http://en.wikipedia.org/wiki/Panhard	PD

6	curved dash Olds	http://commons.wikimedia.org/wiki/File:Oldsmobile_Curved_Dash_Runabout_1903.jpg	CC
7	Model T	http://en.wikipedia.org/wiki/File:1919_Ford_Model_T_Highboy_Coupe.jpg	CC
8	diagram of Porsche 911	author original	AO
9	diagram of Porsche Cayman	author original	AO
10a	1949 Beetle	http://commons.wikimedia.org/wiki/File:1949_VW_Beetle.jpg	PD
10b	1949 Beetle interior	http://en.wikipedia.org/wiki/Volkswagen_Beetle#mediaviewer/File:1949_VW_dash.jpg	PD
11	diagram of BMW 530i	author original	AO
12	universal joint	author original	AO
13	diagram of Corvette	author original	AO
14	Citroen 2CV	http://commons.wikimedia.org/wiki/File:Citroen_2CV_3.jpg	PD
15a	diagrams of front drive	http://commons.wikimedia.org/wiki/File:Automotive_diagrams_07_En.png	CC
15b	diagrams of front drive	http://commons.wikimedia.org/wiki/File:Automotive_diagrams_08_En.png	CC
15c	diagrams of front drive	http://commons.wikimedia.org/wiki/File:Automotive_diagrams_10_En.png	CC
16	Morris Mini 1959	http://upload.wikimedia.org/wikipedia/commons/thumb/b/bd/1959_Morris_Mini-Minor_Heritage_Motor_Centre%2C_Gaydon.jpg/1280px-1959_Morris_Mini-Minor_Heritage_Motor_Centre%2C_Gaydon.jpg	CC
17	Fiat 128 drivetrain	http://www.fiat500usa.com/2009/02/dante-giacosa.html	PM
18	double cardan CV joint	http://en.wikipedia.org/wiki/Constant-velocity_joint#mediaviewer/File:Double_Cardan_Joint_(animated).gif	PD
19	Tracta CV joint	http://en.wikipedia.org/wiki/File:Tracta_Constant_Velocity_Joint.jpg	PD
20a	Rzeppa patent drawing	U.S. Patent and Trademark Office	PD
20b	Rzeppa CV joint photo	http://upload.wikimedia.org/wikipedia/commons/thumb/9/9a/Gelenk-welle.jpg/1280px-Gelenk-welle.jpg	CC
21	4X4 shift knob (adapted)	http://en.wikipedia.org/wiki/Four-wheel_drive#mediaviewer/File:Transfer_lever_4wd_gnangarra.jpg	CC
22	center differential	http://en.wikipedia.org/wiki/Four-wheel_drive#mediaviewer/File:Center_differential_Jeep_Grand_Cherokee_AWD.jpg	PD
23	1903-Spyker-60-HP	http://en.wikipedia.org/wiki/Four-wheel_drive#mediaviewer/File:1903-Spyker-60-HP-1920x1440.jpg	PD
24	Jensen FF in snow	http://en.wikipedia.org/wiki/Four-wheel_drive#mediaviewer/File:Jensen_FF_mk11_in_snow.jpg	CC
25a	all wheel drive	http://commons.wikimedia.org/wiki/File:Automotive_diagrams_02C_En.png	CC
25b	rear engine AWD	http://commons.wikimedia.org/wiki/File:Automotive_diagrams_06_En.png	CC

APPENDIX D LIST OF FIGURES

25c	front longitudinal AWD	http://upload.wikimedia.org/wikipedia/commons/thumb/b/b6/Automotive_diagrams_09_En.png/1280px-Automotive_diagrams_09_En.png	CC
25d	front transverse AWD	http://upload.wikimedia.org/wikipedia/commons/thumb/3/37/Automotive_diagrams_11_En.png/1280px-Automotive_diagrams_11_En.png	CC
26	AMC Eagle	http://en.wikipedia.org/wiki/Four-wheel_drive#mediaviewer/File:1987_AMC_Eagle_wagon_burgundy-woodgrain_NJ.jpg	CC
27	Deux Chevaux Sahara	author photos	AO

CHAPTER 3 Engine Configurations

Fig	Subject	Source	License
1	engine configurations	author original	AO
2	DeDion Bouton engine	http://en.wikipedia.org/wiki/Crankcase#mediaviewer/File:De_Dion-Bouton_engine_(Rankin_Kennedy,_Modern_Engines,_Vol_III).jpg	PD
3	Wolseley 6-cyl cast in pairs	http://upload.wikimedia.org/wikipedia/commons/thumb/8/8e/Wolseley_6-cylinder_marine_oil_engine_%28Rankin_Kennedy%2C_Modern_Engines%2C_Vol_V%29.jpg/1280px-Wolseley_6-cylinder_marine_oil_engine_%28Rankin_Kennedy%2C_Modern_Engines%2C_Vol_V%29.jpg	PD
4	en bloc engine and cylinder head	http://upload.wikimedia.org/wikipedia/commons/3/32/Cylinder_block_and_head_of_sidevalve_engine_%28Autocar_Handbook%2C_Ninth_edition%29.jpg	PD
5	BMW 6-cyl engine	http://commons.wikimedia.org/wiki/File:CarterBMW1.jpg	CC
6	2-cylinder shaking force	author original	AO
7	2-cylinder crank-phase diagram	author original	AO
8	2-cylinder shaking force	author original	AO
9	2-cylinder crank-phase diagram	author original	AO
10	2-cylinder shaking force	author original	AO
11	2-cylinder gas force plot	author original	AO
12	4-cylinder crankshaft and pistons	author original	AO
13	4-cylinder crankshaft	author original	AO
14	4-cylinder gas force plot	author original	AO
15	4-cylinder gas torque plot	author original	AO
16	5-cylinder gas force plot	author original	AO
17	Jaguar E-Type engine	http://en.wikipedia.org/wiki/File:Jaguar_XK6_engine_1.jpg	CC
18	6-cylinder gas force plot	author original	AO
19	6-cylinder gas torque plot	author original	AO
20	3-cylinder gas force plot	author original	AO

21	3-cylinder gas torque plot	author original	AO
22	Duesenberg straight-8 engine	http://commons.wikimedia.org/wiki/File:Model_J_engine.jpg	PD
23	1933 Bugatti straight-8 engine	http://commons.wikimedia.org/wiki/File:1933_Bugatti_Type_59_Grand_Prix_engine.jpg	CC
24	crankshaft from Buick 1937 engine	http://grabcad.com/library/crankshaft-from-1937-buick-straight-8	PM
25	8-cylinder gas force plot	author original	AO
26	8-cylinder gas torque plot	author original	AO
27	1905 inline Wolseley 12-cyl engine	http://commons.wikimedia.org/wiki/File:Wolseley_12_cylinder_360hp_petrol_or_oil_marine_engine_(Rankin_Kennedy,_Modern_Engines,_Vol_III).jpg	PD
28	V-engine geometry	author original	AO
29	connecting rods	http://en.wikipedia.org/wiki/V-twin_engine#mediaviewer/File:Forked_connecting_rods_(Autocar_Handbook,_13th_ed,_1935).jpg	AO, PD
30	Harley gas force plot	author original	AO
31	Harley gas torque plot	author original	AO
32	V8 crankshafts	author original	AO
33	V8 primary moment plot	author original	AO
34	V6 gas torque plot	author original	AO
35	VW VR15 engine	author photo	AO
36	Packard Twin Six	http://upload.wikimedia.org/wikipedia/commons/thumb/9/9a/1916Packard1-35TownCarLimoKimballEngine.jpg/1280px-1916Packard1-35TownCarLimoKimballEngine.jpg	CC
37	Ferrari V12 engine	http://upload.wikimedia.org/wikipedia/commons/thumb/e/ea/1961_Ferrari_250_TR_61_Spyder_Fantuzzi_engine.jpg/1280px-1961_Ferrari_250_TR_61_Spyder_Fantuzzi_engine.jpg	CC
38	Cadillac V16	http://upload.wikimedia.org/wikipedia/commons/thumb/9/91/Cadillac452engine.jpg/1237px-Cadillac452engine.jpg	CC
39	Marmon V16	http://en.wikipedia.org/wiki/V16_engine#mediaviewer/File:1933MarmonV16-engine.jpg	PD
40	boxer engine cranks	author original	AO
41	BMW flat twin	wiki/File:Opposed-cylinders-500.jpg	CC
42	Lanchester 2-cyl engine	author original	
43	Anzani W3 engine	http://en.wikipedia.org/wiki/Anzani#mediaviewer/File:Anzani_Military_Model_Fan_type.jpg	CC
44	Napier W12 enine	http://commons.wikimedia.org/wiki/File:Napier_Lion_W12_@_Brooklands_Museum.jpg	CC
45	VW W12	http://upload.wikimedia.org/wikipedia/commons/thumb/1/14/Volkswagen_W12.jpg/768px-Volkswagen_W12.jpg	CC
46	VW W16	http://upload.wikimedia.org/wikipedia/commons/thumb/7/74/Volkswagen_W16.jpg/1280px-Volkswagen_W16.jpg	CC

APPENDIX D   LIST OF FIGURES

47	rotary engined bicycle	http://upload.wikimedia.org/wikipedia/commons/thumb/c/c9/Felix_Millet.jpg/1280px-Felix_Millet.jpg	CC
48	Gaucher rotary engine	author photo	AO
49	Radial engine	author photo	AO
50	Wasp engine	author photo	AO
51	bearing forces	author original	AO
52	balance shafts	author original	AO
53	floating power	author original	AO

CHAPTER 4 Valve Trains, Induction, and Supercharging

Fig	Subject	Source	License
1	cams and cam followers	author original	AO
2	conjugate cams	author original	AO
3	cam spline functions	author original	AO
4	valve configurations	author original	AO
5	T-head cross section	http://commons.wikimedia.org/wiki/File:T-head_single-cylinder_Otto_engine_(Army_Service_Corps_Training,_Mechanical_Transport,_1911).jpg#filehistory	PD
6	Ricardo L-head	http://upload.wikimedia.org/wikipedia/commons/thumb/5/56/Side_valve_engine_with_Ricardo%27s_turbulent_head_02.png/790px-Side_valve_engine_with_Ricardo%27s_turbulent_head_02.png	CC
7	OHV valve train	author original	AO
7b	OHV valve train	http://en.wikipedia.org/wiki/Overhead_valve#mediaviewer/File:Pushrod2.PNG	CC
8	follower force	author original	AO
9	SOHC cross section	author original	AO
9b	Honda SOHC head	http://upload.wikimedia.org/wikipedia/commons/thumb/f/f2/Head_D15A3.jpg/928px-Head_D15A3.jpg	CC
10	DOHC cross section	http://commons.wikimedia.org/wiki/File:DOHC-Zylinderkopf-Schnitt.jpg	CC
10b	Napier Lion DOHC	http://upload.wikimedia.org/wikipedia/commons/thumb/3/38/Napier_Lion_cambox.jpg/1280px-Napier_Lion_cambox.jpg	CC
11	Neon acceleration	author original	AO
12	bent valves	http://upload.wikimedia.org/wikipedia/commons/thumb/0/0d/Bent_Valves.jpg/1139px-Bent_Valves.jpg	CC
13	Knight cross section	public domain	PD
14	cutaway Knight engine	author photo	AO
15	valve overlap	author original	AO
16	Honda VTEC head	http://commons.wikimedia.org/wiki/File:K20_head.jpg	CC
17	Hyundai vane phasors	http://upload.wikimedia.org/wikipedia/commons/thumb/d/d1/Vane_phasers_T-GDI.jpg/1280px-Vane_phasers_T-GDI.jpg	CC

18	carburetor diagram	http://upload.wikimedia.org/wikipedia/commons/thumb/2/2b/Carburetor.svg/1091px-Carburetor.svg.png	CC
19	transparent float bowls	http://commons.wikimedia.org/wiki/File:Holley_Visiflo.jpg	CC
20	Jaguar carburetors	author photo	
21	roots pump	http://upload.wikimedia.org/wikipedia/commons/thumb/7/7b/Roots_blower_-_2_lobes.svg/2000px-Roots_blower_-_2_lobes.svg.png	CC
22	Lysholm screw pump	http://commons.wikimedia.org/wiki/File:Lysholm_screw_rotors.jpg	CC
23	vane supercharger	http://commons.wikimedia.org/wiki/File:Powerplus_vane-type_supercharger_diagram.png	PD
24	centrifugal supercharger	http://commons.wikimedia.org/wiki/File:ATI_ProCharger_Supercharger_Cutaway.jpg	CC
25	blown Rolls-Royce Merlin	http://upload.wikimedia.org/wikipedia/commons/thumb/7/7d/Rolls-Royce_Merlin.jpg/1280px-Rolls-Royce_Merlin.jpg	CC
26	Blower Bentley	http://upload.wikimedia.org/wikipedia/commons/thumb/a/a0/1929_Bentley_front_34_right.jpg/1280px-1929_Bentley_front_34_right.jpg	CC
27	AMC Marlin with Roots blower	http://upload.wikimedia.org/wikipedia/commons/thumb/7/73/1968_AMX_blown_and_tubbed_e.jpg/1280px-1968_AMX_blown_and_tubbed_e.jpg	PD

CHAPTER 5 Transmissions and Differentials

Fig	Subject	Source	License
1	two-stage gear train	author original	AO
2	crashbox transmission	author original plus public domain	AO, PD
3	4-speed shift mechanisms	http://commons.wikimedia.org/wiki/File:Manual_transmission_clutch_Reverse_gear.PNG	PD
4	18-wheeler trans.	http://commons.wikimedia.org/wiki/File:MeritorColorHiLitesSF.jpg	CC
5	gears	author photos	AO
6	synchromesh	author original	AO
7	synchronizers	author photo	AO
8	friction drive	http://commons.wikimedia.org/wiki/File:1906_Lambert_2-cylinder_chassis.png	PD
9	chain driven CVT	http://commons.wikimedia.org/wiki/File:Pivgetriebe.png	CC
10	Toyota CVT	http://commons.wikimedia.org/wiki/File:Toyota_Super_CVT-i_01.jpg	PD
11	planetary gear train	author original	AO
12	Model T transmission	author original	AO
13	Model T pedals	author photos	AO
14	Warner overdrive	Half-Hour History of Overdrives" Special Interest Autos, Mar–Apr 1974, p.47	PM
15	automotive clutch	author original	AO

APPENDIX D   LIST OF FIGURES

16	Daimler Fluid Flywheel	http://commons.wikimedia.org/wiki/File:Fluid_flywheel,_part_section_(Autocar_Handbook,_13th_ed,_1935).jpg	PD
17	Olds semi-automatic transmission	"Almost Automatic," Special Interest Autos, Jan–Feb 1974, p 27	PM
18	Chrysler Vacamatic trans.	SAE Journal V 47 No 5 Nov 1940, p.455 - SAE has no copyright.	PD
19	Sturtevant Patent Drawing	U.S. Patent and Trademark Office	PD
20	GM Hydramatic trans.	SAE Journal Vol 45 No 5 Nov 1939 p. 462 - SAE has no copyright.	PD
21	ZF torque converter	http://commons.wikimedia.org/wiki/File:Bauma_2007_ZF_Drehmomentwandler.jpg	CC, AO
22	GM Turbohydramatic	Anatomy of a Motorcar, Crescent Books	PD
23	ZF 8-sp auto trans.	http://commons.wikimedia.org/wiki/File:ZF_Stufenautomatgetriebe_8HP70.jpg	CC
24	DCT schematic	http://commons.wikimedia.org/wiki/File:Dual-clutch_transmission.svg	PD
25	VW DSG trans.	http://commons.wikimedia.org/wiki/File:VW_DSG_transmission_DTMB.jpg	CC
26	south-pointing chariot	http://commons.wikimedia.org/wiki/File:South-pointing_chariot_(Science_Museum_model).jpg	CC
27	bevel gear differential	http://commons.wikimedia.org/wiki/File:Differential_(Manual_of_Driving_and_Maintenance).jpg	PD
28	spiral bevel differential	http://commons.wikimedia.org/wiki/File:Transmission_diagram.jpg	CC
29	hypoid differential	http://commons.wikimedia.org/wiki/File:Differentialgetriebe2.jpg	CC
30	Torsen differential	http://en.wikipedia.org/wiki/Torsen#mediaviewer/File:Audi_quattro_AWD_system.jpeg	CC
31	Columbia rear end	Half-Hour History of Overdrives" Special Interest Autos, Mar–Apr 1974, p.47	PM

CHAPTER 6   Suspension and Steering

Fig	Subject	Source	License
1	full elliptic leaf spring	http://commons.wikimedia.org/wiki/File:Spring_3_(PSF).png	PD
2	1904 Stanley	author photo	AO
3	semi elliptic spring	http://commons.wikimedia.org/wiki/File:Leafs1.jpg	PD
4	Alvis spring	author photo	AO
5	Model T	author photo	AO
6a	torsion bar no load	http://en.wikipedia.org/wiki/File:Torsion-Bar_no-load.jpg	PD
6b	torsion bar with load	http://en.wikipedia.org/wiki/File:Torsion-Bar_with-load.jpg	PD
6c	Citroen torsion bar	http://en.wikipedia.org/wiki/Citroën_Traction_Avant#mediaviewer/File:Citroen_front_suspension_(Autocar_Handbook,_13th_ed,_1935).jpg	PD
7	coil spring	author original	AO
8	spring rate	author original	AO

9	Citroen SM	author photo	AO
10	Otto axle	author photo	AO
11	caster, etc.	author original	AO
12	chain drive Simplex	author photo	AO
13	semi and full floating axles	author original	AO, PD
14	De Dion Patent	author scan	AO
15	forces on wheel	author original	AO
16	Hotchkiss driveline	author original	AO
17	leaf spring deflection	http://en.wikipedia.org/wiki/Hotchkiss_drive#mediaviewer/File:Leaf_spring,_showing_torque_reaction_effects_(Manual_of_Driving_and_Maintenance).jpg	PD
18	Torque-tube driveline	author original	AO
19	Watt linkage	author original	AO
20	multi-link suspension	author original	AO
21	generic independent suspension	author original	AO
22	Sizaire 1908 IFS	http://commons.wikimedia.org/wiki/File:Sizaire-Naudin_1908.jpg	PD
23	sliding pillar suspension	http://commons.wikimedia.org/wiki/File:MorganSuspension.jpg	CC
24	Lancia Lambda	author photo	AO
25	Tatra suspension	author photo	AO
26	Alvis suspension	author photo	AO
27	Talbot Lago suspension	author photo	AO
28	Dubonnet suspension	Half-Hour History of Independent Front Suspensions - SIA Nov-Dec 1973 - Hemmings Motor News	PM
29	Swinging axle suspensions	SAE Journal V 33, No 4 1933 p. 326 - SAE has no copyright on this.	PD
30	Ford twin Ibeam suspension	http://www.hemmings.com/hmn/stories/2008/06/01/hmn_feature19.html	PM
31	GM independent suspension	Half-Hour History of Independent Front Suspensions - SIA Nov-Dec 1973 - Hemmings Motor News	PM
32	MacPherson strut patent	U.S. Patent and Trademark Office	PD
33	Fornaca strut patent	U.S. Patent and Trademark Office	PD
34	Morgan cyclecar	author photo	AO
35	Jaguar rear suspension	author photo	AO
36	multilink suspension	http://commons.wikimedia.org/wiki/File:5link3Drear1.gif	PD
37	damping curves	author original	AO
38a	Ackerman steering straight	http://commons.wikimedia.org/wiki/File:Ackermann_simple_design.svg	CC
38b	Ackerman steering turning	http://commons.wikimedia.org/wiki/File:Ackermann_turning.svg	CC
39	steering mechanisms	author original	AO

APPENDIX D   LIST OF FIGURES

40	Fardier de Cugnot steering	John Perodeau	PM
41	roll center	author original	AO
42	roll axis	author original	AO
43	sway bar	http://commons.wikimedia.org/wiki/File:Alfetta_front_suspension_antiroll.jpg	CC
44	ant-dive and anti-squat	author original	AO
45	hopalong Vega	author original	AO

CHAPTER 7   Brakes

Fig	Subject	Source	License
1	wheel cylinder	http://commons.wikimedia.org/wiki/File:Wheel_cylinder.jpg	CC
2	drum brake rendering	http://commons.wikimedia.org/wiki/File:Drum_brake_testrender.jpg	CC
3	self-energizing brake	author original	AO
4	disk brake assembly	http://commons.wikimedia.org/wiki/File:Disk_brake_dsc03680.jpg	CC
5	disk brake caliper	http://commons.wikimedia.org/wiki/File:Callipers_Twin_Pot.jpg	PD
6	Porsche disk brake	http://en.wikipedia.org/wiki/File:PCCB_Wiki_9949.jpg	PD
7	Carbon-Ceramic disk brake	http://commons.wikimedia.org/wiki/File:AMG_Carbon_Ceramic_brake..jpg	PD
8	Hispano-Suiza servo brakes	author photos	AO
9	brake booster	http://commons.wikimedia.org/wiki/File:Servofreno_seccionado.jpg	PD

CHAPTER 8   Body Design

Fig	Subject	Source	License
1	Skene buggy	author photo	AO
2	Stanley runabout	author photo	AO
3	Mercer raceabout	author photo	AO
4	De Dion Bouten vis-a-vis	author photo	AO
5	Pope-Hartford tonneau	author photo	AO
6	Knox surrey	author photo	AO
7	Packard Victoria	author photo	AO
8	Peugeot phaeton	author photo	AO
9	Stevens Duryea tourer	author photo	AO
10	Rickenbacker sedan	author photo	AO
11	Morgan Plus 8	author photo	AO
12	Model T depot hack	author photo	AO
13	drag plots	author original	AO
14	1934 Chrysler airflow front	http://commons.wikimedia.org/wiki/File:1934ChryslerAirflow.jpg	CC
15	1934 Chrysler airflow rear	http://commons.wikimedia.org/wiki/File:%2734_airflow_rear_2.jpg	CC

16	1937 Chrysler Airflow	http://commons.wikimedia.org/wiki/File:1937ChryslerAirflow.jpg	CC
17	Tatra T77 model	http://commons.wikimedia.org/wiki/File:Tatra_77_maquette_1-10_by_Paul_Jarray.jpg	CC
18	Tatra T77 on track	http://commons.wikimedia.org/wiki/File:Tatra_77A_dutch_licence_registration_AM-44-01_pic10.jpg	PD
19	auto profiles	SAE Journal, Vol 45, No 3, p. 367 - SAE has no copyright on it.	PD
20a	Lancia Lambda	author photo plus public domain	AO
20b	Lancia Lambda structure	http://www.dieselpunks.org/profiles/blogs/lord-ks-garage-96-lancia	PD
21	1933 Adler Trumpf	author photo	AO
22	Citroen Traction Avant	http://commons.wikimedia.org/wiki/File:Tractionfr02.jpg	PD
23	Citroen Traction Avant body	http://en.wikipedia.org/wiki/Citroën_Traction_Avant#mediaviewer/File:Citroen_Traction_Avant_body-chassis_unit_(Autocar_Handbook,_13th_ed,_1935).jpg	PD
24	Cord 812	http://commons.wikimedia.org/wiki/File:1937_Cord_812_Phaeton,_Lime_Rock.jpg	CC
25	Lincoln Zephyr	http://commons.wikimedia.org/wiki/File:Lincoln_zephyr_06011701.jpg	CC
26	1942 Tatra T87	courtesy of John Perodeau	PM
27	Nash 600	http://commons.wikimedia.org/wiki/File:1946_Nash_600_gray_2-door_sedan_ny.jpg	PD
28	Nash 600 X-ray	http://commons.wikimedia.org/wiki/File:1942_Nash_Ambassador_X-ray.jpg	PD
29	1915 Pierce Arrow	author photo	AO
30	1937 Peugeot Daarl'Mart	courtesy of John Perodeau	PM
31	1938 Amilcar	author photo	AO
32	1953 Hotchkiss-Gregoire	courtesy of John Perodeau	PM
33	Jag D-type	http://commons.wikimedia.org/wiki/File:1955_Jaguar_XKD_34_left.jpg	CC
34	300SL gullwing	http://commons.wikimedia.org/wiki/File:1955_Mercedes-Benz_300SL_Gullwing_Coupe_34.jpg	CC
35	300 SL roadster	author photo	AO
36	original Acura NSX	http://commons.wikimedia.org/wiki/File:Honda_NSX_red.jpg	CC
37	2015 Acura NSX	http://commons.wikimedia.org/wiki/File:Detroit_NAIAS_2015_2016_Acura_NSX.jpg	CC
38	Audi A8	http://commons.wikimedia.org/wiki/File:A8_white.jpg	CC
39	Audi A2	http://commons.wikimedia.org/wiki/File:Audi_A2_front_20071002.jpg	CC
40	Morgan AeroMax	http://commons.wikimedia.org/wiki/File:MorganAeroMax.jpg	CC
41	Jaguar XJ8	http://commons.wikimedia.org/wiki/File:Jaguar_XJ8_Vanden_Plas.jpg	AO
42	1954 Corvette	http://commons.wikimedia.org/wiki/File:1954_Corvette.jpg	AO

APPENDIX D LIST OF FIGURES

43	1963 Corvette	http://commons.wikimedia.org/wiki/File:1963_Corvette_Sting_Ray.jpg	AO
44	2014 Corvette	http://commons.wikimedia.org/wiki/File:2014_Chevrolet_Corvette.jpg	CC
45	Lotus Elite	http://commons.wikimedia.org/wiki/File:Lotus_Elite_at_Mallory_Park.jpg	CC
46	Lotus Elan backbone	http://en.wikipedia.org/wiki/Lotus_Elan#mediaviewer/File:Lotus_Elan_car_chassis.jpg	CC
47	Lotus Elan	http://commons.wikimedia.org/wiki/File:LotusElan%2B2Side.jpg	CC
48	Pontiac Fiero	http://commons.wikimedia.org/wiki/File:Fiero88.jpg	AO
49	2013 Viper	http://commons.wikimedia.org/wiki/File:2013_SRT_Viper_--_2012_NYIAS.jpg	AO
50	Bugatti Veyron	http://commons.wikimedia.org/wiki/File:Bugatti_Veyron_16.4_–_Frontansicht_(3),_5._April_2012,_Düsseldorf.jpg	CC
51	Lexus LFA	http://commons.wikimedia.org/wiki/File:Lexus_LFA_Yellow_Las_Vegas.jpg	CC
52	Lamborgini Aventador chassis	http://commons.wikimedia.org/wiki/File:Lamborghini_Aventador_LP_700-4_chassis_-_Flickr_-_J.Smith831.jpg	CC
53	Lamborgini Aventador coupe	http://en.wikipedia.org/wiki/Lamborghini_Aventador#/media/File:Musée_Lamborghini_0133.JPG	CC
54	McLaren MP4	http://commons.wikimedia.org/wiki/File:McLaren_MP4-12C_–_Frontansicht_(3),_30._August_2012,_Düsseldorf.jpg	CC
55	Mercedes SLR	http://commons.wikimedia.org/wiki/File:Mercedes-Benz_SLR_McLaren_2_cropped.jpg	PD
56	BMW i3	http://commons.wikimedia.org/wiki/File:BMW_i3_01.jpg	CC
57	FEA simulation of crush	http://en.wikipedia.org/wiki/Crumple_zone#mediaviewer/File:FAE_visualization.jpg	PD
58	crash test at GM	http://commons.wikimedia.org/wiki/File:Strefa_zgniotu_w_tescie_zderzeniowym.jpg	PD
59	pedestrian crash	http://commons.wikimedia.org/wiki/File:PedCrashSequence.png	PD

NOTES

Appendix E

Bibliography

General

Automobile Quarterly. "Automobile Quarterly." 1971–1975.

Beaulieu, Lord Montague. *Jaguar*. New York: A. S. Barnes and Co, 1967.

Bird, A. and Hutton-Stott, F. *Lanchester Motor Cars: A History*. London: Montague, 1965.

Bissell, Thomas A. "Trends in Design of 1940 Cars." *SAE Journal* 45, no. 5 (1939): 457–471.

Bohacz, Ray T. "The Silent Worker—Electricity and the Automobile." *Hemmings Classic Car*, May 2014, 76–78.

Bollee, "http://en.wikipedia.org/wiki/Amédée_Bollée" (accessed Mar 2014).

Brull, Charles B. "Modern European Light Cars." *SAE Journal* 46, no. 4 (1940): 177–183.

Cerf, Alain. *Dimitri Sensaud De Lavaud, an Extraordinary Engineer*. Tampa, FL: Tampa Bay Automobile Museum, 2010.

Dluhy, R. D. *American Automobiles of the Brass Era*. London: McFarland and Co., Inc., 2013.

Duerksen, Menno. "The Walter P. Chrysler Story—Part 1." *Cars and Parts*, Aug 1982, 10–16.

Duerksen, Menno. "The Walter P. Chrysler Story—Part 2." *Cars and Parts*, Sep 1982, 10–16.

Duerksen, Menno. *History of the Great American Classics*. Sidney, OH: Amos Press Inc, 1987.

Georgano, G. N. *Encyclopedia of American Automobiles*. New York: E. P. Dutton, 1968.

Georgano, G. N. *Cars 1886–1930*. New York: Crescent Books, 1990.

Georgano, G. N. *The American Automobile, a Centenary 1893–1993*. New York: Smithmark, 1992.

Hudson, "http://www.Allpar.Com/Cars/Adopted/Hudson-1936.html" (accessed Mar 2014).

Lamm, Michael. "Model A—The Birth of Ford's Interim Car." *Special Interest Autos*, Aug–Oct 1973, 12–39, 56.

Lamm, Michael. "Turbine Cars: Chrysler's Bronze Blowtorch." *Hemmings Classic Car*, Oct 2012, 18–23.

Lancia. "The Lancia Aurelia." *The Automobile Engineer*, February 1951, 43–51.

Litwin, Matthew. "18 Most Innovative American Cars Ever Built." *Hemmings Classic Car*, Feb 2013, 20–25.

Ludvigsen, Karl. *Porsche, Excellence Was Expected*. Princeton, NJ: Princeton Publishing, 1977.

MacIlvain, W. O. "The Stearns Company." *Bulb Horn*, Jan–Feb 1972, 12–21.

Norton, Robert L. *Kinematics and Dynamics of Machinery, 2ed*. Singapore: McGraw-Hill, 2013.

Olley, Maurice. "National Influences on American Passenger Car Design." *Proc. of the Institution of Automobile Engineers*, (1938): 65.

Olley, Maurice. "European Postwar Cars." *SAE Trans.* 61, (1953): 503–528.

Padgett, Martin. *50 Years with Car and Driver*. New York: Fillipachi Publishing, 2005.

Pomeroy, L. H. "What Is Wrong with American Cars?" *SAE Journal* 33, no. 3 (1933): 12.

Renault. "Renault Dauphine." *Sports Car Illustrated*, October 1956, 28.

Report, Drive. "Two Look-Alikes—Ford & Citroën." *Special Interest Autos*, Jan–Mar 1072, 4655.

SAE_Historical_Society. *The Automobile, a Century of Progress*: SAE, 1997.

Schmidt, O. C. *Practical Treatise on Automobiles, 2 Vols*. Philadelphia: Stanley Institute, 1914.

Setright, L. J. K. *The Designers, Great Automobiles and the Men Who Made Them*. Chicago: Follett Publishing Company, 1976.

Setright, L. J. K and Ward, Ian. *Anatomy of the Motor Car*. New York: Crescent Books, 1976.

Stein, Ralph. *The World of the Automobile*. New York: Random House.

Stein, Ralph. *The Treasury of the Automobile*. New York: Ridge Press, 1961.

Stein, Ralph. *The Great Cars*. New York: Grosset and Dunlap, 1967.

Subenrauch, Bob. *The Fun of Old Cars*. New York: Dodd, Meade, and Co., 1967.

Ulmann, Alec. "The WWI Lanchester Armored Car Story." *Bulb Horn*, Nov–Dec 1975, 3440.

Ulmann, Alec. "Fiat Made in U.S.A." *Bulb Horn*, Sep–Oct 1976, 1419.

Walton, Harry. *The How and Why of Mechanical Movements*. New York: Popular Science Publishing Co, 1968.

Wolf, Austin M. "Automobile Engineering Progress." *SAE Journal* 30, no. 1 (1932): 16.

Wolf, Austin M. "Striking Engineering Progress Revealed in 1933 Cars." *SAE Journal* 32, no. 1 (1933): 21.

Wolf, Austin M. "1934 Marks Turn from Conventional in Automobile Design." *SAE Journal* 34, no. 1 (1934): 17.

Wolf, Austin M. "Marked Advances Shown in Designs of 1935 Automobiles." *SAE Journal* 36, no. 1 (1935): 17.

Wolf, Austin M. "Refinements and Super Styling in the 1936 Models." *SAE Journal* 37, no. 5 (1935): 16.

Wolf, Austin M. "Trends in 1937 Car Design." *SAE Journal* 39, no. 5 (1936): 16.

Wolf, Austin M. "Trends in Design of 1939 Cars." *SAE Journal* 43, no. 5 (1938): 12.

Chapter 1 Motive Power

Atkinson_Engine, "http://en.wikipedia.org/wiki/Atkinson_Cycle#Modern_Atkinson_Cycle_Engines" (accessed Mar 2013).

Atkinson_Engine, "http://www.Animatedengines.Com/Atkinson.html" (accessed Mar 2013).

Atkinson/Miller_Engine, "http://www.Curbsideclassic.Com/Blog/the-Atkinson-and-Miller-Cycle-Engines-Not-Exactly-How-They-Started-out-to-Be/" (accessed Mar 2013).

Brayton_cycle, "http://en.wikipedia.org/wiki/Brayton_Cycle" (accessed Jan 2014).

Cerf, A. A. *Nicolas Cugnot and the Chariot of Fire*. Tampa, FL: Tampa Bay Automobile Museum, 2010.

Chevrolet_Volt, "http://www.Plugincars.Com/Exclusive-Chevrolet-Volt-Chief-Engineer-Explains-Volt-Drivetrain-Says-Volt-Electric-Vehicle-90758.Ht" (accessed Jan 2014).

hybrid_vehicle. "http://en.wikipedia.org/wiki/Hybrid_Vehicle_Drivetrain#Full_Hybrids."

hydraulic_hybrid, "http://www.Caranddriver.Com/Columns/a-Hybrid-That-Has-No-Batteries" (accessed Jan 2014).

Lohner-Porsche, "http://en.wikipedia.org/wiki/Lohner-Porsche" (accessed Jan 2015).

Norton, Robert L. *Design of Machinery pp. 660–749*. New York: McGraw-Hill, 2012.

Oliver_Evans, "http://Todayinsci.Com/E/Evans_Oliver/Evansoliver-Sciam(1886).htm" (accessed Feb 2015).

Otto_engine, "http://en.wikipedia.org/wiki/Otto_Engine" (accessed Jan 2014).

Stirling_engine, "http://en.wikipedia.org/wiki/Stirling_Engine" (accessed Feb 2015).

Toyota_Prius, "http://en.wikipedia.org/wiki/Hybrid_Synergy_Drive" (accessed Jan 2014).

Watt_engine, "http://en.wikipedia.org/wiki/Watt_Steam_Engine" (accessed Feb 2015).

APPENDIX E BIBLIOGRAPHY

Chapter 2 Chassis Layouts and Drivelines

2CV, Citroen, "http://en.wikipedia.org/wiki/Citroën_2cv" (accessed Feb 2015).

4-wheel-drive, "http://en.wikipedia.org/wiki/Four-Wheel_Drive" (accessed Feb 2015).

Adolphus, D. T. "The Staver Belt-Drive Automobile." *Hemmings Classic Car*, Jan 2010, 54–61.

Chase, Herbert. "Front-Wheel Drives, Are They Coming or Going?" *SAE Trans.* 23, (1928): 267–290.

Citroen, "http://en.wikipedia.org/wiki/Citroën" (accessed Feb 2015).

Clough, Albert L. "The Sturtevant Touring Car." *The Horseless Age* 1904, 4.

friction_drive, "http://en.wikipedia.org/wiki/Friction_Drive" (accessed Feb 2015).

Gleason, F. H. "Chassis Lubrication." *SAE Trans.*, (1926): 427–443.

Heldt, P. M. "Front Wheel Drives." *SAE Trans.* 25, (1930): 7.

Koch, Jeff. "Tenacious Toronado." *Hemmings Classic Car*, Jan 2011, 30–33.

Ludvigsen, Karl. "Automotive Adventures of Emile Claveau." *Hemmings Sports & Exotic Car*, Nov 2009, 62–67.

Spice, C. W. "Action, Application and Construction of Universal Joints." *SAE Trans.*, (1926): 358–385.

Tracta, "http://en.wikipedia.org/wiki/Tracta" (accessed Feb 2015).

Chapter 3 Engine Configurations

Bohacz, Ray T. "One Size Fits All—the 1981 Cadillac V-8-6-4 Engine." *Hemmings Classic Car*, Apr 2008, 82–85.

Bohacz, Ray T. "Slipping Away—Engine Bearings." *Hemmings Classic Car*, Jul 2008, 70–73.

Bohacz, Ray T. "Silent Knight—How Sleeve-Valve Engines Work." *Hemmings Classic Car*, Dec 2010, 80–82.

Bohacz, Ray T. "Chemistry Lesson—Detroit Reduces Nox with Egr." *Hemmings Classic Car*, Sep 2011, 72–74.

Bohacz, Ray T. "Solid as a Rock—Chevrolet's Highly Successful Stovebolt Six Engine." *Hemmings Classic Car*, Dec 2011, 94–96.

Bohacz, Ray T. "Racing into the Future—Ford DOHC Engine." *Hemmings Classic Car*, Jun 2011, 76–78.

Bohacz, Ray T. "Spin Her Up!—Chrysler's Gas Turbine Engine." *Hemmings Classic Car*, Mar 2011, 72–75.

Bohacz, Ray T. "The Firing Squad—Detroit Explores the Ignition's Firing Order." *Hemmings Classic Car* 2012, 72–74.

Bohacz, Ray T. "Negative Pressure—the Positive Attributes of Engine Vacuum." *Hemmings Classic Car*, Dec 2012, 72–76.

Bohacz, Ray T. "Going the Distance—in-Cylinder Flame Propagation." *Hemmings Classic Car*, Feb 2013, 74–76.

Bohacz, Ray T. "On the Verge—Coolant Phase Change in an Engine." *Hemmings Classic Car*, Aug 2013, 70–72.

Bohacz, Ray T. "Seal It up Good—the Combustion Seal." *Hemmings Classic Car*, Nov 2013, 74–75.

Bohacz, Ray T. "Logjam—the Simply Designed Exhaust Manifold." *Hemmings Classic Car*, Apr 2014, 76–77.

Bohacz, Ray T. "The Real Deal—Quantifying an Engine's Fuel Usage." *Hemmings Classic Car*, Jun 2014, 74–76.

Bohacz, Ray T. "Controlled Chaos—Combustion Chamber Mixture Motion." *Hemmings Classic Car*, Mar 2014, 74–75.

Cameron, Kevin. "Engine Performance: Beyond the Dyno." *Cycle*, Sep 1967, 63–67.

Crane, H. M. "Engine Characteristics as Affected by Cylinder and Crankcase Arrangement." *SAE Paper 260027*, (1926): 1–49.

Cummins, Lyle. *Internal Fire the Internal Combustion Engine 1673–1900*. Warrandale, PA: SAE, 1989.

Daniels, Jeff. *Driving Force, the Evolution of the Car Engine*. Somerset, UK: Haynes Publishing, 2002.

Drive_Report. "Willys-Knight." *Special Interest Autos*, Nov–Dec 1971, 25–29.

Drive_Report. "1932 Pontiac V8." *Special Interest Autos* 1972, 42–47.

Drive_Report. "1929 Viking V8." *Special Interest Autos*, Apr–May 1972, 41–47.

Drive_Report. "Hemi in Full Flite - the 1951 Chrysler." *Special Interest Autos*, 1973.

Ganahi, Pat. "Max Tork Vs. Hi H-Pwr." *Hot Rod*, Apr 1988, 56.

Ganahi, Pat. "Max Tork for Hi H-Pwr." *Hot Rod*, May 1988.

Hartman, Jeff. *Turbocharging Performance Handbook*. St Paul, MN: Motorbooks, 2007.

Hendry, Maurice. "Half-Hour History of V8 Engines." *Special Interest Autos* 1974, 36–39.

Heywood, John B. *Internal Combustion Engine Fundamentals*. New York: McGraw-Hill, 1988.

Hewlett, Van W. "Automobile Bearings." *Bulb Horn*, Sep–Oct 1971, 26–30.

Hudson. "A Test of the Hudson "Super Six" Engine." *The Automobile Engineer*, (1927): 88–89.

Hudson_balanced_crank, "http://Books.Google.Com/Books?Id=Evtmaaaamaaj&Pg=Pa82&Lpg=Pa82&Dq=Hudson+Balanced+Crankshaft&Source=Bl&Ots=Krpq9_Cwlj&Sig=Tqf9gz8iz1nfb3y0lioilv4yb_E&Hl=En&Sa=X&Ei=Q9ydu__Wac7xkqetnicgdg&Ved=0ccwq6aewaq#V=Onepage&Q=Hudson%20balanced%20crankshaft&F=False" (accessed Feb 2015).

Jennings, Gordon. "A Short History of Wonder Engines." *Cycle*, May 1979, 68.

Lamm, Michael. "Milestone Engines." *Special Interest Autos*, May–Jun 1971, 22–27.

Lamm, Michael. "1949 Cadillac Fastback with Ohv V8." *Special Interest Autos*, Jun–Jul 1972, 10–17.

McLean, Ross. "The 1931 Plymouth with Floating Power." *Special Interest Autos*, Jan–Feb 1971, 38–42.

Norton, Robert L. *Design of Machinery, 5ed*. New York: McGraw-Hill, 2012.

Pomeroy, L. H. "Tendencies in Engine Design (Slipper Pistons)." *SAE Trans.* 25, no. 1 (1920): 362–366.

Stone, Richard. *Introduction to Internal Combustion Engines*. Warrandale, PA: SAE, 1992.

Taub, Alex. "Value of Crankshaft Counterweights." *SAE Journal* 31, no. 2 (1933): 316.

Taylor, Charles F. *The Internal Combustion Engine in Theory and Practice, 2 Vols*. Cambridge, MA: MIT Press, 1966.

Ulmann, Alec. "Who Invented the Rotary Engine?" *Bulb Horn*, Jan–Feb 1972, 22–23.

Ulmann, Alec. "Ettore Bugatti's Aircraft Engine Fiasco." *Bulb Horn*, Sep–Oct 1972, 42–46.

Ulmann, Alec. "The FN Four and Single Cylinder Motorcycles." *Bulb Horn*, Mar–Apr 1972, 28–31.

Ulmann, Alec. "The Hispano-Suiza H6B Engine Variants." *Bulb Horn*, Nov–Dec 1973, 12–16.

Ulmann, Alec. "The WWI Peugeot Aircraft Engine." *Bulb Horn*, Jan–Feb 1974, 21–23.

Ulmann, Alec. "The 1915 Chalmers-Weidley OHC Engine." *Bulb Horn*, Nov–Dec 1976, 30–34.

Woodruff, Peter. "Built to Last 100,000 Miles"—The Pontiac Straight-Eight. *Hemmings Classic Car*, Jan 2011, 44–49.

Chapter 4 Valve Trains, Induction, and Supercharging

"Looks Like a Packard-Knight Controversy." *The Automobile*, Feb 1910.

"More on the Knight-Engined Cars." *Bulb Horn*, May–Jun 1972, 14–17.

Bohacz, Ray T. "Bump and Grind—the Automotive Camshaft." *Hemmings Classic Car*, Sep 2008, 70–73.

Bohacz, Ray T. "Half-Breed—GM/Rochester Throttle-Body Injection." *Hemmings Classic Car*, Aug 2011, 68–70.

Bohacz, Ray T. "Enriching the Mixture—the Carburetor Choke." *Hemmings Classic Car*, Jul 2011, 76–78.

Bohacz, Ray T. "Fuel Sipper—Detroit Goes after Fuel Economy." *Hemmings Classic Car*, May 2011, 72–74.

APPENDIX E BIBLIOGRAPHY

Bohacz, Ray T. "Fuel Atomization." *Hemmings Classic Car*, Aug 2012, 70–64.

Bohacz, Ray T. "Manifold Destiny—Intake Manifold Design." *Hemmings Classic Car*, 2013 2013, 80–82.

Bohacz, Ray T. "Conflict Resolution—Correct Carburetor Sizing for Production Engines." *Hemmings Classic Car*, Dec 2013, 76–77.

Bohacz, Ray T. "Intrinsically Linked—the Camshaft Drive Mechanism." *Hemmings Classic Car*, Jun 2013, 70–72.

Brown, W. Ferrier. "Sleeve-Valve Engine Development (Short)." *The Automobile Engineer*, (1926): 18–25.

Brown, W. Ferrier. "Sleeve-Valve Engine Development (Full Article)." *Proc. Inst. of Automobile Engineers*, (1926): 157–263.

Camless_valves, "http://Bioage.Typepad.Com/Greencarcongress/Docs/Lotuseaton.pdf" (accessed Feb 2015).

Fiat_Multiair, "http://www.Autozine.org/Technical_School/Engine/Vvt_6.html" (accessed Feb 2015).

Frederick, W. A. "The Single-Sleeve Valve Engine." *SAE Trans.* 22, (1927): 102–121.

Heldt, P. M. "Sleeve Valve Engines." *SAE Trans.*, (1926): 171–206.

Lamm, Michael. "Spyder Packages, the Corsa, and Turbochargers." *Special Interest Autos*, May–Jun 1974, 16.

Norton, Robert L. "Cams and Cam Followers." In *Modern Kinematics*, edited by A Erdman, 271–331. New York: Wiley and Sons, 1993.

Norton, Robert L. *Cam Design and Manufacturing Handbook*. New York: Industrial Press, 2009.

supercharging, "http://en.wikipedia.org/wiki/Supercharger" (accessed Feb 2015).

Ulmann, Alec. "Turbochargers Are the Newest Thing Out." *Bulb Horn*, Mar–Apr 1976, 30–36.

VVT, "http://www.Autozine.org/Technical_School/Engine/Vvt_5.html" (accessed Feb 2015).

Chapter 5 Transmissions and Differentials

"4-Speeds of the Early 1930s." *Special Interest Autos*, Apr–May 1972, 28–29.

"Packard Invents the Hypoid Rear End." *Special Interest Autos*, Nov–Dec 1973, 54.

"Half-Hour History of Overdrives." *Special Interest Autos*, Mar–Apr 1974, 46–48.

Alison, N. L. "Fluid Transmission of Power." *SAE Journal* 48, no. 1 (1941): 1–9.

Automatic_Transmission, "http://en.wikipedia.org/wiki/Automatic_Transmission" (accessed Jun 2014).

Bohacz, Ray T. "Bringing up the Rear—the Positive-Traction Limited-Slip Differential." *Hemmings Classic Car*, Jan 2011, 82–84.

Bohacz, Ray T. "Smooth Operator—Buick's Dynaflow Automatic Transmission." *Hemmings Classic Car*, Feb 2011, 70–72.

Bohacz, Ray T. "Leave the Shifting to Us—Detroit Perfects the Automatic Transmission." *Hemmings Classic Car*, Feb 2012, 72–74.

Churchill, Harold E. "Mechanical Minds for Motor Cars." *SAE Journal* 49, no. 3 (1941): 11.

Clough, Albert L. "The Sturtevant Automatic Transmission." *Horseless Age* 14, no. 6 (1904): 124–127.

Crouse, W. H., and Anglin, D. L. *Automotive Transmissions and Power Trains, 5ed*. New York: McGraw-Hill, 1976.

CVT, "http://en.wikipedia.org/wiki/Continuously_Variable_Transmission" (accessed November 2014).

DCT, "http://en.wikipedia.org/wiki/Dual-Clutch_Transmission" (accessed November 2014).

Differential, "http://en.wikipedia.org/wiki/Differential_(Mechanical_Device)" (accessed Jul 2014).

Drive_Report. "1937 Olds 6 with Semi-Automatic Transmission." *Special Interest Autos*, Sep–Oct 1971, 22–26.

Drive_Report. "1933 Reo Self Shifter." *Special Interest Autos*, Oct–Nov 1972, 30–35.

Drive_Report. "Almost Automatic: The Oldsmobile Semi-Automatic Transmission." *Special Interest Autos*, Jan–Feb 1974, 24–27.

Fluid_Coupling, "http://en.wikipedia.org/wiki/Fluid_Coupling" (accessed November 2014).

Fluid_Drive, "http://en.wikipedia.org/wiki/Fluid_Drive" (accessed November 2014).

Greenlee, Harry R. "Automatic and Semiautomatic Transmissions." *SAE Journal* 54, no. 8 (1946): 9.

Griswold, W. R. "Hypoid Rear-Axle Design and Lubrication." *SAE Journal* 40, no. 5 (1937): 12.

Heldt, P. M. "Automatic Transmissions." *SAE Journal* 40, no. 5 (1936): 15.

Jeffe, S. D. and Cartwright, B. W. "Chrysler Torqueflite Transmission." *SAE Trans.* 66, (1958): 9.

Manumatic_Transmission, "http://en.wikipedia.org/wiki/Manumatic" (accessed November 2014).

Palazzolo, Joseph. *High Performance Differentials, Axles, and Drivelines*. North Branch, MN: Car Tech, 2009.

Presto-Matic_Transmission, "http://en.wikipedia.org/wiki/Presto-Matic" (accessed November 2014).

Rodger, W. R. and Syrovy, A. J. "The Chrysler Powerflite Transmission." *SAE Trans.* 62, (1954): 15.

Semi-Automatic_Transmission, "http://en.wikipedia.org/wiki/Semi-Automatic_Transmission" (accessed November 2014).

SMG, "http://en.wikipedia.org/wiki/Electrohydraulic_Manual_Transmission" (accessed November 2014).

Sturtevant, T. L. and T. J. "Clutch Device Power Transmitting Mechanism." Patent No. 766551. USA, 1904.

Waclawek, M J. "Torque Converter for Industrial and Commercial Vehicles." *SAE Trans.* 62, (1954): 8.

Winchell, F. J., Route, W. D., and Kelly, O. K. "Chevrolet Turboglide Transmission." *SAE Trans.* 66, (1958): 12.

Wolf, Austin M. "Automatic Transmissions." *SAE Journal* 41, no. 1 (1937): 9.

Wolf, Austin M. "Trends in Design of 1938 Cars: Olds Semiautomatic Transmission." *SAE Journal* 41, no. 2 (1937): 2.

Wolf, Austin M. "Trends in Design of 1940 Cars: Gm Hydramatic Transmission." *SAE Journal* 45, no. 5 (1939): 2.

Wolf, Austin M. "Trends in Design of 1941 Cars: Chrysler Fluid Drive." *SAE Journal* 47, no. 5 (1940).

Chapter 6   Suspension and Steering

Ackerman_Steering, "http://en.wikipedia.org/wiki/Ackermann_Steering_Geometry" (accessed Nov 2014).

Air_Suspension, "http://en.wikipedia.org/wiki/Air_Suspension" (accessed Dec 2014).

Ball_Joint, "http://en.wikipedia.org/wiki/Ball_Joint" (accessed Dec 2014).

Barnes, W B. "Front Axle Movements." *SAE Trans.* 24, (1929): 21–25.

Bastow, Donald. "Steering Problems and Layout." *Proc. of the Institution of Automobile Engineers* 32, (1937): 125–178.

Bastow, Donald. "Independent Rear Suspension." *Proc. of the Institution of Automobile Engineers* 5, no. 1 (1951): 35–58.

Bohacz, Ray T. "Let the Good Times Roll—the Wheel and Tire." *Hemmings Classic Car*, Jan 2010, 66–69.

Bohacz, Ray T. "Steer Me Straight!—the Automotive Steering Gearbox." *Hemmings Classic Car*, Apr 2011, 68–71.

Bohacz, Ray T. "Front End Suspension Geometry." *Hemmings Classic Car*, Feb 2013, 74–76.

Broulheit, Georges "Independent Wheel Suspension." *SAE Trans.* 33, no. 4 (1933): 325–351.

Brown, Roy W. "The Tire Factor in Automobile Riding Quality." *SAE Journal* 30, no. 1 (1932): 16–21.

Brown, Roy W. "Air Springs—Tomorrow's Ride." *SAE Journal* 31, no. 4 (1936): 126–133.

Brull, Charles B. "Modern European Light Cars." *SAE Journal* 46, no. 4 (1940): 177–183.

Csere, Csaba. "A Surprising New Active Suspension." *Car and Driver*, Oct 2004.

Davis, Francis W. "Power Steering for Automotive Vehicles." *SAE Journal* 53, no. 4 (1945): 18.

Dillman, O. D. and Love, R. R. "Chrysler Torsion-Aire Suspension." *SAE Trans.* 66, (1958): 10.

Donald, Bastow. "Independent Rear Suspension." *Proc. of the Institution of Mechanical Engineers* 5, no. 1 (1951): 35–58.

APPENDIX E BIBLIOGRAPHY

Dubonnet_Suspension. "1934 Pontiac 8." *Special Interest Autos*, Nov–Dec 1973, 44–48.

Evans, R. D. "Properties of Tires Affecting Riding, Steering, and Handling." *SAE Journal* 36, no. 2 (1935): 9.

Evans, W.M. "The Case for the Independently Sprung Wheel." *Proc. of the Institution of Automobile Engineers* 23, (1929): 8.

Griffith, Bain. "Design Figures of the New Ford Axle." *SAE Trans.* 66, (1958): 4.

Habibi, H., Shirazi, K., and Shishesaz, M. "Roll Steer Minimization of MacPherson-Strut Suspension Using Genetic Algorithm Method." *Mechanism and Machine Theory* 43, (2008): 57–67.

Hicks, H. A. and Parker, G. H. "Harshness in the Automobile." *SAE Trans.* 34, (1939): 8.

Howard, Bill. "Bose Reimagines Auto Suspension." *PC Magazine*, Aug 26 2004.

James, W. S., Churchill, H. E. and Ullery, F. E. "Sky Hooks for Automobiles." *SAE Journal* 37, no. 3 (1935): 9.

Jansen, Laura M. and Dyke, Shirley. "Semi-Active Control Strategies for Magnetorheological Dampers: A Comparative Study." *ASCE Journal of Engineering Mechanics* 126, no. 8: 795–803.

Lanchester, F. W. "Automobile Steering Gear—Problems and Mechanism." *Proc. of the Institution of Automobile Engineers* 22, (1928): 45.

Lanchester, F. W. "Motor Car Suspension and Independent Springing." *Proc. of the Institution of Automobile Engineers* 30, (1936): 93.

Lanchester, F. W. and Lanchester, G .H. "Independent Springing." *Proc. of the Institution of Automobile Engineers*, (1938).

Lewis, R. P. and O'Brien, L. J. "Rear Axles—Today and Tomorrow." *SAE Trans.* 66, (1958): 19.

Lincoln, C. W. "Hydraulic Steering in General Motors Cars." *SAE Trans.* 62, (1954): 8.

Norby, Jan P. "Half-Hour History of Independent Suspensions." *Special Interest Autos*, Nov–Dec 1973, 40–43,54.

Norbye, Jan P. "A Short History of Air Suspension." *Special Interest Autos,* 1973.

O'Connor, B. E. "Damping in Suspensions." *SAE Journal* 54, no. 8 (1946): 6.

Olley, Maurice. "Independent Wheel Suspension—Its Whys and Wherefores." *SAE Journal* 34, no. 3 (1934): 9.

Olley, Maurice. "National Influences on American Passenger Car Design." *Proc. of the Institution of Automobile Engineers*, (1938): 65.

Olley, Maurice. "Road Manners of the Modern Car." *Proc. of the Institution of Automobile Engineers* 41, (1946): 36.

Platt, Maurice. "Rear Suspension." *The Automobile Engineer* 1939, 37–41.

Polhemus, V. D. "Secondary Vibrations in Rear Suspensions." *SAE Journal*, (1950).

Polhemus, V. D., Kehoe Jr., L. J., Cowin, F. H. and Milliken, S L. "Cadillac's Air Suspension." *SAE Trans.* 66, (1958): 11.

Poynor, James C. "Innovative Designs for Magnetorheological Dampers." MSME, Virginia Polytechnic University, 2011.

Riley, R. Q., "https://Sweetmfg.Biz/Uploads/Files/Tech-04understandingsteering-4.pdf" (accessed Jan 2015).

Simionescu, P. A. and Beale, D. "Synthesis and Analysis of the Five-Link Rear Suspension System Used in Automobiles." *Mechanism and Machine Theory* 37, (2002): 815–832.

Steering_Dynamics, "http://www.Idsc.Ethz.Ch/Courses/Vehicle_Dynamics_and_Design/11_0_0_Steering_Theroy.pdf" (accessed Nov 2014).

Suspension, "http://www.Rqriley.Com/Suspensn.htm" (accessed Dec 2014).

Ulmann, Alec. "The Hotchkiss Drive—What Is It and Why." *Bulb Horn*, Mar–Apr 1974, 16–20.

Vincent, J. G. and Griswold, W. R. "A Cure for Shimmy and Wheel Kick." *SAE Trans.* 24, (1929): 9.

Wolf, Austin M. "Trends in Design of 1941 Cars: Rear Suspensions." *SAE Journal* 47, no. 5 (1940): 5.

Chapter 7   Brakes

Anti_Lock_Brakes, "http://en.wikipedia.org/wiki/Anti-Lock_Braking_System" (accessed Nov 2014).

Beaulieu, Lord Montague *Car Braking.* Vol. 3 Cars and Motorcycles, Edited by Lord Montague and Marcus Bourdon, 1928.

Bohacz, Ray T. "Stopping Power—the Clamping Advantages of Disc Brakes." *Hemmings Classic Car*, Nov 2010, 80–83.

Brake_History, "http://www.Secondchancegarage.Com/Public/History-of-Automotive-Brakes-1.Cfm" (accessed Nov 2014).

Brake_History, "http://www.Motorera.Com/History/Hist07.htm" (accessed Nov 2014).

Brake_History, "http://www.Autoevolution.Com/News/Braking-Systems-History-6933.html" (accessed Jan 2015).

Brakes, "http://en.wikipedia.org/wiki/Vehicle_Brake" (accessed Nov 2014).

Disk_Brake, "http://en.wikipedia.org/wiki/Disc_Brake" (accessed Jan 2015).

Drum_Brake, "http://en.wikipedia.org/wiki/Drum_Brake" (accessed Jan 2015).

Norby, Jan P. "A Short History of Disc Brakes." *Special Interest Autos,* 1973.

SAE_Historical_Society. *The Automobile, a Century of Progress, Chapter 4*: SAE, 1997.

Chapter 8   Body Design

Allen, Edwin L. "Body Engineering—Past, Present, and Conjecture as to Future." *SAE Journal* 45, no. 3 (1939): 365–377.

BMW_i3, "https://en.wikipedia.org/wiki/BMW_i3" (accessed Jan 2015).

Body, Aluminum, "http://www.Paintgages.Com/List-of-Car-Manufacturers-Who-Use-Aluminum-Body-Panels-S/51.htm" (accessed Jan 2015).

Body, Aluminum, "http://www.Supanet.Com/Top-10-Aluminium-Cars-G84p2.html" (accessed Jan 2015).

Body, Aluminum, "http://www.Jflf.org/Pdfs/Wi202/Aluminum.pdf" (accessed Jan 2015).

Body, Aluminum, "http://www.Drivers.Com/Article/245/" (accessed Jan 2015).

Body_History, "http://www.Motorera.Com/History/Hist09.htm" (accessed Jan 2015).

Body-on-Frame, "http://www.Goodcarbadcar.Net/2009/09/Body-on-Frame-Truck-Based-Suvs.html" (accessed Jan 2015).

Brunn, H. A. "Body Comfort and Interior Appointments." *SAE Journal* 30, no. 1 (1932): 21–22.

Budd_Company, "http://en.wikipedia.org/wiki/Edward_G._Budd" (accessed Jan 2015).

Chrysler_Airflow, "http://en.wikipedia.org/wiki/Chrysler_Airflow" (accessed Jan 2015).

Drive_Report. "1954 Corvette." *Special Interest Autos*, Jan–Feb 1971, 20–23.

Drive_Report. "1950 Nash Rambler - America's First Successful Postwar Compact." *Special Interest Autos* 1974, 30–35.

Fitzgerald, Craig. "Bertone: One of the Industry's Longest-Running Coachbuilders." *Hemmings Sports & Exotic Car*, Nov 2009, 69.

Foster, Patrick. "The Travelall: International Harvester's Rugged and Roomy Station Wagon." *Hemmings Classic Car*, Sep 2014, 48–53.

Georgano, G. N. *Cars 1886–1930*. New York: Crescent Books, 1990.

Honda NSX, 2015, "http://en.wikipedia.org/wiki/Acura_Nsx_(2015)" (accessed Jan 2015).

Honda NSX, original "http://en.wikipedia.org/wiki/Honda_Nsx" (accessed Nov 2014).

Lamm, Michael. "Fliptops." *Special Interest Autos*, Jan–Feb 1971, 24–29.

Lamm, Michael. "Magnificent Turkey - The Chrysler Airflow." *Special Interest Autos* 1973.

Lancia_Lambda, "http://en.wikipedia.org/wiki/Lancia_Lambda" (accessed Jan 2015).

APPENDIX E BIBLIOGRAPHY

Ledwinka, Joseph. "Combined Body and Chassis Construction (Excerpts)." *SAE Trans.* 33, (1938): 1.

Ludvigsen, Karl. "Edmund Rumpler's Tropeen-Auto." *Hemmings Sports & Exotic Car,* 2009, 52–55.

Marti, Othmar K. "Streamlining Applied to Automobiles." *SAE Trans.* 26, (1931): 4.

McCain, George L. "Dynamics of the Modern Automobile (Chrysler Airflow)." *SAE Journal* 35, no. 1 (1934): 248–256.

Milliken, S. L. and Parker, J. R. "The Cadillac Frame: A New Design Concept for Lower Cars." *SAE Trans.* 66, (1958): 14.

Murrray_Bodies. "Body by Murray." *Special Interest Autos*, Jan–Feb 1974, 36–39.

Norby, Jan P. "Half-Hour History of Unit Bodies." *Special Interest Autos*, Aug–Oct 1973, 24–54.

SAE_Historical_Society. *The Automobile, a Century of Progress (Chapter 7)*: SAE, 1997.

Shea, Terry. "Fashioned by Function—the Chrysler Airflow." *Hemmings Classic Car*, Sep 2013, 25–31.

Sherman, Roger, "http://www.Pierce-Arrow.org/Features/Feature26/index.php".

Swallow, W. "Unification of Body and Chassis Frame." *Proc. of the Institution of Automobile Engineers* 33, (1939): 431–475.

Tietjens, O. G. "Economy of Streamlining of Automobiles." *SAE Journal* 30, no. 3 (1932): 3.

NOTES

INDEX

A

ABS 210, 211
Acura
   NSX 229
aeolipile 2
air brakes 212
Alfa Romeo 65
   SMG transmission 158
   Variator 106
all wheel drive 46, 47, 48, 49, 50,
   161, 162, 176, 189, 198, 230
aluminum bodies 226
Alvis 41, 129, 143, 179, 183
American Motors 14, 48, 88, 118, 156
   Eagle 48, 161
AMT 139, 156, 157, 158, 159
anti-dive 200
anti-roll bar 199
anti-squat 200
anti-sway bar 199
Anzani 83
Aston Martin 39, 80
Atkinson cycle 21, 22, 23, 24, 29
Auburn 65, 79, 118, 163
Audi 42, 49, 50, 60, 61, 64, 78,
   84, 108, 158, 162, 168, 188,
   189, 198, 230
   A2 230
   A8 230
   Quattro 49, 162, 230
Austin Mini
   39, 43, 44, 111, 134
automatic transmission 148
automobile
   electric 9, 26, 27, 28, 34
   front-mid-engined 38, 237
   hybrid 28
      flywheel 32
      Lohner-Porsche 28, 41, 47
      parallel 28
      series 28, 30
      series-parallel 28, 29, 31
   mid-engined 34, 37, 38, 42, 45,
      50, 230, 234, 236, 237
   rear-engined 33, 38, 49, 83, 188,
      197, 221
axle 201
   beam 170, 171, 172, 173, 174, 189
   dead 174
   De Dion 175
   live 174
   shaft
      full-floating 175
      semi-floating 175
   tramp 177

B

B17 Flying Fortress 119
B24 Liberator 119
balance
   crankshaft 88
   shafts 89
balancer
   Lanchester 82, 89, 90, 111, 135,
      157, 173, 185, 203, 206, 210
   Nakamura 89
Barnes, William 162, 163
BDC 4, 13, 15, 16, 17, 18, 21
Bendix 115, 143, 144, 204, 208, 211
Benelli 63
Benz, Carl 33
Besler steamer 12
Bleriot 83
BMW 27, 30, 37, 39, 43, 45, 49, 52,
   63, 79, 81, 101, 107, 108, 111,
   134, 156, 157, 158, 159, 189,
   191, 196, 198, 211, 219, 231,
   237
   I3 237
   Mini 43
   Valvetronic 107, 108
   VANOS 107
body
   aluminum 216, 226, 227
   boat-tail 215
   by Fisher 213
   closed 216
   crushable 237
   custom built 213
   fabric 217
   open 213, 214
   pedestrian-friendly 238
   station wagon 217
   unit 179, 186, 214, 220, 222, 223,
      224, 225, 226
   wood 213
body-on-frame 214, 215
Borg-Warner 138, 144, 156
Bose active suspension 192
bounce 172
brakes 203
   air 29, 212
   anti-lock 210
   automatic trans. 151
   carbon-composite 207
   ceramic disk 207
   cross-drilled 207
   disk 203, 204, 206, 207, 208, 210,
      212, 216, 228, 233, 240
      floating-caliper 207
   drum 203, 204, 206, 207, 208,
      212, 232
   emergency 212
   four-wheel 204
   hydraulic 108, 204, 205, 210
   hydraulic power 210
   inboard 208
   Lockheed 204
   mechanically operated 204
   mechanical power 209
   parking 212
   planetary trans. 135, 143, 144
   power 209, 210
   self-energizing 204, 205, 206,
      208, 210
   split 205
   vacuum power 210
braking
   regenerative 26, 27, 28, 29, 31, 32
Bristol 86
BSA 41, 54
Budd Company 213, 216, 217, 223,
   224, 225
Bugatti 65, 66, 84, 118, 143,
   213, 235
   Veyron 84, 235
Buick 61, 65, 67, 76, 77, 79, 97, 115,
   129, 135, 145, 150, 153, 154,
   156, 162, 168, 178, 179, 204
bump steer 171, 201

C

Cadillac
   Eldorado 42, 44, 167
   Northstar DOHC V8 101
   V8-6-4 110
CAFE 63, 231
cam
   follower 93, 94, 95, 97, 98, 99, 100
   functions 95
   lobe 93, 106, 107, 108
   Polydyne 99
   tappet 93, 94, 97, 106, 107
camber 171, 172, 179, 180, 181,
   182, 184, 188, 189, 197
camshaft 17, 93, 108
   auxiliary 107
   dual-overhead 61, 66, 76
   F-head engine 97
   hop-up 105
   L-head engine 97
   motion functions 95
   overhead 61, 76, 98, 99, 100, 101
      single 99
   overhead-valve 73, 76
   phase-shifting 107
   T-head engine 96

carbon fiber 231
carburetor 15, 16, 17, 18, 19, 20, 93, 111, 112, 113, 114
   fixed venturi 111
   SU-type 113
   variable venturi 112
Cardano, Gerolamo 45
caster 171, 172, 179, 181, 184, 189, 197
castor oil 85, 86
center of gravity 37, 38, 39, 75, 76, 200, 224
chain
   inverted-tooth 101
   roller 101
   silent 101
chain drive 101
Chandler 65
Chapman struts 186, 188
Chevrolet
   Corvair 39, 83, 188, 226
   Corvette 11, 41, 45, 114, 119, 120, 138, 167, 187, 188, 191, 198, 208, 213, 214, 219, 221, 231, 232, 233, 234
   Stovebolt Six 61, 76
   Tahoe 45, 219
   Volt 29, 30
Christie 164, 179, 181
Chrysler 32, 45, 48, 65, 76, 87, 88, 91, 110, 115, 129, 131, 141, 143, 144, 145, 146, 147, 148, 152, 156, 162, 167, 185, 195, 196, 204, 208, 210, 211, 216, 219, 220, 224, 226, 234, 235
   Airflow 219, 220, 224
   floating power 90
   Fluid Drive 146, 147, 148, 152
   PowerFlite 156
   Torqueflite 156
   Vacamatic 144, 146, 148
Citroën 42, 44, 45, 50, 64, 81, 91, 111, 166, 168, 169, 197, 208, 212, 217, 223, 225
   2CV 42, 50
      Sahara 50
   Maserati SM 169
   Sahara 50
   Traction Avant 42, 45, 166, 168, 223
Citroën, Andre 42
Clerk, Sir Dugald 18, 21, 115
cluster gear 123
clutch 5, 7, 10, 27, 28, 30, 31, 34, 39, 40, 41, 44, 125, 126, 127, 129, 130, 131, 132, 135, 137, 139, 140, 141
   centrifugal 143, 144, 147, 148
   cone 129, 130, 140
   dog 125, 129, 130
   hydraulic 144, 150, 157
   in AMT trans. 158
   in preselector trans. 143
   limited-slip 160
   lock-up 153
   mechanical 139, 140, 142, 153
   multi-plate 157
   one-way 152
   steering-column 197
   synchromesh 125, 129, 146, 158
   vacuum-operated 145
Columbia 26, 104, 163
compression-ignition 20
compression ratio 15, 20, 21, 22
   def. 21
compression stroke 17
connecting rod 2, 3, 4, 16, 21, 22, 51, 56, 59
   fork and blade 70, 73, 74
constant velocity joint 45, 46, 154, 178
Continental 82, 84, 105, 168, 208, 211
Cord 41, 42, 118, 144, 223, 224
   810/812 42, 223, 224
   L29 41
countershaft 123, 124, 125, 127, 129, 131, 158
crankcase 18, 51, 52, 70, 73, 74, 83, 84, 85
crankshaft 2
   balanced 88
      exactly 88
   bearings 88
   boxer engine 81
   Buick straight-eight 67
   cast iron 75
   common 13
   counterweights 61, 65, 73, 74, 75, 76, 77, 88, 90
   cross-plane 73, 74
   dead points 4
   eccentric 25
   even-firing 57
   fatigue failure 67
   five-cylinder 61
   flat plane V8 73
   four-cylinder 58
   Harley-Davidson 71
   inline-twelve 69
   mirror-symmetric 58
   nine-main bearing 66
   offset crankpins 77
   overbalanced 88, 89
   phase angles 54, 60, 61, 87
   primary moment 73
   radial engine 84
   rotary engine 65, 84
   shared 70
   single-plane 73
   six-cylinder 62
   stiffer 70
   straight-eight 66, 67
   stresses in 60
   thirty cylinder 87
   three-cylinder 65
   torque oscillations in 67
   torsional damper 62, 67, 90
   torsional vibration 67, 70, 81, 90
   uneven firing 57
   V4 72

V6 optimum 3-throw 77
V6 six-throw 77
V6 uneven firing 77
V10 78
V12 78
V16 80
V-twin 71
W8 84
W12 83, 84
W16 84
crashbox transmission 127
cross-head 3, 4
Crossley 144
crushable bodies 237
Cugnot, Nicolas 3, 5, 6, 7, 41, 195
Curved Dash Olds 35, 36, 135
CVCC 17
CVT 123, 132, 133, 134, 156, 163
cycle
   Atkinson 21
   Brayton 24
   Diesel 20
   four-stroke 16, 18, 26, 96
   Miller 24
   two-stroke 18
   Wankel 25
cylinder 3
   arrangements 53
   deactivation 109

D

Daimler 17, 34, 35, 54, 65, 70, 71, 79, 103, 104, 110, 111, 113, 115, 132, 141, 143, 144, 146, 157, 203, 226
damper 62, 67, 90, 112, 189, 190
   magnetorheological 190, 191
dashpot 112
dead center 4, 132
De Dion 52, 54, 175, 214
   axle 175
degrees of freedom 134, 159
Delage 118, 143
Delahaye 143, 213
desmodromic 92, 94, 95
DeSoto 65
Deux Chevaux 42, 83, 225
Dewar Trophy 7, 104
Diesel, Rudolf 20, 21, 22, 28, 109, 120, 121, 210
differential 6, 7, 13, 18, 33, 34, 38, 39, 40, 41, 42, 43, 47, 48, 49, 50, 92, 129, 159, 160, 161, 162, 170, 174, 175, 178, 179, 188, 230
   center 47, 48, 49, 50, 161, 162
   limited-slip 160
   open 160
   Torsen 162
Doble Steamer 10
Dodge 45, 48, 49, 65, 78, 162, 167, 175, 213, 216, 224, 226, 234, 235
   Viper 78, 234

INDEX

DOF 134
dog clutch 125, 129, 130
DOHC 51, 99, 100, 101, 106, 234
double-clutching 127
downshift 127, 147, 157
drag coefficient 218, 221, 233
driverless cars 197, 240
driveshaft 7, 38, 39, 40, 41, 43, 49, 50, 75, 129, 137, 138, 159, 161, 162, 174, 177, 178, 179, 212
   open 179
D-Type 61, 114, 208, 222, 228, 229
Dual-Cowl Phaeton 215
Dubonnet 179, 183, 185
Ducati 72, 95
Dudley 99
Duesenberg 65, 67, 74, 101, 118, 204, 210, 213
dwell 93, 94, 99
dynamic balance 57
dynamometer 88, 103, 119
Dynaslow 154
Dynaslush 154

E

ECE 2, 4, 14
efficiency
   thermal 20
electric car 9, 26, 27, 28, 34
   pure 26, 27
Electric Hand 143, 145
electromagnetic
   clutch 143
electronic stability control 211
en-bloc 52
energy 4, 9, 11, 13, 20, 22, 26, 27, 28, 29, 30, 31, 32, 67, 90, 94, 119, 141, 142, 152, 154, 189, 190, 206, 237, 238
   kinetic 26, 28, 29, 31, 32, 141, 206, 238
engine 65, 83
   air-cooled 86
   Atkinson 21, 22, 23, 24
   atmospheric 6
   balanced 53, 54, 61, 63, 71, 77, 78, 80, 81, 82, 88, 90
   balance shaft 57, 61, 65, 77, 90
   balancing 87
   boxer
     four 82, 83
     six 83
     twin 81, 82
   compound expansion 4
   compression-ignition 14
   crank throws 51, 57, 58, 65, 66, 67, 70, 78, 80, 88
   diesel 15, 20, 28, 109, 230, 231
   double-acting 4, 10
   dynamic balance 57
   eight-cylinder
     Inline 65

V8 30, 42, 65, 70, 72, 73, 74, 75, 76, 77, 78, 79, 81, 84, 88, 101, 104, 105, 110, 113, 115, 118, 148, 119, 120, 147, 224, 225, 106, 230, 231, 233, 234, 235, 236, 155
   even-firing 54, 55, 56, 57, 58, 62, 69, 70, 71, 72, 77, 78, 82, 83, 236
   five-cylinder 60, 61, 78, 85
   flat-12 78, 81
   flathead 73, 75, 76, 79, 80, 81, 87, 97, 98, 223, 118, 222, 106
   four-cylinder 36, 51, 52, 57, 58, 59, 60, 69, 73, 77, 82, 89, 91, 101, 110, 227
   four-stroke 18, 20, 25, 26, 33, 34, 51, 53, 54, 55, 58, 60, 65, 85
   gas force 55, 60, 61, 65, 67
   gas torque 60, 62, 65, 67, 90
   harmonics
     first 56, 58, 59
     second 56, 57, 58, 59, 61, 62, 89
     third 61, 62, 65, 66, 77
   hemi 61, 76, 100, 101, 76
   high-compression 76
   inline 54, 57, 60, 61, 63, 65, 67
   inline twin 54, 57
   interference 101
   main bearings 66, 67, 73, 76, 88
   Mazda-Miller 24
   normally aspirated 120
   overhead-valve 75, 76
   perfectly balanced 65, 66, 80, 82
   primary force 57, 71, 72, 83, 88
   primary moment 59, 61, 65, 72, 73, 74, 77, 90
   radial 84, 86, 87
   rotary 25, 65, 84, 85, 86
   secondary force 57, 71, 88
   secondary moment 59, 60, 61, 65
   shaking force 53, 54, 55, 56, 57, 58, 60, 61, 62, 65, 66, 67, 71, 82, 89
   shaking moment 58, 60, 61, 62, 66, 67, 71, 82
   simple expansion 4
   single-acting 4
   single-cylinder 4, 26, 33, 34, 35, 36, 51, 53, 54, 216
   six-cylinder 39, 48, 51, 52, 60, 61, 62, 65, 69, 87, 101, 110
   spark-ignition 14
   Stirling 2, 12, 13, 14
   straight-eight 65, 66, 67, 74, 75, 80, 105
   thirty-cylinder 87
   three-cylinder 63, 85, 91, 101, 115
   torque 60, 121, 130, 132, 150, 151
     gas 60, 62, 65, 67, 90
     inertia 60, 61, 62, 65, 67, 77, 81, 82, 83, 89
     total 62, 65, 67

torsional damper 62, 67
turbojet 105
twelve-cylinder 67
two-cylinder 8, 10, 25, 26, 36, 42, 52, 55, 56, 57, 59, 81, 82, 237
two-stroke 18, 19, 25, 115
uneven-firing 57
V4 70, 72
V6 63, 70, 77, 78, 84, 90, 101, 106, 110, 120, 169, 229, 230, 231, 239
V8 30, 42, 65, 70, 72, 73, 74, 75, 76, 77, 78, 79, 81, 84, 88, 101, 104, 105, 110, 113, 115, 118, 148, 119, 120, 147, 224, 225, 106, 230, 231, 233, 234, 235, 236, 155
V10 78, 235, 236
V12 70, 73, 78, 79, 84, 104, 110, 118, 119, 224, 236
V16 73, 80, 81, 94
V-angle 69, 70, 71, 72, 73, 77, 78, 80, 235
   optimum 70, 73, 77, 78
V-engine 69, 70, 71, 72, 81, 84
vibration 25, 51, 53, 54, 57, 58, 59, 65, 67, 70, 72, 73, 74, 75, 81, 83, 85, 88, 90, 91, 93, 95, 99, 202
V-twin 12, 57, 70, 71, 72, 187
W3 83
W8 84
W12 83, 84, 230
W16 84, 235
W18 83
Wankel 25, 26
W-engine 83, 84
engine start-stop 110
E-Type 51, 61, 76, 79, 100, 113, 114, 166, 188, 208, 222, 229
Evans, Oliver 6, 7, 182
even firing 54, 55, 56, 57, 58, 62, 70, 77, 78, 82, 83
exhaust stroke 17
expansion ratio 21, 22
external combustion 1, 2, 12

F

Fardier de Cugnot 41, 195
fatigue failure 67, 81, 192
Ferrari 37, 74, 79, 157, 158, 176, 191, 198, 229, 234
   Fiorano 158, 191, 198
Fiat
   128 44, 45
   500 54, 111
   Multiair 108
   Uno 134
   X1/9 45
fiberglass 231
floating power 90

fluid coupling 5, 40, 139, 141, 142, 143, 144, 146, 147, 148, 149, 150, 151, 152, 153
    versus clutches 139
Fluid Flywheel 146
flywheel 2, 4, 13, 32, 62, 65, 67, 85, 126, 140, 142, 143, 148
    torque 65, 67
flywheel hybrid 32
Ford
    automated foundry 75
    Expedition 45, 48, 189, 226
    Explorer 45, 226
    F150 231
    Fiesta 44, 45, 63, 134
    Ford-O-Matic 156
Ford, Henry 36, 37, 42, 58, 75, 81, 127, 135, 217, 224
fork and blade 70, 73, 74
Föttinger, Hermann 141, 152
fourbar
    mechanism 194
    slider-crank mechanism 21, 186
four wheel drive 46, 47, 48, 50
Franklin 58, 61, 79
Frazer 12, 150
freewheeling 129, 139, 147
friction drive 131, 132
front-wheel drive 6, 35, 37, 41, 42, 43, 44, 45, 46, 49, 50, 77, 131, 181, 183, 184, 187, 188, 189, 198, 212, 223, 224, 228, 230
    longitudinal engine 42
    transverse engine 42
fuel control 111
fuel injection 15, 23, 109, 111, 113, 114, 115, 210, 229
    direct 114
    electronic 114
    mechanical 114
    port 114
    throttle body 114
    throttle-body 110
fuzzy logic 156

G

Gardner 65
gas
    force 55, 60, 61, 65, 67
    pressure
        curve 17
    torque 60, 62, 65, 67, 90
Gaucher, Roland 86
gear
    bevel 40, 129, 162
    hyperboloid 161
    hypoid 161
    pinion 6, 123, 129, 159, 161, 162, 178, 194, 195, 196, 197, 228
    planet 2, 3, 25, 134, 159
    ring 25, 134, 135, 138, 159, 160, 161

    sun 25, 134, 135, 138, 139, 160
gearbox 157, 159
    semi-automatic 139
    sequential manual 157
    Wilson 143
gear ratio 123, 124, 125, 129, 130, 137, 150, 153, 154, 161, 162
gears
    helical 106, 107, 128
    herringbone 128
    spur 128, 129, 131
    straight-cut 128
gearset 122, 123, 125, 126, 127, 129, 134, 136, 150, 161
gear train 2, 40, 123, 133, 145, 148, 159
    compound 123, 148
    epicyclic 134
    planetary 2, 133, 134, 159
Giacosa, Dante 44
Gregoire, Jean-Albert 41
gyroscopic
    couple 173, 174
    force 174
    sensor 211

H

handling 172, 197
Harley-Davidson 57, 71
harmonic balancer 89
harmonics 57, 60, 82
    first 56, 58, 59
    second 56, 57, 58, 59, 61, 62, 89
    third 61, 62, 65, 66, 77
heat pump 14
helical gear 106, 107, 128
Hispano-Suiza 79, 203, 209
Hitler 38, 82, 221
Honda 17, 23, 44, 54, 57, 72, 82, 83, 106, 107, 110, 111, 134, 176, 229, 230
    Accord 23, 44
    Civic 44, 107
    Gold Wing 82, 83
    Shadow 72, 169
    VTEC 106, 107, 110
Hotchkiss 143, 176, 177, 178, 179, 228
Hotchkiss-Gregoire 228
Hotchkiss suspension 176, 177
Hudson 65, 75, 76, 88, 97, 143, 145, 150, 157, 158, 166, 185, 216, 225
    Drive-Master 145
    Stock Car Races 88
    Super Six 88
hybrid
    def. 28
hybrid vehicle 1, 22, 23, 27, 28, 29, 30, 31, 32, 41, 47, 64, 97, 108, 110, 111, 121, 123, 134, 230, 236, 237, 240
    def. 30

hydraulic
    hybrid 30, 32
        parallel 31
        series 31, 32
    motor 5, 28, 30, 31
    pump 31, 109, 210
hydraulic lifters 94, 104, 110
hypoid gear 161
hypoid rear end 75, 162
Hyundai 23, 107

I

ICE 2, 7, 9, 13, 14, 27, 28, 29, 30
IFS 164, 167, 179, 181, 182, 184, 185, 187, 188
impeller 117, 142, 150, 152, 153, 154
independent suspension 170, 171, 179
    Chapman strut 186, 188
    Dubonnet 183
    leaf-spring linkage 182
    MacPherson strut 186
    sliding pillar 181
    swing-axle 184, 188
    trailing arm 183
    unequal arm 184
Indianapolis 500 41, 65, 114, 118
individual wheel drive 47
induction system 16
inertial force 53, 56, 62, 81, 199
inline engines 54
    eight-cylinder 65
    five-cylinder 60
    four-cylinder 57
    six-cylinder 51, 61, 62, 63, 64, 65, 67, 77, 78, 83, 84, 120, 228, 229, 232
    three-cylinder 63, 65
    twelve-cylinder 67
    two-cylinder 54
instant center 193, 194, 198, 200, 201
intake stroke 16
interference engine 101
IRS 164, 179, 185, 187, 188, 189
Isotta-Fraschini 204
Issigonis, Alec 43

J

jackshaft 37, 123, 131, 132
Jaguar 30, 39, 79, 100, 118, 166, 188, 198, 222
    D-Type 61, 114, 208, 222, 228
    E-Type 51, 61, 76, 79, 100, 114, 166, 188, 208, 222, 229
    engine 61, 113
    Le Mans wins 61
    XJ 231
    engine 61
Jeep
    Grand Cherokee 47
    origin of name 48
    Quadra-Trac 48
Jowett 81

INDEX

K

Kaiser 12, 48, 150
Kettering, Charles 73
kingpins 170, 171, 173, 181, 183
Knight, Charles Yale 101
Knight sleeve-valve engine 103
    RAC test of 103

L

Lamborgini 50, 78, 80, 159, 235, 236
    Aventador 236
    Gallardo 78
Lanchester, F. W. 82, 89, 90, 111, 135, 157, 173, 185, 203, 206, 210
    harmonic balancer 89
    perfectly balanced engine 80, 82
    planetary transmission 135
Lancia 72, 77, 175, 179, 182, 189, 222
    Lambda 182, 222
        praise for 182
LaSalle 162
Laverda 63
Le Mans 32, 41, 61, 64, 76, 208, 222, 227, 228, 229
Levavasseur, Leon 73
Lexus
    LFA 235
Liberty engine 79, 119
Lincoln 23, 74, 79, 150, 163, 168, 211, 224, 226
    Zephyr 224
linkwork 3, 4
Locomobile 7, 8, 9, 58, 65
locomotive
    diesel electric 28, 121
loss-lubrication 85
Lotus 74, 109, 186, 191, 233, 234
    active valve train 109
    Elan 234
    Elite 233
Lycoming 82, 83

M

MacPherson struts 186, 187, 189
Marmon 65, 72, 77, 79, 81
Maserati
    Quattroporte 38, 78, 198, 237
    SMG transmission 158
mass center 197, 198, 199
Maybach 17, 34, 111, 113, 143, 168, 203
Mazda RX-7 26
McLaren
    MP4-12C 236
    P1 236
mechanical advantage 107, 121, 122, 194

Mercedes-Benz 17, 23, 27, 30, 37, 45, 49, 61, 63, 65, 80, 94, 101, 104, 110, 114, 118, 168, 169, 174, 176, 184, 185, 188, 189, 191, 192, 207, 208, 225, 226, 229, 237
    300SL 229
    SLR McLaren 237
Merlin V12 79, 118
metallurgy 149
Milburn Electric 26
Miller, Harry A. 41
Miller, Ralph 24, 41, 118
Mitsubishi 27, 54, 89, 110, 176
Model T 10, 11, 36, 37, 39, 42, 58, 135, 137, 143, 162, 165, 204, 217
momentum 13, 18, 19, 123, 206, 209
monocoque 222, 224, 228, 229, 230, 233, 234, 235, 236
Morgan
    Aero 8 230
    cyclecar 182, 187
    Plus Eight 216
    sliding pillar suspension 181
motive power 1, 28
Moto-Guzzi 72
motor
    electric 14, 22, 28, 29, 30, 32, 196, 197, 121
    hydraulic 5, 28, 30, 31, 121
motor mounts 54, 88, 91, 178, 202
multi-link suspension 178

N

Nader, Ralph 83, 83
Napier 61, 83
    Lion W12 83
Nash 65, 88, 150, 185, 225
    Ambassador 600 225
natural frequency 202
Newcomen 2, 6
Newton's 2nd Law 95
NHTSA 205, 211, 238, 239
Nissan 27, 64, 106, 108, 113, 134, 189, 197
    NVTCS 106
    VVEL 108
NVH 239

O

OHC 92, 98, 99, 101
OHV 76, 95, 97, 98, 99, 100, 101, 110
Oldsmobile 35, 42, 44, 61, 65, 74, 75, 76, 135, 144, 145, 146, 149, 156, 167, 179, 211, 226
    Curved Dash Olds 35, 36, 135
    safety transmission 144
    Toronado 42, 44, 167, 211
Olds. Ransom E. 35, 144, 203
Olley, Maurice 184, 201
OPEC 44
Otto cycle 16, 17, 18, 19, 20, 22
Otto, Nikolaus 15, 16, 17, 18, 19, 20, 21, 22, 23, 34, 35, 51, 170
overdrive 131, 137, 138, 139, 142, 143, 144, 145, 146, 147, 150, 162, 163
overdrive rear axle 162
overhead valve 76, 95, 97, 98, 99, 100, 101, 61
oversteer 41, 172, 180, 188, 197, 199, 200, 225

P

P38 Lightning 119
P47 Thunderbolt 119
P51 Mustang 118
Packard 61, 65, 66, 74, 76, 78, 79, 129, 151, 153, 156, 161, 166, 185, 215, 226
    Twin-Six 78
    Ultramatic 156
Panhard-Levassor 35, 39, 40, 135, 178
Panhard rod 178, 179, 186, 189
parallel motion 3
parallelogram linkage 182, 184, 194, 195
pawl 5, 6, 139, 150
pedestrian-friendly cars 238
Peugeot 35, 64, 99, 104, 215, 227
    Darl'Mart 227
phase 4
phase angle 54, 107
Pierce-Arrow 61, 65, 210, 226, 227
piston
    air brake 212
    brake band 144
    brake shoe 204
    cup seals 204
    disk brake caliper 207
    displacement 2
    fuel injection pump 114
    guiding 2
    hydraulic tappet 94
    IC engine 14
    in pure rotation 85
    master cylinder 204
    Multiair valve train 108
    multiple 13
    power steering 196
    shock absorber 186, 190
    slipper 76
    steam engine 2
    variable venturi carburetor 112
planetary motion 25
plastic bodies 231
Plymouth 45, 91, 162, 226
polar moment of inertia 38, 39, 197
Pontiac 12, 65, 74, 75, 150, 156, 185, 234
    Fiero 234

Porsche
  911   32, 38, 49, 107, 197
  911 GT3R   32
  944   89
  Boxster/Cayman   38
  Carrera GT   78
  Lohner-Porsche   28, 41, 47
  PDK transmission   158
  Variocam   107
Porsche, Ferdinand   28, 38, 221
potato-potato   57, 71
power   1, 90, 118, 121, 122, 126, 142, 195, 197, 209, 210, 236
  stroke   17
power brakes   209, 210
power pulse   55
power steering   195, 196, 210
Pratt & Whitney   86
pumping losses   22, 23, 108, 110, 119
pushrod   97, 98, 99, 100, 101, 110, 120

R

radial engine   84, 86, 87
ratchet and pawl   5, 6
ratchet wrench   6
rear end
  hypoid-gear   161
  two-speed   162
rear-wheel drive   37, 38, 39, 41, 45, 46, 48, 49, 50, 77, 83, 131, 159, 162, 177, 182, 184, 188, 189, 197, 198, 229, 230, 231, 234, 236
redline   92, 94, 100, 107, 157, 159
regenerative braking   26, 27, 28, 31, 32
regenerator   12, 13
Renault   119, 143, 203, 206, 216
Reo
  Royale   144
resonance   92, 94, 99, 100, 173
Ricardo, Harry   76, 97, 105
ride comfort   202
ride frequencies   202
rocker arm   97, 107, 110
roll axis   197, 198, 199
roll center   198
Rolls-Royce   66, 79, 97, 118, 150, 151, 168, 185, 209, 213
rotary engine   25, 65, 84, 85, 86
  Bentley BR2   86
Ruckstell   162
Rumpler   164, 179, 184, 188
Ruxton   41
Rzeppa joint   46

S

Salmson   143
Saturn   234
self-starter   9
sequential manual gearbox   157
Setright, L. J. K.   42

shaft
  axle   174, 175, 188
    full-floating   175
    semi-floating   175
  counter   123, 124, 125, 127, 129, 131, 158
  input   123, 125, 126, 129, 131, 139, 140, 141, 142, 143, 148
  output   124, 125, 127, 129, 130, 131, 137, 138, 140, 148, 158, 159
shaking force   53, 54, 55, 56, 57, 58, 60, 61, 62, 65, 66, 67, 71, 82, 89
shear thickening   161
shimmy   172
shock absorber   184, 186, 187, 189, 190, 191, 197
silent chain   101
Silent Knight, the   102, 103
Simca   44, 45, 143
sixbar mechanism   22
sleeve valves   101, 105
  single   105
slip angle   199
slipper pistons   76
SMG   157, 158, 236
SOHC   99, 100, 101, 237
South-Pointing Chariot   134, 159
spark-ignition   20
speed shift   130
Spitfire   118
spline   95, 124, 125, 130
spring rate   165, 170
springs   43, 92, 103, 140, 164, 165, 219
  air   167, 168, 169
    AMC Rambler   *168*
    Buick   *168*
    Cadillac Eldorado   *168*
    Continental   *168*
    helper   *168*
  change in length   177
  coil   167, 175, 179, 184, 185, 186, 187, 188, 189, 205
  composite-plastic   167
  deflections   179
  force-deflection behavior   168
  full elliptic   165
  half-elliptic   165
  helper   168
  leaf   165, 167, 168, 170, 175, 176, 177, 178, 182, 184, 185, 188
  longitudinal   165, 178
  Model T   165
  nonlinear   168, 170
  power steering valve   196
  quarter elliptic   165
  semi-elliptic   165
  torque reaction force   177
  torsion bar   166
  transverse   165, 167, 168, 178, 182, 185, 188
Spruce Goose   86

sprung-unsprung weight ratio   41, 188, 197, 202
sprung weight   170
spur gear   128
Spyker   48, 61
stabilizer bars   199
Stanley Steamer   7, 8, 165, 214
steam   2
  engine   3, 4, 6, 7, 8, 12, 14, 134
    terminology   *3*
  turbine   2
Stearns-Knight   65, 104
steering   164, 193, 194, 195, 197, 211
  Ackerman   193, 194
  column   110, 129, 143, 147, 150, 157, 194, 195, 196, 197
  parallelogram   194, 196
  power   195, 196, 210
    degressive   *196*
    electric-assist   *197*
    hydraulic-assist   *195*
  rack and pinion   6, 129, 194, 195
Stevens-Duryea   61, 216, 226
streamlining   218, 220, 221
stroke
  compression   15, 17, 21, 22, 23, 24, 113
  exhaust   17
  intake   16, 17, 20, 22, 55, 120
  power   4, 17, 18, 21, 22, 26, 55
Studebaker   65, 75, 76, 118, 144, 156, 162, 185, 205, 208
Sturtevant   110, 148, 149
Stutz   65
Subaru
  BRZ   50
  Justy   63, 134
superchargers
  axial flow compressor   117
  positive displacement   24, 31, 116
  positive pressure   117
supercharging   92, 115, 118
suspension
  A-arm   180, 183, 185
  active   191, 192
  air   168
    AMC Rambler   *168*
    Cadillac Eldorado   *168*
  caster, camber, and toe   171
  Chapman strut   186, 188
  Chrysler TorsionAire   167
  Citroën hydro-pneumatic   168
  double-wishbone   180, 184, 185, 187
  Dubonnet   183
  Hotchkiss   177
  independent   164, 170, 179
    front   *170, 181*
    rear   *187*
  leaf-spring linkage   182
  MacPherson strut   186
  multi-link   178
  non-independent   170, 179, 220
    front   *170*
    rear   *174*
  Packard Torsion-Level   166

INDEX

Panhard rod 178, 179, 186, 189
seven-link 189
shimmy, wobble, and bounce 172
sliding pillar 181
spatial-linkage 188
sprung weight 170
swing axle 184
torque-tube 177
traction bars 177
trailing arm 183
unequal arm 180, 182, 184
unsprung weight 170, 175, 180, 181, 182, 183, 188, 197, 202, 207, 208
Watt linkage 3, 178, 179
Suzuki 43, 63, 72
sway bars 199
swing axle suspension 184
synchromesh 35, 123, 125, 127, 129, 130, 131, 138, 140, 141, 146, 149, 157, 158, 223
Systeme Panhard 35, 36

T

Talbot-Lago 164
tappets
    bucket 100
    hydraulic 94, 139
    roller 94
Tatra 35, 38, 82, 182, 188, 221, 224
    T77 221
    T87 224
TDC 4, 13, 15, 16, 17, 18, 20, 21
Tiger Tank 143
timing belt 93, 101
tires 169
    pneumatic 164
tire scrub 172, 183, 199
toe 171, 172, 179, 184, 189, 197, 202
top gear 67, 83, 125, 137, 138, 143
torque 4
    balance shaft 89
    boxer 4 82
    boxer 6 83
    boxer twin 81, 82
    Corvette Z06 119
    diesel 20
    drops to zero 130
    effect on power 121
    electric motor 5, 27
    flywheel 65, 67
    gearbox 124
    IC engine 4, 11, 39, 40
    in fluid coupling 142, 150
    inline 3-cyl 65
    inline 4-cyl 60
    inline 5-cyl 60, 61
    inline 6-cyl 62
    inline 8-cyl 66, 67
    in top gear 125

locomotive 29
Miller cycle 24
multiplication 123, 152, 154
multiplication ratio 121, 154
oscillations 67
reaction 90, 91, 178, 179
split 47, 160, 162
steam engine 4, 8, 11
two-stroke 18
V4 72
V6 77
V8 73
V10 78
W engine 83
torque converter 40, 133, 139, 141, 148, 149, 152, 153, 154, 155, 156, 158
    acceleration 152
    coupling 152
    lock-up 151, 153
    stall 152
torque ratio 122, 124, 133
    122, 124, 133
torque-tube 176, 177, 178, 201
Torsen differential 162, 230
torsion bar 166, 167, 184, 199, 200
Toyota
    4Runner 50
    Corolla 44
    Prius 23, 28, 29, 30, 111
    Rav4 50
    Valvematic 108
Tracta 41, 45, 46, 166
Traction Avant 42, 45, 166, 168, 223
traction bars 177
traction control systems 211
trailing arm suspension 183
train ratio 123, 124, 125
transaxle 38, 41, 188, 236
transfer case 46, 47, 48
transfer port 18
transmission 5, 7, 40
    18-wheeler 127
    automated manual 139, 140, 157
        dual-clutch 158, 163, 230, 235
        predecessor 157
        single clutch 157, 158, 236
    automatic 40
        Dynaflow 153, 154, 155
        first 148
        Ford-O-Matic 156
        Hydramatic 149, 150, 151, 152, 155
        modern 156
        PowerFlite 156
        Powerglide 154, 155, 232
        Sturtevant 148
        Torqueflite 156
        Turboglide 152, 154
        Turbo Hydramatic 155, 156
        Ultramatic 156
    constant-mesh 125
        shifting 125

constant velocity 46, 154
continuously variable
    CVT 123, 132, 133, 157
    Ford 134
    Nissan 134
countershaft 129
crashbox 127, 128, 129, 135, 137, 149
definition 121
double-clutch 127
downshift 127, 147, 157
Dynaflow 154
fluid drive 146
fluid flywheel 146
freewheeling 138
friction drive 131, 132
helical gears 131
Hydramatic 149, 151
in hybrids 28, 29, 31
manumatic 157
Mini 43
Model T 137
overdrive 131, 137, 138, 163
over front axle 42
planetary 29, 36, 127, 134, 135, 136, 149, 158, 159
    Curved Dash Olds 135
    Lanchester 135
    Model T 36, 134, 135
    Oldsmobile 35
Powerglide 153
preselector 143, 145
    Hudson Drive Master 146
    Hudson Electric Hand 145
purpose of 121
reverse-gear whine 131
semi-automatic 35, 139, 141, 144, 145, 146, 148, 149, 150, 152, 157, 185
    Chrysler Fluid Drive 145
sliding-gear 35, 123, 124, 127, 128, 129, 135
    shifting 125, 127
    speed-matching 126
straight-cut gears 128
synchromesh 35, 129, 130, 141, 223
    shifting 130
    speed shifting 130
Systeme Panhard 39
torque ratio 122, 124, 133
Turbo Hydramatic 155
unfortunate experience with 155
Vacamatic 144, 146
White steamer 10
with transverse engine 44
Trevithick, Richard 6, 7
Triumph 54, 63, 208
Tucker 83, 144
turbine 2, 24, 25, 117, 119, 120, 142, 150, 152, 153
turbocharger 117, 119, 120
two-stroke cycle 18, 20

U

underdrive 144, 147, 148, 162
understeer 41, 171, 172, 197, 198, 199, 200
unit-body 222
universal joint 40, 187, 188
    Cardan 40, 45, 177
    constant velocity 42, 45, 46
    Hooke's coupling 40
Unsafe at Any Speed 83, 188
upshifting 127

V

V4 70, 72
V6 63, 70, 77, 78, 84, 90, 101, 106, 110, 120, 169, 229, 230, 231, 239
V8 30, 42, 65, 70, 72, 73, 74, 75, 76, 77, 78, 79, 80, 81, 84, 88, 101, 104, 105, 110, 113, 115, 118, 148, 119, 120, 225, 147, 221, 224, 225, 106, 230, 231, 232, 233, 234, 235, 236, 237, 155
V10 78, 235, 236
V12 70, 73, 78, 79, 84, 104, 110, 118, 119, 224, 236
V16 73, 80, 81, 94
valve
    actuation 93
    arrangements 95
    cam 18, 92, 93, 94, 95, 97, 99, 100, 101, 102, 103, 104, 105, 106, 107, 108, 114, 143, 228
    clearance 94
    exhaust 17, 92, 93, 94, 96, 105
    in-block *95*
    in head 97
    poppet 103, 104
    seat 94
    sleeve 102, 105
    spring 93, 94, 95
    timing
        variable 94, 105, 106, 235
    train 18, 26, 93, 94, 99, 100, 104, 109
        camless 109
valve float 92, 95, 98, 99
V-engine 69, 70, 71, 72, 81, 84
venturi 111, 112, 113, 210
viscous coupling 160, 161
volatility 14, 15, 20
V-twin 12, 57, 70, 71, 72, 187
VVT 94, 105, 106, 107, 108, 109, 110
VW
    Beetle 38, 39, 44, 82, 83, 92, 166, 184, 188, 219, 224, 225
    Passat 84
    Rabbit 44
    VR15 78, 84

W

W3 83
W8 84
W12 83, 84, 230
W16 84, 235
W18 83
Wankel, Felix 25, 26
Watt, James 2, 3, 4, 6, 7, 25, 134, 178, 179
Watt linkage 3, 178, 179
weight distribution 197
W-engines 78, 83
wheel alignment 171, 172
White Steamer 9, 10
Wills Sainte Claire 101
Willys 48, 104, 135, 150, 204
Willys-Overland 104
Wilson gearbox 143, 146
Wilson-Pilcher 82, 83
Wilson, W. G. 143
wobble 172
Wolseley 52, 67
woodies 217
worm gear 129
WWII 38, 42, 44, 48, 62, 75, 76, 79, 80, 81, 86, 87, 98, 104, 105, 114, 118, 119, 134, 143, 148, 162, 168, 188, 195, 214, 218, 221, 225

Y

Yamaha 54, 57, 63, 72

Z

zero inertial torque 66
zero shaking force 62, 66
zero shaking moment 62, 66